PLACE IN RETURN BOX to remove this checkout from your record.
TO AVOID FINES return on or before date due.
MAY BE RECALLED with earlier due date if requested.

DATE DUE	DATE DUE	DATE DUE
MAY 2 9 2008		

1/98   c:/CIRC/DateDue.p65-p.14

Estuaries

Advances in

ECOLOGICAL RESEARCH

VOLUME 29

Estuaries

Advances in

ECOLOGICAL RESEARCH

Edited by

D.B. NEDWELL

Department of Biological Sciences, University of Essex, UK

D.G. RAFFAELLI

Department of Zoology, University of Aberdeen, UK

Series Editors

A.H. FITTER

Department of Biology, University of York, UK

D.G. RAFFAELLI

Department of Zoology, University of Aberdeen, UK

VOLUME 29

ACADEMIC PRESS

A Harcourt Science and Technology Company

San Diego   San Francisco   New York   Boston
London   Sydney   Tokyo

Academic Press
A Harcourt Science and Technology Company
24–28 Oval Road, London NW1 7DX, UK
http://www.hbuk.co.uk/ap/

Academic Press
525 B Street, Suite 1900, San Diego, California 92101–4495, USA
http://www.apnet.com

ISBN 0–12–013929–4

A catalogue record for this book is available from the British Library

Typeset by Saxon Graphics Ltd, Derby
Printed in Great Britain by MPG Books Limited, Bodmin, Cornwall

99 00 01 02 03 04 MP 9 8 7 6 5 4 3 2 1

Contributors to Volume 29

K.S. BLACK, *Sediment Ecology Research Group, Gatty Marine Laboratory, University of St Andrews, St Andrews KY16 8LB, UK.*

S. CROOKS, *Centre for Social and Economic Research on the Global Environment (CSERGE), School of Environmental Sciences, University of East Anglia, Norwich NR4 7TJ, UK.*

C.H.R. HEIP, *Netherlands Institute of Ecology, PO Box 140, 4400 AC Yerseke, The Netherlands.*

P.M.J. HERMAN, *Netherlands Institute of Ecology, PO Box 140, 4400 AC Yerseke, The Netherlands.*

D.O. HESSEN, *Department of Biology, University of Oslo, PO Box 1027 Blindern, 0316 Oslo, Norway.*

T.D. JICKELLS, *School of Environmental Sciences, University of East Anglia, Norwich NR4 7TJ, UK.*

J. KROMKAMP, *Centre for Estuarine and Coastal Ecology, Netherlands Institute for Ecology, 4400 AC Yerseke, The Netherlands.*

J.J. MIDDELBURG, *Netherlands Institute of Ecology, PO Box 140, 4400 AC Yerseke, The Netherlands.*

D.B. NEDWELL, *Department of Biological Sciences, University of Essex, Colchester CO4 3SQ, UK.*

D.M. PATERSON, *Sediment Ecology Research Group, Gatty Marine Laboratory, University of St Andrews, St Andrews KY16 8LB, UK.*

R. SANDERS, *School of Environmental Sciences, University of East Anglia, Norwich NR4 7TJ, UK.*

M. TRIMMER, *Department of Biological Sciences, University of Essex, Colchester CO4 3SQ, UK.*

R.K. TURNER, *Centre for Social and Economic Research on the Global Environment (CSERGE), School of Environmental Sciences, University of East Anglia, Norwich NR4 7TJ, UK.*

G.J.C. UNDERWOOD, *Department of Biological Sciences, University of Essex, Colchester CO4 3SQ, UK.*

J. VAN DE KOPPEL, *Netherlands Institute of Ecology, PO Box 140, 4400 AC Yerseke, The Netherlands.*

Preface

The importance and ecological significance of estuaries as an interface between the land and sea has long been appreciated, with a considerable history of scientific examination of the physicochemical characteristics of estuaries and of the structure and function of estuarine biological communities. Recently, concern with increased nutrient loads to estuaries, and consequent eutrophication, has focused the attention of both ecologists and legislators. Trends towards a holistic, multidisciplinary approach to the ecology of the coastal zone has emphasized the interaction between the river catchment, the estuary and the coastal sea in terms of element fluxes. This has been epitomized, at least in Europe, by the imminent adoption of the Water Framework Directive by the European Community, which will require integrated management of water resources across the entire system of river basin, estuary and coastal sea.

In the last few years there has been increased research activity focused in estuaries, and there have been significant advances in understanding of at least certain aspects of estuarine ecology. However, much remains to be done before the complete framework of scientific background and understanding necessary for the integrated management strategy envisaged by legislation is in place. Without attempting to cover all aspects of estuarine ecology, this volume aims to provide overviews of a number of the important areas of estuarine research where there have been significant improvements in scientific understanding. This ranges from the macroscale in terms of how differences in river basins influence the loads of nutrients to estuaries (Hessen), through the impacts and fates of nutrient loads in estuaries (Nedwell *et al.*), to how both phytoplanktonic and benthic estuarine primary production are regulated (Underwood and Kromkamp). Paterson considers the influence of benthic microbial communities, while Herman *et al.* discuss the interactions between benthic animals and sediments. Many of the key factors influencing the outcome of management decisions for estuaries may not be ecological at all but result from economic, social or political pressures. Crooks and Turner review these aspects from the point of sustainability of estuarine resources.

We hope that this volume will provide both the specialist in estuarine ecology and the interested newcomer with new insights into the ecology of estuaries.

Dave Nedwell
Dave Raffaelli

Contents

Catchment Properties and the Transport of Major Elements to Estuaries

D.O. HESSEN

Nutrients in Estuaries

D.B. NEDWELL, T.D. JICKELLS, M. TRIMMER AND R. SANDERS

Primary Production by Phytoplankton and Microphytobenthos in Estuaries

G.J.C. UNDERWOOD AND J. KROMKAMP

Water Flow, Sediment Dynamics and Benthic Biology

D.M. PATERSON AND K.S. BLACK

Ecology of Estuarine Macrobenthos

P.M.J. HERMAN, J.J. MIDDELBURG, J. VAN DE KOPPEL AND C.H.R. HEIP

Integrated Coastal Management: Sustaining Estuarine Natural Resources

S. CROOKS AND R.K. TURNER

Catchment Properties and the Transport of Major Elements to Estuaries

D.O. HESSEN

I. SUMMARY

This paper reviews major anthropogenic activities and watershed properties that govern nutrient fluxes, particularly nitrogen (N), from catchments to estuaries. Nitrogen export from most catchments has increased strongly during the past decades, mainly resulting from increased inputs of fertilizers and increased atmospheric deposition of oxidized and reduced N, the first originating mainly from combustion processes, the second mainly from volatilization of manure. While "pristine" watersheds export some 100–200 kg N km^{-2} year^{-1}, agricultural and urbanized watersheds may export more than 10 000 kg N km^{-2} year^{-1}. A major part of total N input to watersheds (typically > 75%) is either denitrified in soils or freshwaters or stored as organic N in soils and biomass, although this N retention may show large spatial and temporal variability. Atmospheric deposition of N seems to correlate well with catchment export and, for many catchments receiving in excess of 2000 kg N km^{-2} year^{-1} from wet and dry deposition, there are signs of a progressing "N saturation" yielding increased area-specific runoff rates. The flux of N is tightly coupled to the pools and the turnover of organic carbon (C) in many temperate watersheds, and a major fraction of total N export is in the form of organic N, associated with organic C (mainly dissolved humic matter). Thus the flux and fate of organic C (which may range from 1 to 8 tonnes km^{-2} year^{-1}) may be a main determinant of annual fluxes of N, although the role of organic N for estuaries is not finally settled.

Similarly, export of phosphorus (P) will show pronounced variability among catchments, depending on population and land use, and range from "pristine" minima of < 4 kg P km^{-2} year^{-1} to maxima of more than 1000 kg P km^{-2} year^{-1}. While fluxes of both N and P increase with catchment disturbancy, there is a strong tendency towards higher N : P ratios in more disturbed watersheds, and N : P ratios may show extreme variability in agriculturally dominated watersheds. Although silicate (Si) fluxes are governed mainly by natural processes (weathering rate), a strong site-specific variability of total Si export is also recorded, typically ranging from < 500 to > −5000 kg SiO$_2$ km^{-2} year^{-1}. For all elements, the presence of larger lakes and reservoirs may be a major determinant to seasonal patterns, and may also strongly decrease export to estuaries. Since most freshwaters are P-limited, retention of P is normally most efficient and can be estimated from basin morphometry and hydraulic regimes by simple empirical models. Retention of N may also to some extent be predicted by corresponding models, yet with far less accuracy. In strictly P-limited systems with low denitrification rates, retention of N may be negligible. The flux of Si to estuaries has probably decreased for many systems, owing to eutrophication and thus increased Si uptake in freshwater systems.

Land use, soil and vegetation, and hydrology are the major factors controlling retention of the various elements in watersheds. Anthropogenic effects such as fertilizer inputs and atmospheric deposition are superimposed on these catchment properties. Also long-term climatic perturbations may add to this altered flux of N from watersheds. Increased winter temperature may shift the annual flux of N (and P) to estuaries from spring to late winter, while long-term climatic warming may increase mineralization of the large stores of organic N in temperate soils with subsequent strong increases in N loads to estuaries.

II. INTRODUCTION

Imbalances in biogeochemical cycles have become a central environmental problem both on the local and global scale. The nitrogen (N) cycle has been subjected to the most severe disturbances with a number of large-scale environmental effects. Anthropogenic N fixation was, for most of the human history, minor relative to bacterial fixation, but this has now been changed dramatically, and the sum of fixation from combustion, fertilizer industry and leguminous crops now exceeds natural biological fixation (Vitousek, 1994;

Vitousek *et al.*, 1997). These large-scale transformation processes imply a major transfer of inert N_2 into more chemically and biologically reactive compounds. N is a key element for primary production, and worldwide production of agricultural crops, forest and marine phytoplankton is primarily limited by the availability of inorganic N (nitrate or ammonia). Consumers are thus indirectly limited by availability of N, but may also to a great extent face a direct N limitation via scarce proteins (White, 1993). In areas where natural N fixation is low, primary production hinges almost entirely on external supplies. The modern fertilizer industry built its success on the ability to fix atmospheric N_2 into forms of nitrogen that could be utilized directly by primary producers. This 90-year-old industrial story gave rise to a vigorous increase in crop harvests worldwide, termed "the green revolution". In western Europe, nitrogen added by mineral fertilizers increased 10-fold after the second world war, while phosphorus doubled over the same period. However, in most industrialized countries the use of fertilizer N has stabilized during the past 15–20 years (World Commission, 1987). The demand for agricultural products will continue to increase, not only as a result of increased population but also due to a rapid economic growth of South-East Asian countries which may be accompanied by a dietary shift towards more animal protein and processed products. One of the most probable ways to meet these challenges is an intensification of agriculture, which generally will imply larger inputs of N. Thus, while the Western use of fertilizers is close to a saturation level with regard to crop yield, a further increase in the use of fertilizers is to be expected in developing countries.

The other side of this coin is an increased bioavailability of nitrogen in forest soils and runoff to freshwater and marine recipients, where its potential stimulating effect on primary production may be detrimental. Together with increased runoff from the increased population in coastal areas, these sources of excess nitrogen are a major hazard to coastal areas worldwide and cause phytoplankton blooms and community shifts (Howarth, 1988; Smayda, 1990; Nixon, 1992; Vitousek, 1994), as well as pronounced changes in the benthic food webs (Raffaelli, 1997). Also, a reduction in coastal fish stocks has been attributed to increased loading of nutrients and reduced levels of oxygen (Johannessen and Dahl, 1996). Riverine export to marine areas has probably increased 5–10-fold in densely populated areas in Europe since the turn of the century (Pacés, 1982; Howarth *et al.*, 1996). Most of this increase is attributed to increased efflux of sewage, agricultural fertilizers and changes in land use.

Another major input of N in terrestrial and aquatic ecosystems results from the combustion of fossil fuel. While some organic N may be released in gaseous form by combustion of organic compounds, the main source of oxidized N is the transformation of atmospheric N_2 into other gaseous components that in a multitude of ways affect the environment. The oxidized forms of N that are produced by combustion processes, NO and

NO_2 (collectively labelled NO_x) have increased dramatically. Over the past 50 years, emissions of NO_x over Europe have increased sevenfold (Pacyna et al., 1991), resulting also in increased deposition (Brimblecombe and Steadman, 1982; Grennfelt and Hultberg, 1986). The present NO_x emissions in Europe are of the same order of magnitude as the use of commercial fertilizers. Correspondingly, atmospheric outputs of reduced N as NH_4^+ and NH_3 from agricultural activity, particularly livestock emissions, has also increased.

The separation of "natural" and anthropogenic N flux is not straightforward, since even remote or "pristine" areas are subject to anthropogenic influence by increased N-deposition. For sparsely populated areas, the contribution from "background" runoff (i.e. from mountains, heathland or forest) may well exceed 75% of total N runoff (Tjomsland and Braathen, 1996; Kaste et al., 1997), and the contribution from increased atmospheric N load to these diffuse sources is not easily assessed. Also, bedrock nitrogen may constitute a major pool of N for most ecosystems, and locally may contribute significantly to nitrate runoff (Holloway et al., 1998). This high leaching rate of geologically derived N cannot be linked to anthropogenic effects.

Nitrogen is the main limiting nutrient for most terrestrial ecosystems (Gutschik, 1981; Schlesinger, 1991), and increased loadings of N would initially increase productivity of terrestrial ecosystems without increased output of N from these ecosystems. Recent data indicate, however, that in addition to the outputs from sewage and agriculture there is also an increased leaching from soils due to "saturation" effects (Aber et al., 1992; Stoddard, 1994), with increased levels of NO_3^- in surface water. In central Europe, with typical annual depositions of N (NO_3^- + NH_4^+) in the range 4000–6000 kg N km^{-2} $year^{-1}$, continuous increases in NO_3^- concentration have been reported for several brooks and rivers (Hauhs et al., 1989).

The impact of increased N export to estuarine ecosystems is determined not only by N itself, but to a large extent also by the fluxes of other major elements such as phosphorus (P), silica (Si) and carbon (C). N, P and Si are major elements regulating aquatic primary production. If the N : P : Si supply ratio is sufficiently skewed towards N, one could expect a predominant P or Si limitation of primary production. This would in particular be the case in areas with low denitrification rates. One effect of increased Si limitation could be a reduced contribution by diatoms in the coastal spring bloom (Smayda, 1990). Such imbalanced nutrient ratios may be linked to net terrestrial photosynthesis and forest growth (Aber et al., 1995; Cronan and Gringal, 1995). In aquatic systems it will not only affect the overall primary productivity, but may also have profound effects on the species composition of phytoplankton and thus the entire food web. It could also influence the physiological properties of the algae, such as their toxicity and biochemistry, and thus nutritive value for grazers (Sterner and Hessen, 1994). Catchment properties and disturbances will affect the fluxes of P, Si and C as well. Moreover, these elements are intimately linked in biological and physicochemical processes of the catchment, so that changes in the flux of one element could significantly change the flux of others.

P has no gaseous phase, and is transported in the atmosphere only by aerosols. Thus the atmospheric flux of P is negligible compared with that of N, but could nevertheless be important due to the comparatively lower demands for P relative to N in plants and bacteria. All populated watersheds have experienced an increase in loadings of P to marine recipients, owing to sewage and agricultural runoff. While P is less mobile in soils than N, there still is a major "diffuse" runoff from agricultural areas, in addition to point sources from livestock and sewage. In general, there is a tendency for increased loads of P relative to N (i.e. decreased N : P ratio) with increased catchment disturbancy (Billen *et al.*, 1991).

For Si there are fewer data available. In general Si concentrations seem to vary less with catchment properties and disturbances, and thus the major effect of human influence in catchments may be expected to be due to changes in bioavailable P and N. This could cause increased retention of Si in freshwaters, and indeed there were early reports of decreasing trends of Si in freshwaters linked to increased load of P (Schelske and Stoermer, 1971). Although the historical trends and causes of Si depletion are arguable, increased loads of N and P would certainly lead to increased N : Si and P : Si ratios entering coastal areas (Smayda, 1990; Conley *et al.*, 1993; Hessen *et al.*, 1997).

Inorganic C rarely controls estuarine productivity, but allochthonous organic C could be a major determinant to heterotrophic productivity and optical properties of recipient waters. Natural watershed properties such as drainage patterns, hydrology, soil and forest coverage, and not least the proportion of wetlands, are the main determinants to the load of organic C to estuaries. However, catchment disturbances such as deforestation and drainage will also impact the C balance and the export of organic C. The annual increased releases of CO_2 due to anthropogenic effects constitute only a minor fraction of natural C mineralization of ecosystems. Nevertheless, the continuous accumulation of CO_2 in the atmosphere and the consequent changes in the global climate could entail severe feedback effects not only on the carbon budget of catchments, but also on the other major elements concerned. For example, increased CO_2 may entail increased uptake capacities of N, and vice versa.

This account will discuss major watershed properties and anthropogenic factors affecting the export of these major elements to estuarine ecosystems, focusing on nitrogen. In particular, attention will be devoted to the large-scale patterns, the relation between elements and the potential effect of global climate changes.

III. NUTRIENT LOADING TO ESTUARIES

A. Nitrogen Loading and Leakage

1. Large-scale Patterns

Increased sewage efflux to marine areas has been the consequence of both changed sanitation systems and a considerable increase in populations in coastal regions. Howarth *et al.* (1996) estimated that the share of sewage and wastewater in total N load to the North Atlantic was 11%, ranging from 7% to 34%. The highest share was found for the North Sea. Since point sources like sewage output do not involve catchment processes, this will not be considered further here.

Present atmospheric deposition of oxidized and reduced N shows strong regional patterns (Prospero *et al.*, 1996). The major sources are combustion processes (mainly fossil fuel) and livestock emissions (mainly manure), and although concentrations peak in densely populated areas like central Europe increased deposition will occur over great distances due to atmospheric transport. This is particularly the case for the NO_x. Likewise the relative share of these compounds will vary with distance to urban or agricultural areas. Prospero *et al.* (1996) estimated a total deposition of 5.0 Tg N year^{-1} of NO_x over the entire North Atlantic Basin and the Caribbean, while the corresponding value for NH_4^+ is 3.6 Tg N year^{-1}. Particularly over southern latitudes, mineral dust may be the most important source of N. Over the entire North Atlantic region, it was estimated that deposition of oxidized N has increased 4–5-fold since pre-industrial times, while emissions of reduced N from animal waste may have increased 10-fold (Prospero *et al.*, 1996). Over central Europe, a 7-fold increase in NO_x emissions has been estimated over the years 1950–1980 (Pacyna *et al.*, 1991). For the most perturbated regions, atmospheric deposition of $NO_x + NH_4$ may exceed 6000 kg N km^{-2} year^{-1}, while the more "pristine" areas receive less than 200 kg N km^{-2} year^{-1} (Tørseth and Semb, 1997).

Similarly the use of fertilizer has increased dramatically, and is clearly the dominant source of N inputs to most of the North Atlantic regions (Howarth *et al.*, 1996). Again, the regional differences are major, ranging from less than 100 kg N km^{-2} year^{-1} in sparsely populated areas, to almost 6000 kg N km^{-2} year^{-1} in the North Sea region. By far the highest total input of N to watersheds were found in this region, of which approximately 80% was fertilizer inputs. This may be reversed in less populated areas, however. For large areas of the North American, Canadian and North European coast, atmospheric deposition dominates over fertilizer inputs. These regions have, in general, low total N inputs, however.

Long-term records (more than three decades) of N fluxes in rivers are rare, but most of the existing series do indicate an increase in N export from watersheds to estuaries. Also comparisons of N fluxes from "pristine" catchments with those of more populated catchments clearly indicate increased N fluxes (Howarth *et al.*, 1996). In support of this, Henriksen and Brakke (1988) and Henriksen *et al.* (1988) found that nitrate levels in lakes of southern Norway in general doubled over the period from 1976 to 1986, while there has been no further increase since 1988. Similar observations were made by Jassby *et al.* (1995) for northern California. "Background" export of N from undisturbed catchments with low population density may range over an order of magnitude from 84 to 998 kg km^{-2} year^{-1} (Lewis, 1986), but for most temperate watersheds pristine runoff will be in the range of 80–200 kg N km^{-2} year^{-1} (Meybeck, 1992, 1995; Howarth *et al.*, 1996); moderately influenced watersheds typically export some 500–2000 kg N km^{-2} year^{-1}, while heavily influenced catchments may exceed 10 000 kg N km^{-2} year^{-1}. For major river systems feeding the North Atlantic basin, Howarth *et al.* (1996) found the ratio for "pristine" to current river N export to range from less than 1 to almost 20, with an average range of 1.7–5.3 for North American rivers and 3.5–10.6 for European rivers. On a smaller scale and for individual rivers, this "pristine" to current N flux ratio may clearly be far higher.

The causes of the increased N flux are numerous, but nevertheless some large-scale patterns may be predicted from fertilizer inputs and atmospheric deposition. By compiling data from riverine efflux to the North Atlantic Ocean from the major European and North American regions, Howarth *et al.* (1996) demonstrated that the area-specific riverine flux of N could be decribed as a simple linear function of both anthropogenic deposition of oxidized N and fertilizer application (Figure 1). Equations for the two regressions were

$$N \text{ flux} = -89.3 + 1.08 \text{ N dep } (r^2 = 0.79, p = 0.0006)$$

and

$$N \text{ flux} = 411.0 + 0.15 \text{ N fert } (r^2 = 0.39, p = 0.05)$$

(all units in kg N km^{-2} year^{-1}). These equations should imply that riverine export to estuaries could be estimated from deposition data, and this would be valid across different ecosystems. And, somewhat surprisingly, the effect of N deposition would be superimposed on fertilizer application, although inputs of fertilizer greatly exceed atmospheric deposition. Recent data for 33 rivers of the northern USA confirm the relationship between N deposition and riverine export of N (Jaworski *et al.*, 1997). Still, these large-scale regressions mask a pronounced local variability among individual catchments (see below).

A major share of N import to ecosystems will never show up in runoff, i.e. most watersheds show a fairly high retention of N, either by denitrification or

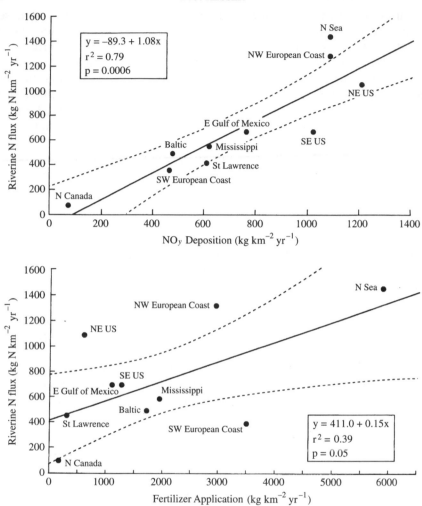

Fig. 1. Regression curves of river nitrogen export per area of catchment versus anthropogenic deposition of total oxidized N (NO$_y$) and fertilizer application for major temperate regions draining to the North Atlantic Ocean. From Howarth *et al.* (1996).

storage as organic N in biomass or soils. Large watershed studies indicate that no more than 20% of N input is exported from watersheds (Lowrance and Leonard, 1988; Jaworski *et al.*, 1992). In support of these values, Howarth *et al.* (1996) estimated an average retention of 75% of total anthropogenic input to watersheds. Their linear relationship between total input and total runoff would indicate that there is a constant proportion of loss, and thus that relative retention is independent of actual watershed load. If N "saturation" of soils and vegetation occurs, one would expect an exponentially increasing loss with increased watershed load. The large share of deposited N that is allocated to organic forms (stored) or lost as denitrification clearly emphasizes the importance of these processes. Any kind of catchment disturbance that significantly alters these processes (acidification, deforestation, drainage, climate change) will have profound effects on the N budget of catchments, and thus on the net export of N to estuaries.

The relative contribution to retention of denitrification or of biomass or soil storage will again depend on site and season. A general estimate shows that no more than 26% of net anthropogenic N input is retained in forests in soils or biomass (Howarth *et al.*, 1996) and since this is only one third of the net anthropogenic inputs that do *not* show up in rivers, this would leave a potential major share of the retention loss for denitrification. Thus, factors that are governing mineralization and denitrification will also be the main determinants of watershed export of N to estuaries.

2. N Fluxes on a Regional Scale

The spatial differences in area-specific gross runoff as well as retention is well illustrated by a detailed Swedish study. A high-resolution study of gross and net (i.e. retention subtracted) runoff from different area types to coastal waters revealed area-specific gross runoff ranging from < 100 to > 5000 kg N km^{-2} year^{-1}, and retention ranging from < 20% to > 80% for non-point sources (Swedish Environmental Protection Agency, 1997). For major forested areas even gross load was less than 100 kg, and these areas also typically had retention efficiencies of anthropogenic N exceeding 80%. These low rates of N export for the less affected parts of the watersheds were greatly offset by the coastal parts of the catchments, which typically had 100-fold higher area-specific export of N, and also frequently had retention rates < 40%. Agricultural runoff contributed some 50% of gross load in this study.

A "test" on the relationship between load and loss from a restricted geographical area was performed by use of data from the Norwegian Paris Convention (PARCOM) sampling programme (Holtan *et al.*, 1996), as well as from major rivers included in the Norwegian acidification monitoring programme. Grouping the rivers and watersheds into 16 large areas of southern Norway gave area-specific runoff ranging from 306 to 2200 kg N km^{-2} year^{-1} with an average of 725 kg N km^{-2} year^{-1} (calculated from Tjomsland and Braathen, 1996). By calculating direct anthropogenic catchment inputs from sewage and fertilizers, the "background" runoff was estimated as the difference between total flux (measured) and anthropogenic flux (calculated). These background fluxes ranged from 125 to 744 kg N km^{-2} year^{-1}, with an average of 327 kg N km^{-2} year^{-1}. Thus, anthropogenic inputs contributed from 23% to 79% (average 42%) of total N export to sea. The actual contribution from human-made inputs are higher, however, since increased N runoff due to increased N deposition was not included in the anthropogenic load.

Although these catchments may show a great variability in N deposition due to elevation and precipitation gradients, an approximation on area-specific N deposition was made for each area, based on Tørseth and Semb (1997). The "background" export of N from the 16 areas correlated well with estimated N deposition, while total runoff was poorly matched with N deposition (Figure 2).

Subcatchment studies in two large watersheds in southern Norway (cf. Henriksen and Hessen, 1997) clearly demonstrated on a smaller scale the role of various catchment properties for N export to estuaries. In the sparsely populated Bjerkreim watershed, dominated by atmospheric inputs (1500–2300 kg N km^{-2} $year^{-1}$), total N export was 230–420, 420–500 and 630–870 kg N km^{-2} $year^{-1}$ from forested, heathlands and mountainous subcatchments respectively (Kaste *et al.*, 1997). This corresponds to a retention of 70–90%, 65–70% and 55–70% respectively, and clearly demonstrates that N export from equally "pristine" catchments may differ significantly, even within a small region. Total annual N export from this watershed ranged from 810 to 1070 kg N km^{-2} $year^{-1}$ over the years 1993–1995. Agricultural land covered < 5% of this catchment, but still contributed 30% to total catchment export of N. The apparent close correlation between atmospheric deposition of N and total catchment export of N that is seen in regional comparisons (Howarth *et al.*, 1996 and Figure 1) may not be valid on a smaller scale. Subcatchments have quite variable N load and export, but there was no correlation between deposition and efflux among the subcatchments of the Bjerkreim watershed. On a small scale, variability in land use, soil and vegetation cover overrules the differences in atmospheric N load.

The other catchment in this study, the Auli watershed, which is dominated by forest and agricultural land, gave an even clearer example of differences on a local scale, related to land use. Total export from this catchment was 2200 kg N km^{-2} $year^{-1}$ (average 1991–1995), ranging from 230 kg N km^{-2} $year^{-1}$ from an old forest stand to 6400 kg N km^{-2} $year^{-1}$ from cereal land (Høyaas *et al.*, 1997). This field also proved an excellent example of the dynamics of N export related to deforestation, where clear-cutting of a forest stand yielded a strong increase in N export, gradually declining over the next 5 years (Mulder *et al.*, 1997).

Whether the export of N from watersheds occurs as organic N, nitrate or ammonia is a major determinant of primary production in the recipient system. Significant fluxes of ammonia to estuaries will occur only where there are major sewage outputs, since reduced N is normally rapidly oxidized, and the relative contribution from inorganic or organic N may show tremendous variability among catchments. Data from the Norwegian PARCOM programme

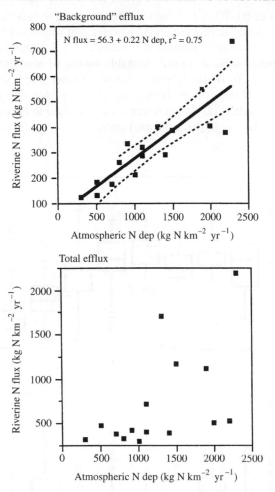

Fig. 2. Regression of river nitrogen export from 16 major watersheds of southern Norway related to calculated atmospheric nitrogen deposition. Upper panel: estimated "background" efflux, i.e. total N flux minus calculated sewage and fertilizer inputs. Lower panel: total measured flux of N. Data from Tjomsland and Braathen (1996) and Tørseth and Semb (1997).

were used to assess regional differences in inorganic : organic N as related to N deposition. Pooling all major rivers, and a number of smaller watersheds, into geographical sub-areas along the coast yielded a remarkable pattern with regard to the share of NO_3^- in total N. Areas receiving high deposition of atmospheric N also yielded a higher share of NO_3^-. The southern and south-western regions, which are most heavily influenced by atmospheric N, had annual medians of 60–70 % NO_3^- of total N, declining to less than 20% in the northernmost region. There were no such consistent trends for PO_4^{2-} : total P, except for the most heavily populated regions 2 and 3 where inorganic P on average contributed 40% of total P, probably owing to agricultural activities and sewage (Figure 3). While the relative share of inorganic N correlated well with atmospheric N deposition, such a pattern may also be caused by low levels of organic carbon (and thus lower levels of organic N) over the south-western regions, due to poorly developed soils.

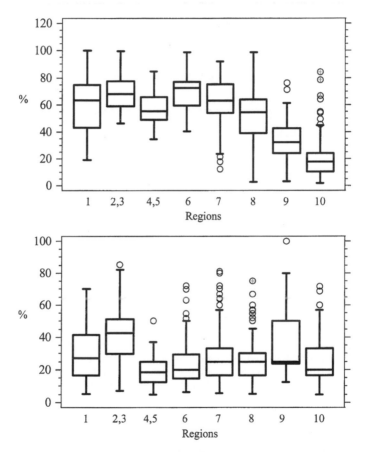

Fig. 3. Box-and-whisker displays of percentages of NO_3 of total N and PO_4 of total P for rivers in major regions along the Norwegian coastline from the Swedish border (region 1) to the Russian border (region 10). Regions 4–6 are the most heavily acidified, receiving the highest load of atmospheric N. Values are medians with quartiles in boxes, 90% confidence limits as bars. Observations outside 90% confidence limits appear as open circles. From Hessen *et al.* (1997a).

3. The "N Saturation" Concept

During the growth season there will be very restricted output of inorganic N from non-point sources, due to efficient retention in soils, uptake by soil biota and vegetation, and high denitrification rates. With increasing N deposition, uptake capacities may gradually be exceeded, leading to increased leakage of inorganic N also during the summer (Aber et al., 1989; Stoddard, 1994). Such a progressing "N saturation" may be distinguished by four stages of increasing NO_3^- concentrations in runoff and decreased retention during the growth season. (Note that the concept "saturation" in this context simply implies reduced retention, and that the causes for the reduced retention may be manifold.) The "pre-saturation" stage is characterized by distinct peaks of NO_3^- in late autumn and winter, with decreasing concentrations as the growing season proceeds, to virtually no NO_3^- in runoff during the summer. At this stage vegetation is still strictly N limited and all inorganic N from deposition and mineralization is consumed by primary producers or soil bacteria. With progressive excess of N above biological demands, uptake of nitrate decreases and a larger fraction and amount of reduced N is nitrified. Some part of this is denitrified in soils, but this will still result in pronounced losses of NO_3^- during the summer. The end-point of such a saturation process represents a situation where not only the concentrations and total efflux may have increased more than 5-fold, but there is also a constant supply of N to estuaries during the entire growth season. Over central Europe, with typical annual depositions of N in the range 4000–6000 kg N km^{-2}, signs of advanced N saturation (continuously raised NO_3^-) have been reported for several brooks and rivers (see Gundersen, 1992, for a review). Representative examples of such scenarios are given in Figure 4. The first (A) is representative of a forested catchment of south-eastern Norway with a low ratio of organic N : NO_3^-, annual precipitation near 1000 mm and an area-specific deposition < 500 kg N km^{-2} $year^{-1}$. The next two (B and C) represent a forested and a heathland catchment respectively in south-western Norway where levels of organic N are extremely low, annual precipitation exceeds 2500 mm and there is a deposition of 2500 kg N km^{-2} $year^{-1}$. The last (D) is from a large river from the same area, downstream of a major lake. This station illustrates the effect of larger lakes on annual N dynamics. While upstream brooks and rivers normally exhibit pronounced seasonality, an apparent saturation scenario may be achieved downstream simply by the dampening effect of lakes with long flushing times within the catchment. Annual N retention ranged from 70–90% in the forested watershed to frequently less than 60% in the heathland watersheds. Most clearly, this illustrates the overall importance of soil and vegetation on apparent N saturation. Although the mountainous heathland catchments showed clear signs of advanced saturation according to the criteria of Stoddard (1994), the cause

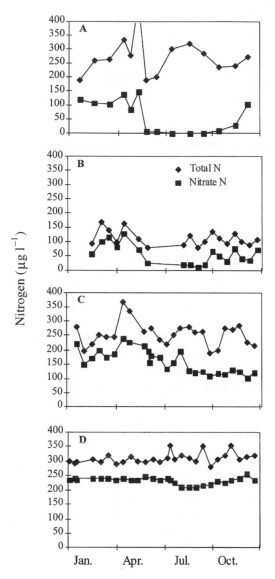

Fig. 4. Annual patterns of total N and nitrate from different subcatchments. (A) is representative of a forested catchment of south-eastern Norway with low ratio of organic N : NO_3^-, annual precipitation near 1000 mm and an area-specific deposition < 500 kg N km^{-2} year^{-1}. (B) and (C) represent forested and heathland catchments respectively of south-western Norway with annual precipitation exceeding 2500 mm and a deposition of 2500 kg N km^{-2} year^{-1}. (D) is from a large river from the same area, downstream of a major lake, illustrating the effect of larger lakes on annual N dynamics.

for this apparent N leakage may be a combination of decades with soil acidification and increased N deposition, poorly developed soils and hydrological dynamics with a high flushing rate.

Shifts from N limitation to P limitation could add to the effects of high flushing rate and thin soils. As indicated from the emission estimates of nitrogen, and assuming nearly stable atmospheric P deposition, the N : P atmospheric loading ratio may have increased up to 7-fold during the past decades. Excess N in the terrestrial parts of the catchment will be interpreted as a N saturation effect (Stoddard, 1994). The potential for terrestrial P limitation is controversial, yet this may be a common property at least of wet heathlands and bogs (Verhoeven et al., 1996). Note that apparent saturation may result from an elemental imbalance relative to the needs for autotrophic production; i.e. the apparent N saturation may be a consequence of a limited availability of P, and as such N saturation could be seen as a phosphorus limitation in disguise. It should also be stressed that any factor affecting (reducing) N storage in soils or biota or denitrification rates would increase the leakage of N.

Studies in several sites with highly different N inputs, combined with experimental removal or additions of N in precipitation, invariably reveal increased losses (lower retention) with increased deposition, and give some insight into the dynamics and actual levels of N deposition required for inducing saturation symptoms. Gundersen (1995) recorded enhanced export of N from catchments receiving more than 1500 kg N km^{-2} year^{-1}. An examination of 65 forested plots and catchments throughout Europe indicated a three-stage response to increasing N inputs (Dise and Wright, 1995). At deposition rates < 1000 kg N km^{-2} year^{-1} no significant leaching occurred. At intermediate levels (1000–2500 kg N km^{-2} year^{-1}) leaching occurred at some sites, while at deposition > 2500 kg N km^{-2} year^{-1} a significant leaching of N was seen at all sites, still with pronounced site-specific variability in N export. Since these data apply to forested catchments with well developed soils, it is reasonable to assume that saturation thresholds would be lower over non-forested areas with less developed soils. Leaching of N from forests may also depend strongly on the C : N ratio in soils (Cole et al., 1993; Dise et al., 1998).

4. N Retention in Aquatic Systems

For most temperate lakes there is strong evidence for a predominant P limitation, and a number of large-scale fertilization experiments have verified the overall importance of P relative to N (Schindler, 1977; Hecky and Kilham, 1988). This is confirmed by the models linking P load to phytoplankton biomass (e.g. Vollenweider, 1975; OECD, 1982), which are among the best examples of management-adapted empirical models in biology. The low N : P ratios in sewage may, however, promote periods of (secondary) N limitation in freshwaters. Also, lakes in regions with low N deposition may experience N rather than P limitation. A fundamental difference between fresh and marine waters is the pronounced capacity for N fixation in the freshwater environment. In fact N limitation judged from sestonic N : P ratios may be an apparently transient situation in many eutrophic lakes, as N deficiency may soon be equilibrated by N_2 fixation by cyanobacteria ("blue-green algae"). The presence of a high proportion of N_2-fixing blue-greens may in itself indicate some level of N limitation.

Nutrient-saturated phytoplankton exhibit in general C : N : P ratios close to the Redfield ratio (106 : 16 : 1 by atoms, 42 : 7 : 1 by weight), although there is a large interspecific variability. Systematic deviations from the Redfield ratio with regard to N or P in phytoplankton communities could then indicate a potential N or P limitation of biomass production. If applying this approach, lakes with an atomic N : P ratio < 16 in the seston would indicate potential N limitation. A survey of 355 Norwegian lakes (Faafeng and Hessen, 1993), covering a wide span in geographical distribution and productivity, indicated that N limitation during the summer would be unlikely in the vast majority of lakes, based on these criteria (N and P as total fractions). A follow-up study of 47 lakes, where particulate P and N was also analysed, confirmed this pattern, with particulate N : P < 16 only in three south-western lakes. There are, however, several problems with using crude seston as an indicator of nutrient status in algae, since both detritus, zooplankton and bacteria will contribute. In particular, a large proportion of bacteria in the sestonic compartment would skew the N : P ratio towards an apparent P limitation. Based on nutrient-enrichment assays from mountain lakes in Colorado, Morris and Lewis (1988) suggested the ratio between dissolved inorganic nitrogen (DIN) and total phosphorus (TP) as an appropriate measure of N versus P limitation. In lakes with a DIN : TP ratio < 9 (molar ratio) phytoplankton would be stimulated by additions of both N and P, while at ratios < 2 there was a strict N limitation. Applying this threshold, N limitation would not be an uncommon phenomenon in lakes in regions with low N concentrations, and in fact the majority of these mountain lakes were limited by N alone or by N and P combined. The frequent close balance between N and P limitation was supported by a study of Elser et al. (1990), who in a literature review of 62 nutrient limitation studies in lakes found that in 82% of these the largest growth stimulation was found when both N and P were added. Since the majority of N-limited lakes also have low P levels, it may be concluded that a biased input of N in general would give small eutrophication effects. In eutrophied, chronically N-limited lakes, increased N could increase algal biomass. Since these lakes often are dominated by blue-greens, however, increased N could in fact promote the competitive ability for other more preferable algae, and as such be advantageous for the whole food web.

The role of lakes (and wetlands) as sinks for N will thus be regulated strongly by their P content, in both absolute and relative terms, and by their flushing rate. N-limited lakes with high productivity and long renewal times will most likely be highly efficient sinks for N, whereas strictly P-limited systems and/or systems with short renewal times would be expected to retain less N. Provided there are sufficient capacities for N uptake by aquatic plants or phytoplankton and for denitrification, significant amounts (20–40%) of N may be retained by lakes (Jansson et al., 1994a,b). The role of both lake productivity and P : N ratios for total N retention was seen clearly in a study covering a gradient of lakes from ultraoligotrophic to eutrophic (Berge et al., 1997). While eutrophic lakes yielded annual N retention rates as high as 20–30%, the rate in oligotrophic lakes was less than 5%, in line with predictions (Jensen et al., 1990). In spite of renewal times of several years, the ultra-oligotrophic lakes in this study showed virtually no N retention at all. These lakes had stable NO_3^- concentrations of around 300 µg l^{-1}, but total P concentrations close to the detection limits for routine analysis (< 1 µg total P l^{-1}) and thus there was simply no potential either for biological uptake or for denitrification (these lakes are also almost devoid of organic carbon). Single lakes may even be net sources for N to the system, i.e. yielding a negative net retention, due to massive N fixation by blue-green bacteria that may occur in N-deficient eutrophic lakes (Berge et al., 1997).

Retention of P is a clear function of residence time, expressed simply as the input : output ratio (R) related to renewal time $(R = [1 + \sqrt{(T_w^{-1})}]^{-1}$ where T_w is the theoretical renewal time in years (e.g. Larsen and Mercier, 1976), but N retention is by far more fluctuating and unpredictable. Still there is a general trend for increased N retention as a function of the ratio of mean depth to residence time for lakes (Kelly et al., 1987). General models for N retention as a function of hydraulic regimes have been predicted, analogous to the P retention model of Larsen and Mercier (1976), acknowledging the fact that this will depend on the P content of the systems. While the model of Kelly et al. (1987) relates N retention to the ratio of mean depth over residence time (to account for denitrification), Bratli et al. (1995) proposed retention of N according to: $R = [0.2/(1 + \sqrt{(1 * T_w^{-1})}] + X$, where $X = 0, 0.1$ and 0.2 for oligotrophic, mesotrophic and eutrophic lakes respectively. The general applicability of these retention factors may be questioned, however. In shallow lakes with long residence times (meaning a high potential for denitrification and algal uptake of N), N retention may be significant (> 50%) for lakes and reservoirs (Howarth et al.,1996, their Figure 7). To the extent that N retention is governed by algal uptake, productivity and the degree of P limitation of primary production may be major determinants of N export from lakes and reservoirs. As an example, Berge et al. (1997) found a mean N retention of 28% for a eutrophic lake with a renewal time of no more than 2.5 days, but less than 5% in an oligotrophic lake with a renewal time of 2.6 years. This clearly emphasizes that general model predictions of N retention based simply on hydrology may not be valid, owing to the fact that few lakes are strictly N limited and that denitrification losses may show tremendous variability among localities.

While phytoplankton uptake of N and subsequent burial in sediments may be a major sink for N in lakes (Ahlgren *et al.*, 1994), denitrification is the second major loss route (Seitzinger, 1988; Ahlgren *et al.*, 1994) and is probably the dominant loss route for N in wetlands. Since denitrification presupposes some minimum levels of phosphorus and organic carbon and P as well as (almost) anaerobic conditions, it normally occurs at the sediment surface. In general, denitrification losses are quite restricted in oxygenated, running waters, but can be strongly enhanced by additions of P or organic C (Faafeng and Roseth, 1993). Jansson *et al.* (1994a) found that no more than 1–2% of total N flux was removed by denitrification in a stream channel draining an agricutural area. For smaller rivers and brooks that are highly P deficient, an increased N uptake by benthic biota may be faciliated by additions of P (Hessen *et al.*, 1997). Such additions may also induce strong diurnal cycling of N with minima during daytime, again underlining the potential uptake of N by primary production once P is provided. For larger rivers, there appears to be only negligible N retention with a possible exception for slow-flowings systems rich in organic matter.

5. Seasonal Runoff Patterns

For the marine recipient, the seasonal runoff patterns as well as the more short-term dynamics may be more relevant than the total annual load. For most catchments, hydrology will be a main determinant to the flow of N and, typically, periods with high precipitation and floods may contribute a major output of nutrients within a short period of time. For northern and alpine areas, snow cover and spring melt normally cause fairly predictable annual dynamics with a pronounced spring flow of N (and other nutrients) associated with snow melt, fuelling the estuarine spring bloom. A second period of N export takes place in autumn, commonly with high levels of precipitation and waterlogged soils, causing episodes of pronounced surface runoff. Agriculturally dominated catchment may have far more dynamic patterns relative to more "pristine" catchments, however. Extreme outputs of N from agricultural land may be caused by crop failure in combination with hydrological events, especially heavy rainfall succeeding periods of drought (Figure 5). Flooding of arable land may also, at irregular intervals, provide major outputs of N.

The presence or absence of large lakes within the catchment will profoundly affect annual patterns. As already stated, lakes or reservoirs with renewal times of months to years could almost completely mask short-term fluctuations and annual patterns from the upstream catchment. Annual

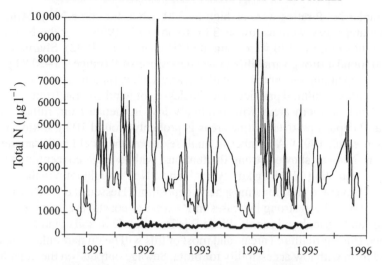

Fig. 5. Annual variability in total N concentrations (thick line) in an agriculturally dominated watershed of southern Norway with annual inputs of fertilizer (animal manure + mineral fertilizers) of 5000–6000 kg N year^{-1} and atmospheric N deposition of 600–1500 kg N year^{-1}. For comparison, values are shown for export from a heathland-dominated watershed of southwestern Norway with atmospheric N deposition ranging from 1500 to 2300 kg N year^{-1}, but with minor agricultural activities. From Hessen *et al.* (1997a).

patterns will also be markedly influenced by climatic factors (see below). Prevailing mild winters may completely shift annual runoff patterns, and moist, mild summers may strongly promote mineralization of organic N stores (Hessen and Wright, 1993; Høyaas *et al.*, 1997).

IV. ELEMENTAL RATIOS

1. N : P Ratios

Stoichiometry of estuarine loads will be a strong determinant of the entire ecosystem, and elemental ratios are strongly dependent on the different sources. For both atmospheric and riverine loadings, there have been systematic trends towards increased N : P ratios. Since P is transported in the atmosphere associated with particles only, atmospheric transport and deposition of P is minor relative to N. Depositions may nevertheless have increased as a result of increased use of fertilizers, and thus increased P associated with particles released from arable land. Data on atmospheric deposition of P are few, and most estimates point to a consistently high N : P ratio in atmospheric deposition, i.e. indicating low contributions of atmospheric P relative to N. For the Baltic

Sea, molar N : P ratios were as high as 340, while for some North American coastal areas they were as low as 33 (Nixon et al., 1996). For the North Sea, N : P ratios close to 130 were estimated (Hessen et al., 1992). Sharpley et al. (1995) found a strong variability in wet deposition of P (range 4.1–20.8 kg km^{-2} year^{-1}), depending on annual rainfall. Clearly such variations may also be associated with agricultural practices and the degree of wind erosion from cultivated land. Over northern areas with relatively low influence from mineral dust, annual P deposition over marine areas is probably around 10 kg P km^{-2} year^{-1} (Hessen et al., 1992). A number of analyses from open-field bulk samplers in southern Norway showed strong fluctuations, with average concentrations of 4 µg P l^{-1}, (equalling 11.2 P kg km^{-2} year^{-1}) and a N : P weight ratio in the order of 130–200 (average 160) for atmospheric deposition (Hessen et al., 1997a).

There have been strong increases in P export to coastal areas in the form of sewage and agricultural runoff. Since P is less mobile in soils than N, less P is leached from agricultural areas, and most of this will be on particulate mineral-bound form with low accessibility for biota. Sewage outputs, on the other hand, have low N : P ratios and a dominance of bioavailable P. The flux of P from catchments will, even more than for N fluxes, depend on catchment properties and land use. Since loadings of N and P do not necessarily co-vary, tremendous oscillations in N : P ratios may be found, especially for disturbed watersheds.

The pronounced variability in N : P ratios from various major river systems was reviewed by Howarth et al. (1996), showing a range from 4.8 to 271 (atomic ratio), but with most temperate regions clustered around a N : P atomic ratio of 20–30, or 9–14 by weight. Billen et al. (1991) examined rivers with different levels of disturbance in their catchments, and found that N : P ratios of forested watersheds in general exceeded 200 (atomic ratio), those from agricultural areas ranged from 30 : 1 to 300 : 1, while urbanized watersheds had ratios between 5 : 1 and 40 : 1.

Export of P to estuaries also shows a tremendous variability between watersheds. For the main rivers systems draining to the North Atlantic basin, Howarth et al. (1996) estimated fluxes ranging from 4.5 to 236 kg P km^{-2} year^{-1}, with the highest flux from the Amazonian basin.

Similar to the losses for N, the area-specific losses of P show a strong variability with land use and various anthropogenic loadings. A US survey of P export from 928 non-point source watersheds (Omernik 1977) gave losses typically ranging from 10 kg P km^{-2} year^{-1} for forested catchments, increasing with decreased forest cover, to losses approaching 30 kg P km^{-2} year^{-1} for agricultural watersheds. Clearly P losses from arable land may greatly exceed these values, and will also depend strongly on fertilizer application, land use and crop types (Sharpley et al., 1995). The concentrations of organic P in soils, and the transformation of this pool may, analogous to N, be a major determinant to area-specific runoff of P (Vaithiyanathan and Correll, 1992).

From the regional set of catchment data pooled into 16 different regions, Tjomsland and Braathen (1996) estimated total fluxes ranging from 8.4 to 909 kg P km^{-2} year^{-1}, while "background" (i.e. non-anthropogenic) fluxes ranged from 3.7 to 47 kg P km^{-2} year^{-1}. On average, anthropogenic P contributed 87% of total P runoff. The frequent decoupling of riverine fluxes of N and P may promote strong variability among catchments and strong seasonal fluctuations in N : P ratio. For the Norwegian regions, N : P atomic ratios of the riverine efflux ranged from 18 to 165 (average 40) for total export and from 63 to 227 (average 128) for background export. Thus, in general, anthropogenic influence yielded lower N : P ratios. The observation of Billen *et al.* (1991) that N : P ratios tend to decrease as total nutrient load increases in response to disturbance, fits in general for Norwegian rivers on a local scale. The lowered N : P ratios for the northernmost watersheds, which also are the most pristine, do, however, underline the potential importance of atmospheric load for riverine N : P ratios.

Data from the Norwegian PARCOM programme again can be used for illustrating regional differences in stoichiometry in riverine loadings. There was a clear geographical pattern in stoichiometry of riverine runoff along the coast. The NO_3^- : PO_4^{2-} ratio shows a pronounced peak in south-western rivers with a median weight ratio above 200 declining to less than 50 in the eastern and northern rivers. The same pattern is found for total N : total P (Figure 6).

A word of caution is needed when it comes to the biological relevance of total export of N and P, as well as the role of elemental ratios. At least for boreal watersheds, a major part of the N flux may be in organic form, so that it will not be directly accessible for estuarine primary production. The potential role of this (often dominant) pool of N will depend primarily on mixing processes and and photo-oxidation, and is still a matter of controversy (see below). This proviso is even more pronounced for P, where a major part of P from agricultural lands, as well as from mountainous and forested areas, may be dominated by mineral-bound P that will rapidly enter sediments of estuaries without participating in biological cycling. Thus the biologically relevant fractions of the total may often be minor, implying also that the elemental ratios of total pools of N and P could be quite irrelevant from a biological point of view. Typically, some 50% or more of total P export from watersheds is in particulate form (Sharpley *et al.*, 1995; Sharpley and Rekolainen, 1997), and this share may exceed 90% under periods of high erosion of arable land. Most often, the bioavailable fraction of the particulate fraction is low. Typically, only 10–40% of total particulate P may be utilized by primary producers (Sharpley *et al.*, 1995). Bioassays (growth of freshwater algae) revealed a wide range of bioavailability of P from various sources as related to PO_4^{2-} (Berge and Källqvist, 1988). While less than 10% of total P from glacial erosion was bioavailable, 35% of P from soil erosion of arable land was available for algal growth. This increased to almost 60% bioavailability for P from flooded land after spread of livestock manure. Such

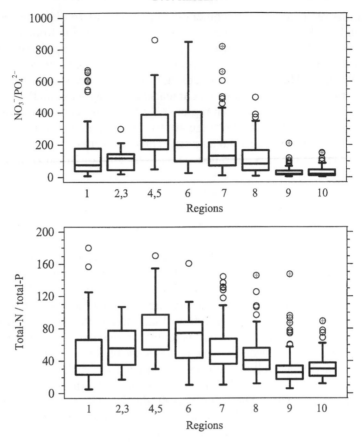

Fig. 6. Box-and-whisker displays of NO_3^- : PO_4^{2-} ratios and total N : total P ratios (by weight) for rivers in major regions along the Norwegian coastline. Values are medians with quartiles in boxes, 90% confidence limits as bars. Observations outside 90% confidence limits appear as open circles. From Hessen *et al.* (1997a).

bioassays will always indicate maximum yields due to continuous mixing and long contact times in culture vessels. In nature, the frequent rapid sedimentation of particulate P out of the euphotic zone will invariably yield lower fractions of bioavailable P. A major determinant to bioavailable P from watersheds may thus be losses of soluble, reactive P from plant leachate, a factor that again will depend strongly on hydrology, crop harvest and crop type (Sharpley *et al.*, 1995). These pronounced variabilities in the fractions of particulate organic or inorganic N and P will, of course, have strong bearings on the ratios of bioavailable N : P. Hessen *et al.* (1997a) found that average ratios of NO_3^- : total N ranged from 0.2 to 0.7 for a number of watersheds, with most rivers giving a ratio above 0.6. In contrast, PO_4^{2-} : total P ranged from 0.2 to 0.4, with most rivers clustered around 0.2. This again would mean that ratios of total N : P could differ markedly from the biologically more relevant PO_4^{2-} : $[NO_3^- + NH_4^+]$ ratios.

2. N : Si Ratios

Compared with N and P, silica (Si) is only modestly influenced by anthropogenic activities. Since Si is primarily a product of weathering, factors such as anthropogenic acidification could possibly change fluxes of Si, but there are no consistent data on this. There are, however, strong indications that eutrophication of lakes over the past few decades may *indirectly* have caused significant depletions of dissolved Si (Schelske and Stoermer, 1971; Schelske, 1988; Conley *et al.*, 1993). The reason for this is that increased inputs of P caused increased production and biomass of phytoplankton, including diatoms, which subsequently depleted the reservoirs of dissolved Si within lakes. Correlations between dissolved P concentrations and the rate of Si depletion during spring bloom of lakes have also been demonstrated (Talling and Heaney, 1988). Similarly, the use of P fertilizers in the catchment correlated strongly with declines in dissolved Si in the Mississippi River (Turner and Rabalais, 1991). The construction of large, human-made reservoirs has also been linked to declines in Si efflux from rivers (Conley *et al.*, 1993), since such reservoirs may act as efficient Si traps owing to uptake by planktonic diatoms. Thus, eutrophication, and to some extent dam constructions, may reduce Si concentrations and thus output of Si to marine systems. In particular, increased loadings of N may strongly skew the N : Si ratios to estuaries, with severe impact on primary productivity and algal community composition (i.e. reduced competitive abilities for diatoms).

Although Si outputs to freshwater systems are governed mainly by slow weathering rates, area-specific efflux may nevertheless vary due to catchment properties. Again referring to the south Norwegian watersheds, efflux of SiO_2 ranged from 502 to 5543 (average 2993) kg SiO_2 km^{-2} year^{-1}. The reason for the strong, area-specific difference in Si runoff from various catchments is not clearcut, but may partly be linked to the geological properties and soil texture of the catchments. Availability and concentrations of P, as well as presence or absence of lakes in the watersheds, could add to this. With the exception of two (heavily regulated) watersheds, a strong correlation was obtained between Si export and P export (Figure 7). By omitting the two outlier watersheds, the regression of catchment export of Si versus P could be described as: $SiO_2 = 5020.61 - 31.60 P$, $r^2 = 0.67$ (where both SiO_2 and P are expressed as kg km^{-2} year^{-1}). An increase in the N : Si ratio could also be seen with increasing total P, from 0.2 to approximately 0.8 (by weight), but again with the same two outliers. Again, the causality for the apparent negative correlation between Si and P is not clearcut. While higher levels of P certainly could induce increased uptake of P in the lakes of these watersheds (which are all strictly P limited),

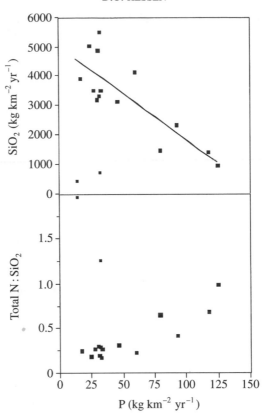

Fig. 7. Annual area-specific fluxes of SiO_2 and the area-specific ratio of total N : SiO_2 related to annual P export for 16 major watersheds of Norway. Two outliers are rejected (see text).

most of the P originated from settlement in the lower part of the watersheds where an eventually increased Si uptake would be associated with epibiotic or epilithic diatoms. The capacity for such uptake should probably be small in major rivers. Nevertheless, data from a highly P-deficient watershed of southern Norway show close co-variation between N and Si, most pronounced in major rivers (Figure 8), less so in the small brooks where there was frequently a delay of some days between minima in N and Si, often yielding poor correlations. These variations were not associated with hydrological patterns, and could indicate simultaneous uptake of Si and N in biota.

Previous analysis of a larger data set covering a larger gradient of N precipitation gave indications of skewed elemental ratios, with excess N relative to P and Si in regions receiving high atmospheric inputs of N (Hessen *et al.*, 1997a). It should be stressed, however, that these patterns give no proof of

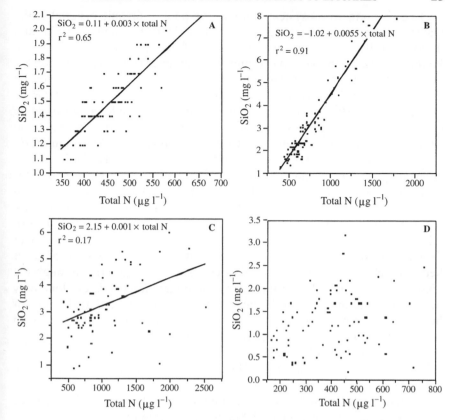

Fig. 8. Examples of regressions of SiO_2 concentrations on total N concentrations from different monitoring stations of the Bjerkreim catchment, Southern Norway. Data for all subcatchments cover three annual cycles: A, main outlet; B, major river from subcatchment; C, forested catchment; D, heathland catchment.

causality. Populations patterns, vegetation, land use and hydrology could also contribute to these variations in elemental ratios.

V. THE COUPLING OF NITROGEN AND ORGANIC CARBON

1. The Flux and Fate of Organic C

Northern ecosystems typically have large stores of C and N in the soil. This decreases southward, to the opposite extreme in tropical soils which are almost devoid of such organic stores. Thus southern and northern ecosystems will behave essentially differently with regard to C and N cycling. In northern ecosystems, the largest pool of organic material in the water column is dissolved organic material (DOM) of humic origin, and dissolved fulvic acid is the largest fraction of the humic matter (40–60 %). Pettersson *et al.* (1997) estimated humic compounds to account for approximately 80% of the organic matter in 24 Swedish and Finnish rivers entering the Gulf of Bothnia. In such northern boreal catchments, dynamics of humus DOM is also instrumental to catchment export of organic N and to some extent of inorganic N. Potentially limiting metals such as iron (Fe) may also to a large extent be associated with fluxes of organic C (Pettersson *et al.*, 1997). The high amounts of dissolved organic carbon (DOC) in humus DOM may also be a major determinant of heterotrophic production in estuarine ecosystems. The microbial degradation of DOC, mainly dissolved humic substances, can be an important process in carbon cycling, directly influencing ecosystem dynamics (Hessen *et al.*, 1990; Wetzel, 1995). Although humus C, which is normally by far the greatest component of DOC, is considered recalcitrant, it is nevertheless a major source of C for heterotrophic bacteria in humic lakes (Hessen *et al.*, 1990). Although the conventional microbial loop fuelled by phytoplankton exudates alone may be a minor link of carbon to higher trophic levels, substantial additions of humus DOC may strongly enhance the role of bacteria in the food web energy flux. One important pathway for microbial utilization of humic carbon is photolysis of the humic material, releasing smaller more labile substrates as well as nutrients (Wetzel *et al.*, 1995). Dissolved humic substances can influence chemical and physical characteristics of the ecosystem, being important in light attenuation and complexation of trace metals (see Hessen and Tranvik, 1998, for an overview).

Some general patterns of DOC concentration across climatic zones have been reported (Meybeck, 1982), although there is a strong within-zone variation due to different altitude, soil and vegetation cover and hydrology. For rivers, Meybeck (1982) concluded that concentrations of DOC were greatest in taiga rivers followed by rivers in the wet tropics, temperate and tundra zones respectively. Boreal catchments show strong gradients in their export of DOC. Heathland and alpine areas may have DOC concentrations < 1 mg C l^{-1}, while catchments dominated by wetlands, bogs and coniferous forests may yield more than 50 mg C l^{-1}. The single largest variable affecting the yield of DOC from most catchments is the proportion of the catchment that is wetland (Urban *et al.*, 1989; Dillon and Molot, 1997). Precipitation has a diluting effect on DOC export. This can also be seen as a regional gradient with generally increasing DOC concentration in freshwater lakes in Canada from humid south-central Ontario (1–4 mg C l^{-1}; Dillon and Molot, 1997), increasing to the west through north-western Ontario (2–12 mg C l^{-1}) to the semiarid western great plains of Alberta (e.g. 20–40 mg C l^{-1}; Curtis and Adams, 1995). It is also consistent with observed decreases in DOC in response to increased precipitation (Tate and Meyer, 1983), and increases in response to drought (Schindler et al., 1992).

There is a strong seasonality in the export of DOC, with typically high values during spring flood, summer and winter minima, and a maximum in late autumn owing to frequent rainfalls and fresh DOC originating from decaying litter (Hessen et al., 1997c; Schiff et al., 1997). Hydrology is a major determinant to DOC dynamics, and hydrological changes invariably cause large changes in DOC concentrations and export (see below). The quality (and thus the biological impact) of the DOC will be determined by catchment properties, and thus the age and source of DOC. While wetlands export mainly recently fixed C (determined from the $^{13}C : {}^{14}C$ ratio) with little seasonal variation, forested areas may export a larger fraction of old and recalcitrant DOC, and show more pronounced seasonal fluctuations (Schiff et al., 1997).

While there is extreme variability in concentrations of DOC among and within catchments, the annual yield appears to be less variable. DOC yield measurements and empirical analyses suggest a range of $1-8$ g C m^{-2} year^{-1} (Dillon and Molot, 1997; Schiff et al., 1997), but with an average yield from terrestrial catchments between 3 and 4.5 g C m^{-2} year^{-1} (Moeller et al., 1979; Rasmussen et al., 1989; Schindler et al., 1992).

After entering the marine environment, some part of the DOC is transformed to particulate matter by flocculation, entering the sediments without contributing to the planktonic food web. Some is photo-oxidized, and an unknown, and probably highly variable, portion enters the food web via heterotrophic bacteria and heterotrophic flagellates. The fate of DOC in estuaries will depend largely on the season and the quality of DOC. While the more consistent supply of old, recalcitrant DOC is poorly utilized by bacteria, the more dynamic, bioavailable pool of "fresh" DOC could to a great extent fuel the microbial loop. It will also control light attenuation, particularly in the ultraviolet (UV) region, and it will to a great extent regulate the bioavailability of metals as nutrients or toxicants by complexation of trace metals by fulvic and humic acids. While these ecological and biogeochemical aspects of DOC have been well studied in freshwater systems (cf. Hessen and Tranvik, 1998), the role of DOC, and particularly allochthonous, humus DOC, is poorly understood in estuarine ecosystems. Nevertheless, the basic properties should not differ markedly between freshwater and low-salinity marine systems. While estuaries and coastal areas with high influx from rivers (like the Baltic) are certainly are influenced by riverine C, the terrestrial contribution of organic C to the open ocean accounts for only about 1% of the total input (Benner, 1998). Stable C isotope ratios of DOC ($\delta^{13}C$) in the ocean are typically in the range of -20 to $-22‰$, close to that found in marine plankton, and considerably different to that of riverine organic C (-27 to $-29‰$) (Hedges et al., 1994; Benner, 1998). This clearly suggests a minor signal from terrestrial C.

2. Linkages of C and N in Aquatic Humus

For lakes and rivers in boreal areas, there is often a strong correlation between organic N and organic C, both being governed by the humus content. Non-anthropogenic total flux of nitrogen to estuaries in temperate areas is often correlated with the flux of organic C, since a major part of N is in organic form. This is most pronounced in coniferous catchments with high total organic carbon (TOC), less so in catchments with low TOC (Figure 9). Particularly in summer, when inorganic N is depleted due to uptake in terrestrial and freshwater vegetation and bacteria, the nitrogen pool is composed chiefly of humus-bound organic N. Humus is normally rather poor in nitrogen, with typical values of 2–6% in humic acids and less than 3% in fulvic acids (Schnitzer, 1985). The molar carbon : nitrogen (C : N) ratio is 18–30 : 1 for humic acids and 45–55 : 1 for fulvic acids (Thurman, 1985). For 451 Norwegian lakes molar C : N ratios ranged from 11 to 20 over a wide span of humic content, with no systematic difference between forest, heathlands or mountainous areas (Wright et al., 1997).

The proportion of aromatic compounds will be a major determinant to C : N ratios of humic DOC. An increased share of lignin derivatives would yield increased C : N ratios, as lignin contains many aromatic rings and no nitrogen. It can be reasoned that the presence of lignin within the precursor pool of organic material would serve as a source of aromatic moieties and as a diluent with respect to nitrogen. Regressing the C : N ratios against the aromatic to aliphatic ratio of water samples yielded a strong positive correlation, and the samples from streams and lakes receiving DOM from a surrounding forested watershed were found to have a higher aromatic carbon content and lower nitrogen content than those of other watersheds (McKnight and Aiken, 1998).

Comparison of 20 small catchments clearly demonstrated that the C : N ratio in runoff decreased with increasing N deposition, and within-catchment studies also indicated a generally lower C : N ratio of heathland areas relative to forested areas (Kaste et al., 1997). This was attributed to increased leaching of inorganic N with increased deposition, since organic N in runoff was largely unaffected by total N deposition (Wright et al., 1997). Since a major portion of inorganic N deposition is actively exchanged with the organic N in soils and biota, the key to the understanding of N dynamics and N runoff lies in the mineralization processes and the fate of the organic N pool. Owing to the intimate interplay between organic N and organic C, the factors governing the runoff of DOC will also be a major determinant to N runoff.

While inorganic N is rapidly allocated to phytoplankton biomass in sea water, the fate of the large, and frequently dominating, pool of organic N is

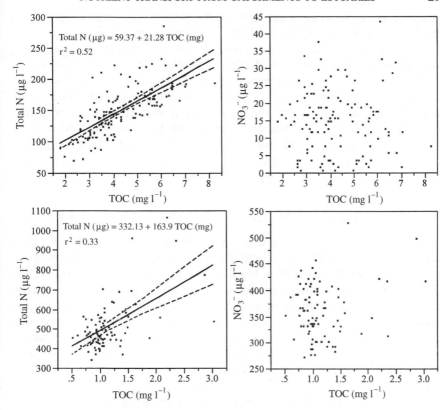

Fig. 9. Correlation between concentrations of TOC and total N (left panels) and TOC and NO_3^- (right panels). Upper panel: region with low deposition of N and moderate to high TOC (River Øyeså, central Norway, 1988–1997). Lower panel: region with low TOC but high atmospheric deposition of N (River Bjerkreim, southern Norway, 1988–1993).

more obscure. Part of the humus-bound N may be adsorbed or chelated to humic aggregates, while there is also some N present in the recalcitrant macromolecules. Amino acids, peptides and remains of nucleic acids make up between 30% and 50% of humus N (Khan and Sowden, 1972; Schnitzer, 1985); the rest is unknown organic residues. The extent to which this large pool of organic N is made available to estuarine phytoplankton could be a major determinant to productivity. From bioassays in the Baltic Sea, Carlsson and Granéli (1993) concluded that a considerable fraction of this organic N could be bioavailable. The Baltic Sea also shows a strong gradient from predominant P limitation of the northern part receiving high inputs of riverine, humic substances, to a predominant N limitation in the southern, more oceanic, areas (Tamminen and Kivi, 1996). Algal assays with organic N in freshwater demonstrated that even NO_3^- concentrations as low as 10 µg NO_3^- l^{-1} were sufficient to depress utilization of organic N. In the absence of NO_3 (i.e. below detection limits), algal cell yield indicated that 5–8% of organic N could be utilized for algal growth (Hessen and Källqvist, 1994). In nature, however, a considerably greater proportion could be bioavailable. Bacteria are able to utilize humus carbon, and they certainly degrade humic compounds (Hessen et al., 1990). Solar radiation, particularly UV radiation, oxidizes both humus C and N, and would promote the breakdown and enhance the availability of recalcitrant C and N compounds (Cooper et al., 1989; Hessen and van Donk, 1994; Miller and Zepp, 1995; Moran and Zepp, 1997).

VI. ANTHROPOGENIC EFFECTS ON NUTRIENT FLUXES

There are a number of reports demonstrating increased concentrations of N and P in riverine runoff over the past decades, owing to various anthropogenic activities (Howarth et al., 1996). For Si and C, there are fewer consistent data, yet for some areas the flux of organic C has decreased as a result of long-term drought and increased temperature (Schindler et al., 1996). Effects of climate changes and long-term increases in N deposition will be major determinants to the N cycle in the long run. There are, however, a number of direct land-use effects that may modify and (most often) increase nutrient export from catchments to estuaries, the most prominent of these being agricultural practices and forestry.

With the exception of direct input of animal or human sewage in densely populated watersheds, fertilizer application is the single largest input of N to most non-pristine watersheds. Only some 50% of applied N fertilizer is removed by crop harvest (Nelson, 1985; Høyaas et al., 1997). The rest is stored as organic N, denitrified, or enters groundwater or surface runoff. The ratio of applied fertilizer to plant uptake may show strong variation related to climatic variables and crop yield. Significant correlations have been shown between total N runoff and crop yield, as well as between N runoff and soil mineral content of N (Vagstad et al., 1997), and during years with crop failure severe losses of N from fertilizers to water courses are commonly reported (Høyaas et al., 1997; Vagstad et al., 1997).

Some correlation between fertilizer input and N concentrations in water leached from agricultural land is commonly observed, yet this relationship depends strongly on soil texture and type of vegetation (see Howarth et al., 1996, for a review). Yet experimental fertilization often fails to demonstrate strong correlations between the amount of N added by applied fertilizers and N runoff, indicating that mineralization of organic N may exceed fertilizer inputs as a determinant for N export even for arable lands (MacDonald et al., 1989; Vagstad et al., 1997). Northern and temperate soils have large stores of organic N (typically 2–7 tonnes ha⁻¹) for boreal, coniferous forests, and even larger for agricultural lands (Hessen and Wright, 1994; Vagstad *et al.*, 1997), and the fate and turnover of organic N may be instrumental to total N flux from catchments. The role of fertilizers versus mineralization of organic N stores for annual runoff is not settled, yet the strong increase in riverine N fluxes related to increased fertilizer inputs (cf. Howarth *et al.*, 1996) is clearly indicative of the role of fertilizers for increased N runoff in the long term. Agricultural activities will also be a strong determinant to gaseous outputs. First, increased levels of denitrified N from soils may entail feedbacks to the entire N cycle. Second, there may be pronounced gaseous emissions of ammonia from volatilized manure. As stated previously, such ammonia emissions equal emissions of oxidized N over most of Europe.

Other types of change in land use certainly also affect fluxes of nutrients from catchment to estuaries. Drainage and forestry both affect total nutrient fluxes as well as seasonal runoff patterns, but again probably with most notable effects for N. In most cases, forest stands are capable of withholding almost all the mineralized N during growing seasons, while clearcutting and deforestation always yield increased leakage of N (Likens *et al.*, 1977). Recovery and increased N retention may be a slow process (Likens *et al.*, 1978; Mulder *et al.*, 1997), as illustrated by studies of N export from various forest stands of southern Norway where both younger and old forest stands had almost complete retention of N during the growing season. Clearcutting of half of a forested subcatchment took place in 1990. In 1992–1993, N loss was 1450 kg N km⁻² year⁻¹ for this catchment (no data were available for 1991). For the next 3 years, the loss was gradually reduced to 1070, 580 and 450 kg N km⁻² year⁻¹ respectively (Høyaas *et al.*, 1997), and the NO_3^- concentration in soil solution (lysimeters) decreased from 15 mg l⁻¹ to virtually zero over the same period (Mulder *et al.*, 1997).

As for N, losses of P have clearly also increased from most watersheds over recent decades, for a number of reasons. Area-specific losses were found to be approximately 10, 20 and 45 kg P km⁻² year⁻¹ for forested and low and medium agricultural activity catchments respectively during 1965–1974, while during 1981–1985 the corresponding losses had increased to approximately 12, 30 and almost 100 kg P km⁻² year⁻¹ respectively (Sharpley and Rekolainen, 1997). These increases can be attributed to changes in fertilizer application, soil drainage and crops. Changes in forestry will also have impact on P export. Following clearcutting and herbicide treatment, specific losses from a forested system increased 10-fold from 2 to 20 kg P km⁻² year⁻¹ (Hobbie and Likens, 1973). However, since agricultural activities are by far the dominant non-point source of both N and P for most watersheds, anthropogenic effects on other parts of the catchments will be less important in the context of estuarine loadings.

VII. CLIMATIC EFFECTS ON NUTRIENT FLUXES

Short-term, or long-term, climatic changes will profoundly affect annual patterns of nutrient export to estuaries, as well as total loading and nutrient stoichiometry (Meybeck, 1993). Again using boreal areas as an example, the role of snow cover is pronounced. In mild winters with reduced or no snow cover, most of the annual export of N occurs during late winter, while during cold and "normal" winters, more of the N export to marine areas takes place during snow-melt in spring (Figure 10). This pattern was seen for all rivers monitored in southern Norway for these years. Thus prevailing mild winters in northern ecosystems would store less N for periods of high marine productivity in spring. More systematic shifts in climate may also have a strong effect on nitrogen budgets of catchments by increased soil mineralization. Typical northern, boreal soils contains 2000–7000 tonnes N km^{-2}, while soils further south commonly have far less. This would imply a potential for increased runoff from northern soils due to increased mineralization rates that could far exceed the present export of N to estuaries. By use of models for soil mineralization of N, using sulfur and N deposition scenarios, and various degrees of N saturation and climatic variables (precipitation and temperature) as dynamic variables, a marked increase in mineralization and watershed

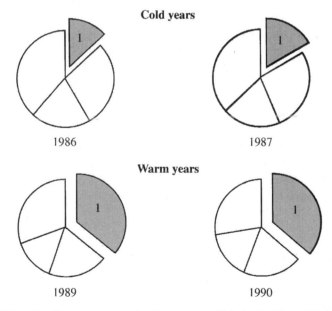

Fig. 10. Climatic effects on quarterly nitrogen runoff from the River Bjerkreimselva, southern Norway for 2 years with cold (normal) winters with snow cover (upper), and for warmer years with reduced or lacking snow cover (lower). The shaded sections represent January to March. From Hessen *et al.* (1997a).

export of inorganic N was predicted for a Norwegian watershed (Hessen and Wright, 1993). These predictions were supported by large-scale experimental manipulations with raised CO_2 and temperature (Wright, 1998). An entire "mini-catchment" (860 m^2), including trees and natural vegetation, was enclosed within a transparent greenhouse. The uppermost 20% of the catchment was partitioned off, serving an as untreated reference. Climate in the greenhouse was manipulated to 560 ppmv CO_2 and temperature was increased 3–5°C above ambient. Precipitation and N deposition were controlled by a sprinkler system, and were identical for both the control and the treated catchment. While the entire catchment had a consistent positive net retention of NO_3^- and NH_4^+ during the pre-treatment years 1985–1994, the treated catchment yielded a strong negative net retention (net loss) of NO_3^- and NH_4^+ during treatment in 1995–1997.

Climate has also effects on concentrations and runoff of DOC, and thus also on organic N. Long-term monitoring at the Experimental Lakes Area, north-western Ontario, provided data on reference lakes over a period of 20 years, coincident with increased drought and a warming of 2°C (Schindler *et al.*, 1992, 1996, 1997). Following these climatic changes, concentrations of DOC in stream runoff increased because the flux of water decreased more than the flux of DOC. The concentration of DOC in lakes decreased, however, even though concentrations in runoff increased. This effect was attributed to increased water residence times in lakes, allowing more DOC to enter sediments and more DOC to be photo-oxidized and degraded by biological processes.

The sensitivity of aquatic ecosystems to climatic perturbation of DOM concentration or quality clearly depends on the complex interaction between climate, surficial geology and topography (cf. Curtis, 1998). Nevertheless, the sensitivity to climatic perturbation can be estimated from loading and flushing rate relationships. In humid regions, export of DOC to marine recipients will likely decrease with climatic warming. Although decreased precipitation could yield increased concentrations, this will be obscured by decreased water flow and increased retention in catchments and lakes. In arid and semiarid regions, climatic warming will probably increase the concentration of DOC of lakes and rivers by evapoconcentration, but again the total export to marine ecosystems will probably decrease. A decrease in the C : N ratio of dissolved inorganic matter has also been reported during climatic warming (Curtis, 1998), for which there was no clear mechanism.

Humus DOC is a major agent for blocking short-wave radiation (Scully and Lean, 1994), and for lakes the loss of DOC in response to climatic warming has been linked to enhanced penetration of underwater UV radiation (Schindler *et al.*, 1997). The same would hold for estuaries receiving less DOC. UV radiation is harmful to all kinds of aquatic biota, and changes in levels of coloured (humic) DOC by climate warming will probably far exceed the effects of any stratospheric ozone decrease. Moreover, while ozone layer depletion has a strong seasonality, with minima in late winter and early spring, the effects of decreased DOC will be pronounced over the entire summer season.

For similar reasons as for C, the export of P, N and Si to estuaries will also decrease with warming, increased evaporation and reduced precipitation. This will be most pronounced for watersheds with lakes or reservoirs, where increased retention times will cause increased retention of all elements. These effects will be most pronounced for changes in water bodies with short renewal times (< 1 year), since the retention equations have the form of asymptotic saturation curves approaching maximum retention for renewal times > 1–3 years.

Finally, there are tight couplings between ambient CO_2 and N uptake of terrestrial vegetation. Increased deposition of N may sequester more CO_2 and raised CO_2 may allow for increased uptake of N (Schindler and Bayley, 1993). Model predictions of increased CO_2 indicated an increase in terrestrial net primary productivity and an increase in the vegetation C : N ratio (Rastetter et al., 1997). Long-term predictions included increased movement from soil to vegetation and increased storage of N.

ACKNOWLEDGEMENTS

The author is indebted to colleagues in the project "Nitrogen from Mountains to Fjords" for co-operation and collection of data presented in this paper, and in particular to Arne Henriksen, Gjertrud Holtan, Brit-Lisa Skjelkvåle and Richard F. Wright for discussions and access to data.

REFERENCES

Aber, J.D., Magill, A., McNulty, S.G., Boone, R.D., Nadelhoffer, K.J., Downs, M. and Hallett, R. (1995). Forest biogeochemistry and primary production altered by nitrogen saturation. *Water Air Soil Pollut.* **85**, 1665–1670.

Aber, J.D., Nadelhoffer, K.J., Steudler, P. and Melillo, J.M (1989). Nitrogen saturation in northern forest ecosystems. *Bioscience* **39**, 378–386.

Ahlgren, I., Sörensson, F., Waara, T. and Vrede, K. (1994). Nitrogen budgets in relation to microbial transformation in lakes. *Ambio* **23**, 367–377.

Benner, R.H. (1998). Cycling of dissolved organic matter in the ocean. In: *Aquatic Humic Substances; Ecology and Biogeochemistry* (Ed. by D.O. Hessen and L. Tranvik), pp. 317–332. Springer, Heidelberg.

Berge, D. and Källqvist, T. (1988). *Bioavailability of Phosphorus in Agricultural Runoff, Compared with Other Sources of Pollution*. Report O-87079. *Norwegian Institute for Water Research, NIVA*. (in Norwegian).

Berge, D., Fjeld, E., Hindar, A. and Kaste, Ø. (1997). Nitrogen retention in two Norwegian watercourses of different trophic status. *Ambio* **26**, 282–288.

Billen, G. and Servais, P. (1991). Modélisation du transport de pollutants par l'estuarie de l'Escant, Cas du Phosphore. In: *Unité de gestiaton du Modèle Mathématique de la Mer du Nord et de L'Estuaire de l'Escaut, Ministère de la Santé Publique et de l'Environment,* Brussels.

Billen, G., Lancelot, C. and Maybeck, M. (1991). N, P, and Si retention along the aquatic continuum from land to ocean. In: *Ocean Margin Processes in Global Change* (Ed. by R.F.C. Matoura, J.M. Martin and R. Wollast), pp. 19–44. Wiley, Chichester, UK.

Bratli, J.L., Holtan, H. and Åstebøl, S.O. (1995). Environmental objectives for surface waters; input calculations. *SFT/ Norwegian Pollution Control Authorities, Report TA-1139/1995* (in Norwegian).

Brimblecombe, P. and Steadman, D.H. (1982). Historical evidence for a dramatic increase in the nitrate component of acid rain. *Nature* **298**, 460–461.

Carlsson, P. and Granéli, E. (1993). Availability of humic bound nitrogen for coastal phytoplankton. *Estuar. Coastal Shelf Sci.* **36**, 433–447.

Cole, J.J., Peierls, B.L., Caraco, N.F. and Pace, M.L. (1993). Nitrogen loading of rivers as a human-driven process. In: *Human as Components of Ecosystems: The Ecology of Subtle Human Effects and Populated areas* (Ed. by J.M. McDonnell and S.T.A. Pickett), pp. 141–157. Springer, New York.

Conley, D.J., Schelske, C.L. and Stoermer, E.F. (1993). Modification of the biogeochemical cycle of silica with eutrophication. *Mar. Ecol. Progr. Ser.* **101**, 179–192.

Cooper, W.J., Zika, R.G., Petasne, R.G. and Fisher, A.M. (1989) Sunlight-induced photochemistry of humic substances in natural waters: major reactive species. *Adv. Chem. Ser.* **219**, 333–362.

Cronan, C.S. and Crigal, D.F. (1995). Use of calcium/aluminium ratios as indicators of stress in forest ecosystems. *J. Envir. Qual.* **24**, 209–226.

Curtis, P.J. (1998). Climatic and hydrologic control of DOM concentrations and quality in lakes. In: *Aquatic Humic Substances; Ecology and Biogeochemistry* (Ed. by D.O. Hessen and L. Tranvik), pp. 93–106. Springer, Heidelberg.

Curtis, P.J. and Adams, H.E. (1995). Dissolved organic matter quantity and quality from freshwater and saline lakes in eastcentral Alberta (Canada). *Biogeochemistry* **30**, 59–76.

Dillon, P.J. and Molot, L.A. (1997). Dissolved organic and inorganic carbon mass balance in central Ontario lakes. *Biogeochemistry* **36**, 29–42.

Dise, N.B. and Wright, R.F. (1995). Nitrogen leaching from European forests in relation to nitrogen deposition. *Forest Ecol. Manag.* **71**, 153–161.

Dise, N.B. Matzner, E. and Gundersen, P. (1998). Synthesis of nitrogen pools and fluxes from European forest ecosystems. *Water Air Soil Poll.* **105**, 143–154.

Elser, J.J., Marzolf, E.R. and Goldman, C.R. (1990). Phosphorus and nitrogen limitation of phytoplankton growth in the freshwaters of North America: a review and critique of experimental enrichments. *Can. J. Fish. Aquat. Sci.* **47**, 1468–1477.

Faafeng, B. and Hessen, D.O. (1993). Nitrogen and phosphorus concentrations and N : P ratios in Norwegian lakes: perspectives on nutrient limitation. *Verh. Int. Verein. Limnol.* **25**, 465–469.

Faafeng, B. and Roseth, R. (1993). Retention of nitrogen in small streams artificially polluted with nitrate. *Hydrobiologia* **251**, 113–122.

Grennfelt, P. and Hultberg, H. (1986). Effects of nitrogen deposition on the acidification of terrestrial and aquatic ecosystems. *Water Air Soil Pollut.* **30**, 945–963.

Gundersen, P. (1992). Mass balance approaches for establishing critical loads for nitrogen in terrestrial ecosystems. In: *Critical Loads for Nitrogen – A Workshop*

Report (Ed. by P. Grennfelt and E. Thörnelöf), pp. 55–110. Environment Canada and Nordic Council of Ministers, NORD.

Gundersen, P. (1995). Nitrogen deposition and leaching in European forests — preliminary results from a data compilation. *Water Air Soil Pollut.* **85**, 1179–1184.

Gutschik, V.P. (1981). Evolved strategies in nitrogen aquisition by plants. *Am. Nat.* **118**, 607–637.

Hauhs, M., Rost-Siebert, K., Raben, G., Paces, T. and Vigerust, B. (1989). Summary of European data. In: *The Role of Nitrogen in the Acidification of Soils and Surface Waters* (Ed. by J.L. Malanchuk and J. Nilsson), Ch. 5. Nordic Council of Ministers, Miljørapport.

Hecky, R.E. and Kilham, P. (1988). Nutrient limitation of phytoplankton in freshwater and marine environments: a review of recent evidence on the effect of enrichments. *Limnol. Oceanogr.* **33**, 796–822.

Hedges, J.I., Cowie, G.L., Richey, J.E., Quay, P.D., Benner, R., Strom, M. and Forsberg, B.R. (1994). Origins and processing of organic matter in the Amazon River as indicated by carbohydrates and amino acids. *Limnol. Oceanogr.* **39**, 743–761.

Henriksen, A. and Brakke, D.F. (1988). Increasing contributions of nitrogen to the acidity of surface waters in Norway. *Water Air Soil Pollut.* **42**, 183–202.

Henriksen, A. and Hessen, D.O. (1997). Whole catchment studies on nitrogen cycling: nitrogen from mountains to fjords. *Ambio* **26**, 264–257.

Henriksen, A., Lien, L., Traaen, T., Sevaldrud, I.S. and Brakke, D.F. (1988). Lake acidification in Norway—present and predicted status. *Ambio* **17**, 259–266.

Hessen, D.O. and Källqvist, T. (1994). Bioavailability of humus bound organic nitrogen for freshwater algae. Nitrogen from mountains to fjords, *Newsletter* 1/94, **4–5**.

Hessen, D.O. and Tranvik, L. (Ed.) (1998). *Aquatic Humic Substances; Ecology and Biogeochemistry*. Springer, Heidelberg.

Hessen, D.O. and Van Donk, E. (1994). UV-radiation of humic water; effects on primary and secondary production. *Water Air Soil Pollut.* **74**, 1–14.

Hessen, D.O. and Wright, R.F. (1993). Climatic effects on freshwaters; model predictions on acidification, nutrient loading and eutrophication. In: *Impact of Climatic Change on Natural Ecosystems With Emphasis on Boreal and Arctic/alpine Areas* (Ed. by W. Oechel, J. Holten and G. Paulsen), pp. 154–167. NINA/DN report, Trondheim, Norway.

Hessen, D.O., Andersen, T. and Lyche, A. (1990). Carbon metabolism in a humic lake; pool sizes and cycling trough zooplankton. *Limnol. Oceanogr.* **34**, 84–99.

Hessen, D.O., Vadstein, O.and Magnusson, J. (1992). *Nitrogen to Marine Areas; On the Application of a Critical Load Concept*. Background document for workshop on critical loads, Lökeberg, Sweden, 1992. Environment Canada and Nordic Council of Ministers, NORD.

Hessen, D.O., Henriksen, A., Hindar, A, Mulder, J., Tørset, K. and Vagstad, N. (1997a). Human impacts on the nitrogen cycle; a global problem judged from a local perspective. *Ambio* **26**, 321–325.

Hessen, D.O., Henriksen, A. and Smelhus, A.M. (1997b). Seasonal fluctuations and diurnal oscillations in nitrate of a heathland brook. *Water Res.* **31**, 1813–1817.

Hessen, D.O., Gjessing, E.T., Knulst, J. and Fjeld, E. (1997c). DOC-fluctuations in a humic lake as related to catchment acidification, season and climate. *Biogeochemistry* **36**, 139–151.

Hobbie, J.E. and Likens, G.E. (1973). Output of phosphorus, organic carbon, and fine particulate carbon from Hubbard Brook watersheds. *Limnol. Oceanogr.* **18**, 734–742.

Holloway, J.M., Dahlgren, R.A., Hansen, B. and Casey, W.H. (1998). Contribution of bedrock nitrogen to high nitrate concentrations in stream water. *Nature* **395**, 785–788.

Holtan, G, Berge, D., Holtan, H. and Hopen, T. (1996). *Paris Convention. Annual Report on Direct and Riverine Inputs to Norwegian Coastal Waters During the Year 1995. A: Principles, Results and Discussions. B: Data Report.* SFT report 674/96. NIVA report O-90001.

Howarth, R.W. (1988). Nutrient limitation of net primary production in marine ecosystems. *Annu. Rev. Ecol. Syst.* **19**, 89–110.

Howarth, R.W., Billen, G., Swaney, D., Townsend, A., Jaworski, N., Lajtha, K., Downing, J.A., Elmgren, R., Caraco, N., Jordan, T, Berendse, F., Freney, J., Kudeyarov, V., Murdoch, P. and Zhao-Liang, Z. (1996). Regional nitrogen budgets and riverine N & P fluxes for the drainages to the North Atlantic Ocean: natural and human influences. *Biogeochemistry* **35**, 75–139.

Høyaas, T.R, Vagstad, N., Bechmann, M. and Eggestad, H.O. (1997). Nitrogen budget in the river Auli catchment, a catchment dominated by agriculture, in southeastern Norway. *Ambio* **26**, 289–295.

Jansson, M., Andersson, H., Berggren, H. and Leonardson, L. (1994a). Wetlands and lakes and nitrogen traps. *Ambio* **23**, 320–325.

Jansson, M, Leonardson, L. and Fejes, J. (1994b). Denitrification and nitrogen retention in a farmland stream in southern Sweden. *Ambio* **23**, 326–331.

Jassby, A.D., Goldman, C.R. and Reuter, J.E. (1995). Long-term change in Lake Tahoe (California–Nevada, USA) and its relation to atmospheric deposition of algal nutrients. *Arch. Hydrobiol.* **135**, 1–21.

Jaworski, N.A., Groffman, P.M., Keller, A.A. and Prager, J.C. (1992). A watershed nitrogen and phosphorus balance: the upper Potomac River basin. *Estuaries* **15**, 83–95.

Jaworski, N.A., Howarth, R.W. and Hetling, L.J. (1997). Atmospheric deposition of nitrogen oxides onto the landscape contributes to coastal eutrophication in the northeast US. *Environ. Sci. Technol.* **31**, 1995–2004

Jensen, J.P., Kristensen, P. and Jeppesen, E. (1990). Relationship between nitrogen loading and in-lake nitrogen concentrations in shallow, Danish lakes. *Verh. Int. Verein. Limnol.* **24**, 201–204.

Johannessen, T. and Dahl, E. (1996). Declines in oxygen concentrations along the Norwegian Skagerrak coast, 1927–1993: a signal of ecosystem changes due to eutrophication? *Limnol. Oceanogr.* **41**, 766–778.

Kaste, Ø., Henriksen, A. and Hindar, A. (1997). Retention of atmospherically-derived nitrogen in subcatchments of River Bjerkreim in southwestern Norway. *Ambio* **26**, 296–303.

Kelly, C.A., Rudd, J.W.M., Hesslein, R.H., Schindler D.W., Dillon P.J., Driscoll, C.T., Gherini, S.A. and Hecky R.E. (1987). Prediction of biological acid neutralization in acid-sensitive lakes. *Biogeochemistry* **3**, 129–149.

Khan, S.U. and Sowden, F.J. (1972). Distribution of nitrogen in fulvic acid fractions extracted from the Black Solonetzic and Black Chernozemic soils of Alberta. *Can. J. Soil Sci.* **52**, 116–118.

Larsen, D.P. and Mercier, H.T. (1976). Phosphorus retention capacity of lakes. *J. Fish. Res. Board. Can.* **33**, 1742–1750.

Lewis, W.M. (1986). Nitrogen and phosphorus runoff losses from a nutrient-poor tropical moist forest. *Ecology* **67**, 1275–1282.

Likens, G.E., Bormann, F.H., Pierce, R.S., Eaton, J.S. and Johnson, N.M. (1977). *Biogeochemistry of a Forested Ecosystem*. Springer, New-York.

Lowrance, R. and Leonard, R.A. (1988). Streamflow nutrient dynamics on coastal plain watersheds. *J. Environ. Qual.* **17**: 734–740.

MacDonald, A.J., Powlson, D., Poulton, P. and Jenkinson, D. (1989). Unused fertilizer nitrogen in arable lands—its contribution to nitrate leaching. *J. Sci. Food Agric.* **46**, 407–419.

McKnight, D.M. and Aiken, G.R. (1998). Sources and age of aquatic humus. In: *Aquatic Humic Substances; Ecology and Biogeochemistry* (Ed. by D.O. Hessen and L. Tranvik), pp. 9–39. Springer, Heidelberg.

Meybeck, M. (1982). Carbon, nitrogen and phosphorus transport by world rivers. *Am. J. Sci.* **282**, 401–450.

Meybeck, M. (1993). C, N, P and S in rivers: from sources to global inputs. In: *Interactions of C, N, P and S Biogeochemical Cycles and Global Change* (Ed. by R. Wollast, F.T. Mackenzie and L. Chou), pp. 163–193. Springer, Berlin.

Miller, W.L. and Zepp, R.G. (1995). Photochemical production of dissolved inorganic carbon from terrestrial organic matter: significance to the oceanic organic carbon cycle. *Geophys. Res. Lett.* **22**, 417–420.

Moeller, R.J., Minshall, G.W., Cummins, K.W., Petersen, R.C., Cushing, C.E., Sedell, J.R., Larson R.A. and Vannote, R.L. (1979). Transport of dissolved organic carbon in streams of differing physiographic characteristics. *Org. Geochem.* **1**, 139–150.

Moran, M.A. and Zepp, R.G. (1997). Role of photochemistry in the formation of biologically labile compounds from dissolved organic matter. *Limnol. Oceanogr.* **42**, 1307–1316.

Morris, D.P. and Lewis, W.M., Jr. (1988). Phytoplankton nutrient limitation in Colorado mountain lakes. *Freshw. Biol.* **20**, 315–327.

Mulder, J., Nilsen, P., Stuanes, A. and Huse, M. (1997). Nitrogen pools and transformation in Norwegian forest ecosystems with different atmospheric inputs. *Ambio* **26**, 273–281.

Nelson, D. (1985). Minimizing nitrogen losses in non-irrigated eastern areas. In : *Plan for Nutrient Use and the Environment*, pp. 173–209. Fertilizer Institute, Washington, DC.

Nixon, S.W. (1992). Quantifying the relationship between nitrogen input and productivity of marine ecosystems. *Adv. Mar. Technol. Conf.* **5**, 57–83.

Nixon, S.W., Ammerman, J., Atkinson, L., Berounsky, V, Billen, G., Boicourt, W., Boynton, W., Church, T., Ditoro, D.M., Elmgren, R., Garber, J., Giblin, A., Jahnke, R., Owens, N., Pilson, M.E.Q. and Seitzinger, S. (1996). The fate of nitrogen and phosphorus at the land–sea marigin of the North Atlantic Ocean. *Biogeochemistry* **35**, 75–139.

OECD (1982). *Eutrophication of Waters: Monitoring, Assessment and Control*. OECD Eutrophication Programme. Final Report, Paris, France.

Omernik, J.M. (1977). *Nonpoint Source Stream Nutrient Relationships: A Nationwide Survey*. EPA-600/3-77-105. US EPA, Washington, DC.

Pacés, T. (1982). Natural and anthropogenic flux of major elements from central Europe. *Ambio* **11**, 206–208.

Pacyna, J.M., Larssen, S. and Semb, A. (1991). European survey for NO_x emissions with emphasis on Eastern Europe. *Atmos. Environ.* **25A**, 425–439.

Pettersson, C., Allard, B. and Boren, H. (1997). River discharge of humic substances and humic-bound metals to the Gulf of Bothnia. *Estuar. Coast Shelf Sci.* **44**, 533–541.

Prospero, J.M., Barrett, K., Church, T., Dentener, F., Duce, R.A., Galloway, J.N., Levy, H., II, Moody J., and Quinn, P. (1996). Atmospheric deposition of nutrients to the North Atlantic basin. *Biogeochemistry* **35**, 75–139.

Raffaelli, D. (1997). Impact of catchment land-use on an estuarine benthic food web. In: *Biogeochemical Cycling and Sediment Ecology* (Ed. by J.S. Gray, W. Ambrose and A. Szaniawska), pp. 147–152. NATO ARW, Kluwer, Dordrecht.

Rasmussen, J.B., Godbout, L. and Schallenberg, M. (1989). The humic content of lake water and its relationship to watershed and lake morphometry. *Limnol. Oceanogr.* **34**, 1336–1343.

Rastetter, E.B., Agren, G.I. and Shaver, G.R. (1997). Responses of N-limited ecosystems to increased CO_2: a balanced-nutrition, coupled-element-cycles model. *Ecol. Appl.* **7**, 444–460.

Schelske, C.L. (1988). Historical trends in Lake Michigan silica concentrations. *Int. Rev. Ges. Hydrobiol.* **73**, 559–591.

Schelske, C.L. and Stoermer, E.F. (1971). Eutrophication, silica depletion, and predicted changes in algal quality in Lake Michigan. *Science* **173**, 423–424.

Schiff, S.L., Arvena, R., Trumbore, S.E., Hinton, M.J., Elgood, R. and Dillon, P.J. (1997). Export of DOC from forested catchments on the precambrian shield of central Ontario: clues from ^{13}C and [14]C. *Biogeochemistry* **36**, 43–65.

Schindler, D.W. (1977). Evolution of phosphorus limitation in lakes. *Science* **195**, 260–262.

Schindler, D.W. and Bayley, S.E. (1993). The biosphere as an increasing sink for atmospheric carbon: estimates from increased nitrogen deposition. *Global Biogeochem. Cycles* **7**, 717–733.

Schindler, D.W., Bayley, S.E., Curtis, P.J., Parker, B., Stainton, M.P. and Kelly, C.A. (1992). Natural and man-caused factors affecting the abundance and cycling of dissolved organic substances in precambrian shield lakes. *Hydrobiologia* **229**, 1–21.

Schindler, D.W., Bayley S.E., Parker, B.R., Beaty, K.B., Cruikshank, D.R., Fee, E.J., Schindler, E.U. and Stainton, M.P. (1996). The effects of climatic warming on the properties of boreal lakes and streams at the Experimental Lakes Area, northwestern Ontario. *Limnol. Oceanogr.* **41**, 1004–1017.

Schindler, D.W., Curtis, P.J., Bayley, S.E., Parker, B., Beaty, K.G. and Stainton, M.P. (1997). Climate-induced changes in the dissolved organic carbon budgets of boreal lakes. *Biogeochemistry* **36**, 9–28.

Schlesinger, W.H. (1991). *Biogeochemistry: An Analysis of Global Change.* Academic Press, San Diego.

Schnitzer, M. (1985). Nature of nitrogen in humic substances. In: *Humic Substances in Soil, Sediment, and Water* (Ed. by G.R. Aiken, D.M. McKnight, and R.L. Wershaw), pp. 303–328. John Wiley, New York.

Scully, N.M. and Lean, D.R.S. (1994). The attenuation of UV radiation in temperate lakes. *Arch. Hydrobiol.* **43**, 135–144.

Seitzinger, S.P. (1988). Denitrification in freshwater and coastal marine ecosystems: ecological and geochemical implications. *Limnol. Oceanogr.* **33**, 702–724.

Sharpley, A.N. and Rekolainen, S. (1997). Phosphorus in agriculture and its environmental implications. In: *Phosphorus Loss from Soil to Water* (Ed. by H. Tunney, O.T. Carton, P.C. Brookes, and A.E. Johnston), pp. 29–40. CAB International, New York.

Sharpley, A.N, Hedley, M.J., Sibbesen, E., Hillbricht-Ilkowska, A., House, A. and Ryszkowski, L. (1995). Phosphorus transfers from terrestrial to aquatic ecosystems. In: *Phosphorus in the Global Environment* (Ed. by H. Tiessen), pp. 171–199. SCOPE, John Wiley, New York.

Smayda, T.J. (1990). Novel and nuisance phytoplankton blooms in the sea: evidence for a global epidemic. In: *Toxic Marine Phytoplankton* (Ed. by E. Graneli). Elsevier.

Sterner, R.W. and Hessen, D.O. (1994). Algal nutrient limitation and the nutrition of aquatic herbivores. *Annu. Rev. Ecol. Syst.* **25**, 1–29.

Stoddard, J.L. (1994). Long term changes in watershed retention of nitrogen. Its causes and aquatic consequences. In: *Environmental Chemistry of Lakes and Reservoirs*. (Ed. by L.A. Baker), Advances in Chemistry Series no. 237. American Chemical Society, Washington, DC.

Swedish Environmental Protection Agency (1997). *Nitrogen from Land to Sea*. Report 4801.

Talling, J.F. and Heaney, S.I. (1988). Long term changes in some English (Cumbrian) lakes subjected to increased nutrient inputs.In: *Algae and the Aquatic Environment* (Ed. by F.E. Round), pp. 1–29. Titus Wilson, Kendal.

Tamminen, T. and Kivi, K. (Ed.) (1996). *Nitrogen Discharges, Pelagic Nutrient Cycles, and Eutrophication of the Northern Baltic Coastal Environments*. PELAG III Final Report (In Finnish).

Tate, C.M. and Meyer, J.L. (1983). The influence of hydrologic conditions and succession state on dissolved organic carbon export from forested watersheds. *Ecology* **64**, 25–32.

Thurman, E.M. (1985). *Organic Geochemistry of Natural Waters*. Nijhoff/Junk, The Hague.

Tjomsland, T. and Braathen, B. (1996). *Tilførsler av næringsstoffer til kysten mellom svenskegrensen og Stad*. Report LNR 3548–96. Norwegian Institute for Water Research.

Turner, R.E. and Rabalais, N.N. (1991). Changes in Mississippi river water quality this century. *BioScience* **41**, 140–147.

Tørseth, K. and Semb, A. (1997). Atmospheric deposition of nitrogen, sulphur and chloride in two watersheds located in southern Norway. *Ambio* **26**, 258–265.

Urban, N.R., Bayley, S.E. and Eisenreich, S.J. (1989). Export of dissolved organic carbon and acidity from peatlands. *Water Resource Res.* **25**, 1619–1628.

Vagstad, N., Eggestad, H.O. and Høyås, T.R. (1997). Mineral nitrogen in agricultural soils and nitrogen losses: relation to soil properties, weather conditions and farm practices. *Ambio* **26**, 266–272.

Vaithiyanathan, P. and Correll, D.L. (1992) The Rode River watershed: phosphorus distribution and export in forest and agricultural soils. *J. Environ. Qual.* **21**, 280–288.

Verhoeven, J.T.A., Koerselman, W. and Meuleman, A.F.M. (1996). Nitrogen- or phosphorus-limited growth in herbaceous, wet vegetation: relations with atmospheric inputs and management regimes. *TREE* **11**, 494–496.

Vitousek, P.M. (1994). Beyond global warming: ecology and global change. *Ecology* **75**, 1861–1876.

Vitousek, P.M., Aber, J., Howarth, R.W., Likens, G.E., Matson, P.A., Schindler, D.W., Schlesinger, W.H. and Tilman, G.D. (1997). Human alterations of the global nitrogen cycle: sources and consequences. *Ecol. Appl.* **7**, 737–750.

Vollenweider, R.A. (1975). Input–output models with special reference to phosphorus loading in limnology. *Schw. Zeitschr. Hydrol.* **37**, 53–84.

Wetzel, R.G. (1995). Death, detritus, and energy flow in aquatic ecosystems. *Freshw. Biol.* **33**, 83–89.

Wetzel, R.G., Hatcher, P.G. and Bianchi, T.S. (1995). Natural photolysis by ultraviolet irradiance of recalcitrant dissolved organic matter to simple substrates for rapid bacterial metabolism. *Limnol. Oceanogr.* **40**: 1369–1380.

White, T.C. (1993). *The Inadequate Environment. Nitrogen and the Abundance of Animals*. Springer, Berlin.

World Commission on Environment and Development (1987). *Our Common Future.* Oxford University Press, Oxford.

Wright, R.F. (1998). Effect of increased carbon dioxide and temperature on runoff chemistry at a forested catchment in southern Norway (CLIMEX project). *Ecosystems* **1**, 216–225.

Wright, R.F., Raastad, I.A and Kaste, Ø. (1997). Atmospheric deposition of nitrogen, runoff of organic nitrogen, and critical loads for soils and waters. *NIVA/Norwegian Institute for Water Research Report SNO 3592-97.*

Nutrients in Estuaries

D.B. NEDWELL, T.D. JICKELLS, M. TRIMMER AND R. SANDERS

I. SUMMARY

Nitrogen and phosphorus loading in rivers has increased considerably as a result of human activity. Silicon loads have been less impacted. However, extrapolating these increased loads in rivers through to coastal waters is not straightforward because of the intense nutrient cycling that can take place in estuaries. In this article we review the inputs to estuaries, methods of calculating fluxes through estuaries to coastal waters, and the processes that give rise to the intense cycling within estuaries. Although all the nutrients are intimately linked through their role in primary production, their cycling processes within estuaries are markedly different and hence estuarine cycling can not only change total nutrient loads, but also modify the ratios of one nutrient to another. These modifications of nutrient ratios may have important implications for both the extent and the number of species involved in primary productivity in coastal waters.

Many estuaries are rather turbid, which limits the extent of primary productivity and hence the impact of this process on nutrient cycling. Primary

ADVANCES IN ECOLOGICAL RESEARCH VOL. 29
ISBN 0–12–013929–4

productivity is probably the dominant process affecting dissolved silicon (Si) fluxes, and hence the modification of Si fluxes may be limited. The high turbidity will promote particle–water exchange reactions, which are particularly important for phosphorus (P) cycling. The high turbidity is also associated with net sedimentation in most estuaries and the sedimented material is rich in organic matter derived from riverine, marine and estuarine sources. The bacterial degradation of this organic matter drives a series of redox reactions which have a major impact on nitrogen (N) and P cycling. Denitrification can be a major sink for nitrate in estuaries, helping to attenuate its impact on coastal ecology, but the concomitant production of nitrous oxide may have a deleterious effect on the atmosphere. Iron(III) reduction in sediments can mobilize P bound to ferric oxyhydroxides in sediments and release this back into the water column.

Considerable progress has been made on understanding nutrient cycling processes in individual estuaries, but we are still some time away from being able to generalize these results for other estuaries and hence effectively predict the present and future nutrient cycling in unstudied systems.

II. INTRODUCTION

Estuaries are the major conduits between land and sea, through which flow the loads of soluble and particulate materials derived from the catchment area of each estuary. These loads may be derived from both leaching and runoff from the land and from atmospheric deposition within the catchment, each of which may have anthropogenic components. On a global basis, the estuarine input of N to the sea has been estimated as between 10 and 200 Tg N year^{-1} with a preferred estimate of 40.6 Tg N year^{-1} (Devol *et al.*, 1991; Galloway *et al.*, 1996; Howarth *et al.*, 1996; see also Hessen, this volume), and phosphorus loads as 21.4 Tg P year^{-1} (Howarth *et al.*, 1996). Historically, nutrient loads to estuaries have increased in parallel with increases in human populations in their catchments, as a result of both raised inputs of human and animal wastes to water courses, and increased rates of application of fertilizers to arable land, with their subsequent leaching to water courses. Overall, about 15% of the fertilizer N applied to arable land is lost (Addiscott *et al.*, 1991), via *in situ* denitrification and leaching from land into waterways and ultimately to estuaries. About 2500 Gmol N year^{-1} is introduced anthropogenically from sewage or fertilizer into the watersheds of the temperate region of the North Atlantic Ocean, of which about 35% is exported to the coastal seas (Galloway *et al.*, 1996), but human inputs can vary enormously between watersheds. For a number of major rivers a log–log relationship between the export of nitrate and dissolved inorganic phosphate (DIP) per km^2 of watershed and the human population density has been shown (Wollast, 1983; Peierls *et al.*, 1991; Howarth *et al.*, 1996) (Figure 1). In addition to riverine inputs, groundwater (e.g. Moore, 1996) and atmospheric inputs (e.g.

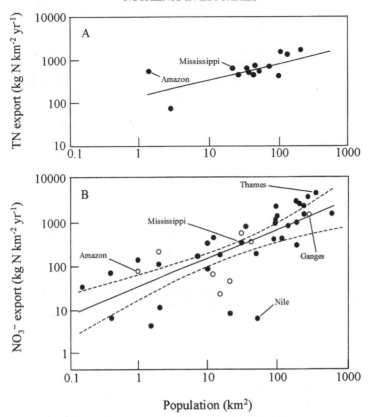

Fig. 1. Relationship between log of population density and (A) log of total nitrogen (TN) export from regions of the North Atlantic basin (with kind permission from Kluwer Academic Publishers) and (B) nitrate in major world rivers (redrawn and reprinted by permission from Peierls *et al.* (1991)). (A) log TN = 2.2 + 0.35 log (population density); $r^2 = 0.45$, $p = 0.01$. (B) log $NO_3^- = 1.15 + 0.62$ log (population density); $r^2 = 0.53$, $p = 0.00001$.

Scudlark and Church, 1993) have enhanced nutrient fluxes through estuaries. However, in most cases the direct atmospheric inputs to an estuary are small compared with the fluvial component, although the picture can change if the whole catchment is considered. The importance of groundwater inputs is uncertain and probably variable.

There is no doubt that long-term data sets illustrate increases in nutrient concentrations during this century in a variety of European and North American rivers and coastal seas (Figure 2) (DOE/CDEP, 1986; see also Vannekom and Salomons, 1980). Turner and Rabalais (1991) showed that nitrate concentrations in the Mississippi had doubled since 1965, and there

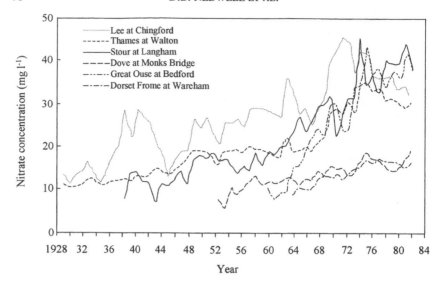

Fig. 2. Time series of nitrate concentrations in a variety of UK rivers. From DOE/CDEP (1986), with kind permission from HMSO.

have been similar or greater increases this century in some European rivers (Pacés, 1982; Balls *et al.*, 1995; Justic *et al.*, 1995). Generally, DIP loads stabilized then decreased after the 1970s, coinciding with the change to low-phosphate detergents (e.g. Klein and van Buuren, 1992). For example, total phosphate and DIP loads through the Rhine/Meuse decreased by about 50% between 1981 and 1989. In some estuaries the nitrogen loads have also stabilized since the 1980s (Klein and van Buuren, 1992; Parr and Wheeler, 1996). However, nitrate and DIP inputs in the Rhine/Meuse were still, respectively, about four and seven times the inputs in 1935. Moreover, these different trends in N and P inputs have also resulted in increases in the N : P ratios of nutrient inputs: for example, in the Rhine/Meuse increasing from about 20 : 1 in 1979 to about 30 : 1 in 1990 (Klein and van Buuren, 1992), which will increase any tendency to P limitation in receiving waters. In contrast, increased discharges of sewage treatment works (STW) effluents, which are relatively high in DIP, have also been suggested to decrease estuarine N : P ratios, increasing N limitation (Howarth, 1988).

These historically increased nutrient loads to estuaries have been reflected in similar trends of increased nutrient concentrations in at least some coastal seawater (Hickel *et al.*, 1993; Allen *et al.*, 1998), raising concerns about nutrification stimulating both primary production and nuisance blooms of algae, not only within estuaries but also in adjacent coastal waters. However, discernible trends of nutrient increase in coastal seas have usually been confined to localized areas subjected to severe impact by large local inputs, such as in the German Bight by the Rhine and Scheldt plumes (Gieskes and Kraay, 1977; de Jonge and Essink, 1991; Jickells, 1998).

In this chapter we intend to review recent advances in understanding of how nutrient loads through estuaries vary, and how these loads influence the processes operating in estuaries. Furthermore, the biological and physico-chemical processes in estuaries also act on the fluxing nutrients, inducing changes. Much of the research in the past 20 years has been directed towards furthering understanding of the effects of increased nutrient loads in estuaries, with consequent focus on high-nutrient, eutrophic estuaries; while there is a relative lack of information on low-nutrient, oligotrophic estuaries.

III. FACTORS AFFECTING ESTUARINE NUTRIENT FLUXES

A. Nature of the Loads

The nutrient loads through individual estuaries reflect the nature of their catchment's soils and also any anthropogenic inputs (e.g. Balls, 1994; Raffaelli *et al.*, 1989; Hessen, this volume). The North Atlantic Ocean is surrounded by the greatest density of industrialized communities in the world, with a relatively good environmental data set, and the input loads to this region have been summarized by Nixon *et al.* (1996). Nutrient loads into estuaries, and potentially coastal seas, in other areas of the globe, particularly in the tropics, are less well understood. Gross inputs to the North Atlantic Ocean are dominated by the nutrient loads from the large rivers; the Amazon contributing 25% of the N input and 68% of the P input, and the Mississippi 14% of N and 5% of P. However, when the different areas of the watersheds drained by each river are taken into account, the N loads per unit area of catchment into the North Sea basin are the greatest (average 1450 kg N km^{-2} catchment year^{-1}), with high loads also along the north-west coast of North America (average 1300 kg N km^{-2} year^{-1}). The smallest loads are from the Canadian rivers (average 76 kg N km^{-2} year^{-1}) which drain relatively less populated catch-ments. While loads of P to the North Sea are also high (117 kgP km^{-2} year^{-1}), the highest input is from the Amazon Basin (236 kgP km^{-2} year^{-1}), probably because of increased rates of erosion and weathering in this watershed compared with others (Devol *et al.*, 1991).

Geographical differences in estuarine nutrient loads are well illustrated by the inputs to the major estuarine systems around the UK. Of the 155 UK estu-aries (Davidson *et al.*, 1991) there are 96 recognized estuarine systems. For example, the Wash estuarine system is made up of the Great Ouse, the Nene, the Welland and the Witham. Riverine inputs to these estuaries are gauged at regular intervals for both water flow and concentrations of nutrients for the Harmonized Monitoring Scheme initiated by the Oslo and Paris Commission. For example, Figure 3 shows the annual nutrient loads to the estuaries for total oxidized nitrogen (TOxN, which is nitrate + nitrite), ammonium, orthophosphate and silicate. TOxN was by far the greatest input of N through the rivers,

Fig. 3. Nutrient loads (mol year^{-1}) to British estuarine systems and nutrient loads normalized for estuary area (mol m^{-2} year^{-1}). Data are the mean values for 1995 and 1996 (Harmonized Monitoring Scheme, unpublished data). Estuarine areas are from Davidson *et al.* (1991).

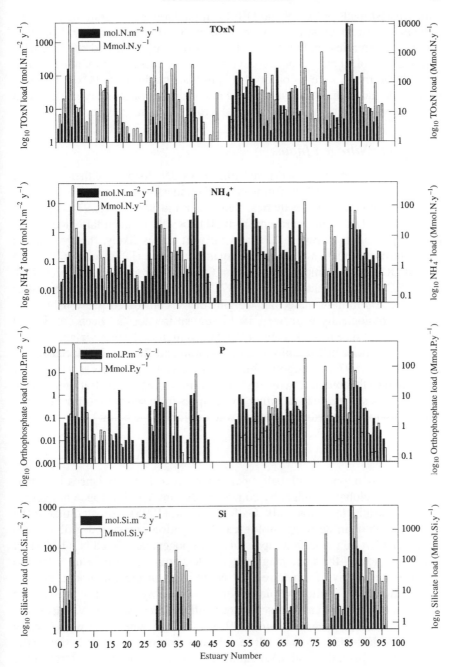

Fig. 3 continued.

with greatest loads of TOxN and DIP from the large rivers draining intensive agricultural catchments; the Severn (estuary 5), Humber (estuary 72), other east coast estuaries including the Thames (estuary 78), and the estuaries entering the Solent on the south coast (estuaries 85 and 86). Ammonium inputs through the rivers were an order of magnitude smaller than TOxN, but may have ecological impacts on algal production disproportionately greater than their size would imply (Underwood and Kromkamp, this volume).

1. Organic Nitrogen Inputs

The magnitude and role of organic nitrogen (ON) loads in either dissolved (DON) or particulate (PON) forms is poorly understood, although it may be a significant input in some estuaries. Seitzinger and Sanders (1997) suggested that DON + PON could represent from 20 to > 90% of the total N load, with DON representing between 20% and 50%. It seems likely, though, that in hypernutrified estuaries dissolved inorganic nitrogen (DIN) load is likely to exceed DON. For example, in the Great Ouse DON represented only 3–20% of the TON load (Rendell *et al.*, 1997). As most dissolved organic matter (DOM) exists in high molecular weight forms, considered previously to be relatively biologically refractory, its biological impact had been considered negligible, but recent work suggests that even high molecular weight organic molecules may be utilized biologically (Amon and Benner, 1994; Seitzinger and Sanders, 1997). In oligotrophic estuaries, DON may represent a larger proportion of the total nitrogen input. In addition, urea concentrations may be significant in some estuaries, either from inputs from sewage treatment works or by emission from sediments to the water column (see below).

2. N_2 Fixation

N_2 fixation is important globally, being the process that balances denitrification in the global N budget by converting N_2 into fixed nitrogen molecules which are available biologically. Galloway *et al.* (1995) estimated that in the absence of human activity, biological N_2 fixation accounted globally for 90–130 Tg N year^{-1}, but that inputs of anthropogenically fixed N_2 (140 Tg N year^{-1}) are now of the same magnitude. At least part of the riverine N load to estuaries may derive from terrestrial N_2 fixation in the catchment, but the contribution of N_2 fixation within estuaries is probably small. The most active N_2 fixers in estuarine environments are benthic cyanobacteria (Vitousek and Howarth, 1991; Howarth *et al.*, 1995), but even where there are heavy cyanobacterial mats such as in salt-marsh pans their overall contribution to the N budget is apparently small (Azni and Nedwell, 1986). Benthic heterotrophic N_2 fixation (Nedwell and Azni, 1980) is also likely to be negligible. Levels of *nitrogenase* are regulated by intracellular levels of ammonium adenylate: high intracellular levels of ammonium adenylate stimulate *glutamine synthetase*, which acts as a repressor of the nitrogenase complex. Many estuarine sediments, particularly those with significant organic matter contents, have high ammonium concentrations in the pore waters which are likely to repress N_2 fixation. It has also been argued that nitrate respiration inhibits N_2 fixation by competing for reductants (Dicker and Smith, 1980). The contribution of N_2 fixation in estuaries is likely to decrease, therefore, as N loads increase. Capone (1983) reported rates of estuarine benthic N_2 fixation averaging 0.4 g N m^{-2} year^{-1}, compared with an average of 11.2 and 23 g N m^{-2} year^{-1} in mangrove sediments and salt-marsh communities respectively. Total estuarine N_2 fixation was estimated at 0.43 Tg N year^{-1}, which is only about 1% of the global N load to estuaries of 40.6 Tg N year^{-1}.

3. Inputs of Phosphorus

Riverine inputs of phosphorus to estuaries contain both dissolved and particulate phosphorus. On a global basis particulate phosphorus dominates (Table 1), although the fluvial particulate flux is uncertain (Milliman and Syvitski, 1992). In a highly contaminated river system such as the Rhine particulate P still dominates, although dissolved P makes up a greater proportion of the total (de Jonge, 1997). Hence, in estimating P fluxes through estuaries, it is necessary to consider both the dissolved and particulate forms. Particulate phosphorus is operationally defined here as that phosphorus which is retained on a filter (usually 0.7 or 0.45 μm pore size). It is poorly characterized chemically but will include organic phosphorus, iron and calcium phosphates and some phosphorus bound within the aluminosilicate lattice of clays. The relative importance of these different components will vary depending on the catchment. However, it is clear that much of this particulate phosphorus is readily exchangeable with the dissolved phase (e.g. Froelich, 1988). Dissolved phosphorus consists of dissolved inorganic phosphorus (DIP—the dissociation products of phosphoric acid whose relative proportions are pH dependent; House *et al.*, 1988) and dissolved organic phosphorus (DOP), which is also poorly characterized. In the Great Ouse and Delaware estuaries (both draining populous catchments and

Table 1

Estimates of riverine inputs of phosphorus to the global ocean (based on Howarth *et al.*, 1995)

	Dissolved P	Particulate P	Total P
Natural state	32	226	258
Modern	65	645	710

Values are 10^9 mol year^{-1}.

subject to significant anthropogenic perturbation), DOP is a minor species in the water column in comparison to DIP (Lebo and Sharp, 1992; Rendell *et al.*, 1997), but in the more pristine Amazon River, where DIP levels are much lower, DOP makes up a much greater proportion of the total dissolved P load (Richey *et al.*, 1991). Little is known about the amounts and role of DOP, and more research in this area would be valuable.

4. Silicate

Silicate is derived from rock weathering, and loads depend on the balance of weathering rates and dilution (Figure 3). Thus, catchment geology plays a key role, and silicate loads are less influenced by human activity. An exception to this is where rivers are dammed and silicate is deposited behind the dam, reducing the silicate load downstream. Damming the Danube greatly reduced the silicate load to the Black Sea (Humborg *et al.*, 1997), reducing the proportion of diatom production in the estuary in favour of microflagellates.

5. Seasonal Variation in Nutrient Loads

While annual loads give comparative information between estuaries, there may be large seasonal differences in the inputs of nutrients, with consequent ecological effects. Figure 4 shows the seasonal changes of inputs of TOxN to the Thames estuary, with peak loads during winter coincident with winter rains flushing nitrate from the catchment. During this time concentrations of nitrate in river water can be extremely high (> 1 mmol l^{-1}). In summer, reduced runoff, NO_3^- uptake by crops and biological uptake in the river combine to lower nitrate concentrations. Hence, the riverine inputs are lowest to this estuary during the summer period when biological activity in the estuary is most rapid and most likely to be nutrient limited. During this period efflux of nutrients from estuarine sediments may be important in maintaining coastal primary production (Trimmer *et al.*, 1998). In contrast to nitrate, DIP inputs are often dominated by point sources such as STW discharges, and hence concentrations fall in winter as river water flows increase, although STW discharges do not (e.g. Fichez *et al.*, 1992). Silicate concentrations in estuaries may also show seasonal variation, with maximum concentrations during the winter and minima during summer, but these changes may be more the result of changes in biological activity in the estuary rather than seasonal changes in silicate loads (Fichez *et al.*, 1992; Balls *et al.*, 1995).

B. Impact of Nutrient Loads in Estuaries

The ecological impact of these nutrient loads will depend upon both the area of the estuary that receives the load and the residence time of the nutrients

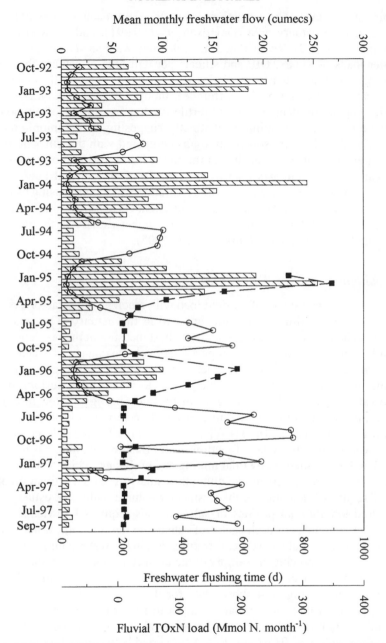

Fig. 4. Mean monthly freshwater flow (bars; sum of rivers Thames, Lee and Roding) and corresponding freshwater flushing times (○) and total oxidized nitrogen (TOxN) loads (■) for the Thames estuary. (Flow and TOxN load data, Environment Agency, unpublished data.)

within the estuary. The annual nutrient load for each UK estuary was normalized for estuary area (Davidson *et al.*, 1991), and showed distinct groupings (Figure 3). Rivers that drain cultivated watersheds (see above) had heavier loads of TOxN, DIP and silicate per unit area of estuary than other more oligotrophic estuaries in Wales and Scotland which drain essentially rocky catchments. Part of this difference, though, may also be attributed to heavier population densities in the fertile catchments. Inputs of ammonium did not reflect these groupings so distinctly, suggesting that ammonium loads were related to different source strengths compared with the other nutrients. One interpretation may be that ammonium inputs were more related to upstream STW discharges than to leaching from agricultural land. As the freshwater load passes down the estuary, dilution and the increasing estuarine area reduces its impact. For example, in the Great Ouse estuary the TOxN load decreases from 455 mol N m^{-2} year^{-1} in the upper estuary to 20 mol N m^{-2} year^{-1} in the lower estuary (Trimmer *et al.*, 1998).

1. Residence Time of Water in Estuaries

The impact on the estuary of nutrient load will, amongst other factors, depend on the residence time of the nutrient within the estuary. There is some confusion in the literature as to what is the best measure of the residence time. Tidal flushing will express the proportionate volume change of water within an estuary, derived from the tidal water volume exchange relative to total water volume in the estuary. However, the degree to which tidal flushing dilutes freshwater inputs will also depend on the morphology of the individual estuary. Freshwater dilution (and its reciprocal, retention) is best described by the Freshwater Flushing Time (FWFT), which expresses the volume of freshwater within the estuary as a function of the freshwater input (Dyer, 1973). FWFT will be influenced by tidal mixing and exchange, which will affect the volume of freshwater remaining within the estuary. Accurate estimation of the FWFT requires calculation of the freshwater volume within the estuary, from measured longitudinal profiles of salinity. This requires knowledge of the cross-sectional areas of each sector of estuary so that the total water volume can be estimated, and any vertical stratification in the water column must also be allowed for. Knowing the salinity of the local coastal seawater, such salinity profiles can be converted into equivalent volumes of freshwater in each sector along an estuary. Figure 4 shows data for the River Thames estuary, illustrating seasonal and interannual variation in flow rates and nutrient loads, dependent upon rainfall in the catchment. Peak water flows occur in winter, coincident with peak TOxN loads. FWFT varied between as little as 15 days during winter peak flows to as much as 765 days in late summer when river flow was least and the freshwater volume in the estuary was replaced only slowly by river flow. When the TOxN load is greatest, the reduction in FWFT decreases the biological impact of the load on an estuary because of fast flushing, but also diminishes the effect of biological processes in the estuary on the TOxN load itself (Trimmer *et al.*, 1998).

C. Estimating Nutrient Fluxes Through Estuaries

Calculating nutrient fluxes through estuaries is not a trivial task. Here we will discuss only the estimations for fluxes derived from riverine sources. Additional inputs from the atmosphere (e.g. Scudlark and Church, 1993) and via groundwater (e.g. Moore, 1996; Nixon *et al.*, 1996) can further complicate the picture. Land Ocean Interactions in the Coastal Zone have published some guidelines for flux calculations in estuaries (Gordon *et al.*, 1995).

Riverine inputs to an estuary can be calculated by multiplying concentrations and flow data, but both can vary on short time scales, and the resultant undersampling of this natural variability needs to be considered (e.g. Walling and Webb, 1985; deVries and Klavers, 1994). The impact of estuarine processes then needs to be assessed in order to estimate nutrient fluxes through the estuary into the coastal sea. Table 2 illustrates the methods used to assess nutrient retention and fluxes within estuaries. In theory it is possible to determine nutrient fluxes at any point in an estuary from an estimate of net seaward water flux and water column concentrations corrected for seawater dilution. However, such an approach becomes progressively more difficult and uncertain moving seaward, although it has been used successfully in some circumstances (e.g. Lebo and Sharp, 1992; Sanders *et al.*, 1997a). Modelling residual water fluxes can offer an alternative to measuring or calculating them directly, and these can then be combined with nutrient measurements to derive nutrient fluxes (e.g. Morris *et al.*, 1995).

Such a flux calculation provides no direct information on processes causing changes in nutrient fluxes. An alternative approach is directly to measure source and sink terms within an estuary and to integrate these to derive flux estimates. Such an approach requires that all relevant processes be identified and quantified, and this is clearly very difficult. Furthermore, in multiplying up measured process rates to whole-estuary estimates, it is necessary to know the area of the estuary. The area physically within the estuary's mouth may be defined relatively easily, but the influence of estuarine processes may extend beyond this into the estuarine plume, and have a marked effect on the aggregated process rate if included (e.g. Trimmer *et al.*, 1998). However, in relatively small and simple estuaries where a few processes dominate, this can be a useful approach (e.g. Nielsen *et al.*, 1995; Ogilvie *et al.*, 1997; Sanders *et al.*, 1997b). In some estuaries it has been possible to combine a hydrodynamic model to estimate water fluxes with a process model to derive nutrient fluxes (e.g. Soetart and Herman, 1995) and such an approach clearly has advantages of allowing the fluxes to be predicted under different conditions. However, the resources needed to develop and validate such models are significant.

Table 2

Methods used to estimate nutrient retention and fluxes within and through mesoscale estuaries

Method	Advantages	Disadvantages	References
Comparison of mean fluxes at various points within the estuary	Conceptually simple	Difficult to apply at high salinity, large data sets required for statistical validity. Gives limited information on processes	Sanders et al. (1997a) Lebo and Sharp (1992)
Chemical mass balances derived by integrating process measurements	Gives information on magnitude, location and nature of removal processes	Requires process measurements. Requires some physical parameters (river flow/estuarine morphology). Requires load estimate to get fractional removal	Nielsen et al. (1995) Ogilvie et al. (1997) Sanders et al. (1997b) Trimmer et al. (1998)
Property : salinity plot analysis (PSPA)	Conceptually simple. Simple sampling strategy. Requires river flow	Requires steady state in inputs, processes and exports over freshwater residence time of estuary. Inappropriate in areas of vertical/lateral inhomogeneity. Only gives fractional removal	Billen et al. (1985) Balls (1994) Sanders et al. (1997a)
Process models	Suitable in complex systems. Can be used as a prognostic tool	Requires hydrodynamic and process models; both require validation	Nielsen et al. (1995) Soetaert and Herman (1995)
Hydrodynamic model coupled to nutrient data set	Suitable in high-salinity environments	Requires hydrodynamic model and large data set	Morris et al. (1995)
Salt balance calculations	Useful in systems with complex freshwater inputs	No information on processes. Only gives export	Simpson et al. (1997)
Regression of removal against residence time	Conceptually simple	Accuracy is dubious. Purely empirical. No theoretical foundations	Nixon et al.(1997)
K_d type model for particle-reactive species	Requires few parameters. Founded in well accepted theory	Works only if particle water interactions are dominant control and for particle-reactive species	Morris (1986) Prastka et al. (1998)

One of the most widely used methods to derive fluxes of dissolved material in estuaries involves "property salinity plots" (Boyle *et al.*, 1974; Liss, 1976; Officer, 1979). This method (Figure 5) involves plotting the parameter of interest (e.g. nitrate) against a conservative tracer of water mixing, usually salinity. Assuming the relationship becomes linear at high salinity, extrapolation of this linear portion back to zero salinity yields an "effective zero salinity end member". Multiplication of this value by the river flow yields the flux of nitrate, in this case out of the estuary after estuarine modification. This approach can therefore yield fluxes, but not any direct information on processes within the estuary. This approach is also sensitive to two key assumptions: (1) that the estuary is at steady state and that variations in input, loss or export terms are negligible over time scales comparable to the water residence time in the estuary; and (2) that the estuary is well mixed.

These criteria are rarely perfectly satisfied, and this can introduce curvature into property : salinity plots, leading to erroneous flux estimates (e.g. Lebo *et al.*, 1994; Shiller, 1996). This method also has problems in dealing with estuaries with more than one major nutrient source, e.g. two rivers, although these can be circumvented. The method also fails to provide any information on processes operating in an estuary. Although widely used, there are significant limitations to the property : salinity plot approach. Firstly for DIP, "phosphorus

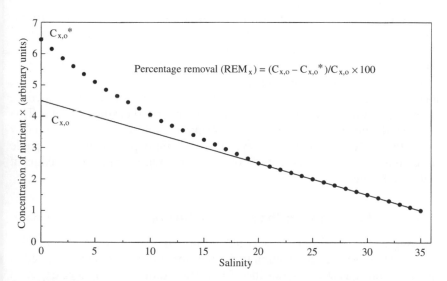

Fig. 5. Methodology conventionally used to estimate percentage removal from a property salinity plot. Extrapolation of the high salinity property : salinity regression back to the y axis yields an effective freshwater end-member concentration ($C_{x,o}$). This is compared with the measured freshwater end-member concentration ($C_{x,o*}$) to derive an estimate of removal.

buffering" at high salinity can complicate the interpretation of property: salinity plots (Sanders *et al.*, 1997a) and denitrification on large intertidal areas at the head of an estuary will be difficult to estimate with this approach because of the lack of a significant linear region of the plot beyond this area.

The nutrient dynamics of particularly complex estuarine systems may not be amenable to analysis using a single methodology. Under these circumstances it is possible to derive an overall budget by combining the results obtained from applying different methodologies to individual subsections of a single system. Using this approach Sanders *et al.* (1997b) derived an overall nutrient budget for the Humber estuarine system by combined flux analysis in the low-salinity freshwater region and property salinity plot analysis in the saline portion of the estuary. The accuracy or validity of estimates of nutrient retention generated by each of the individual methods outlined is rarely assessed. The easiest way to assess the accuracy of individual methods is to compare estimates of nutrient retention in a single system derived from a variety of methods. For this comparison to be valid, the various methods must be applied simultaneously. We are aware of relatively few studies that have attempted this. These include that of Nielsen *et al.* (1995), which compared the results from scaled up sediment incubation results with hydrodynamic model output, and that of Sanders *et al.* (1997b), who compared integrated nutrient uptake derived from primary production measurements with estimates of removal from property : salinity plot analyses.

It is also possible to derive estimates of area- or volume-specific process rates from property : salinity plots if river flow, salinity field and channel morphology are well known, using salt and freshwater conservation considerations. Figure 6 shows estimates of nitrate removal rate from such an analysis of a nitrate : salinity plot from the muddy hypereutrophic Colne estuary, together with measured benthic nitrate removal rates from the same period. Both measurements yielded similar rates and integrated to similar values. This suggested that the two methods of estimating removal were reasonably accurate, and further suggested that property : salinity plot analysis is capable of reproducing the general features observed in estuarine profiles of sediment water exchange.

D. Interaction Between Primary Production and Nutrient Loads

As long as algal primary production is not light limited, it is generally accepted that primary production in fresh water is P limited, but N limited in the sea. Along an estuary a transition from P to N limitation might be expected to occur, therefore, and nutrient availability will play a key role in regulating primary production. The interaction between estuarine primary production and the physicochemical estuarine environment, including nutrient availability, is considered in detail by Underwood and Kromkamp (this volume),

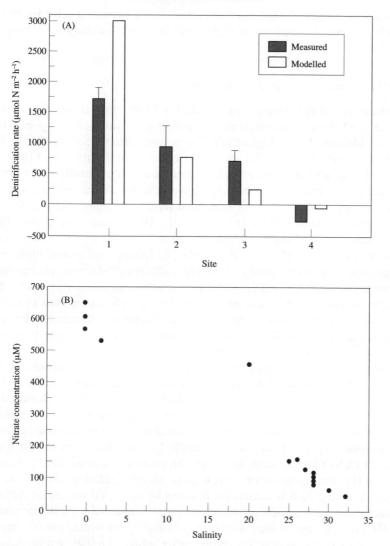

Fig. 6. (A) Measured rates of nitrate uptake (denitrification) by the sediment at four sites down the Colne estuary and (B) derived rates from nitrate : salinity plots for samples collected at the same time (see text). Denitrification data are from Ogilvie *et al.* (1997). Bars indicate standard error, $n = 3$.

and here we confine ourselves to a brief consideration of the overall signifi-cance of nutrient uptake during primary production relative to estuarine nutrient loads. Historical trends of increase of nutrients in perhaps hitherto low-nutrient estuaries have inevitably resulted in increased primary production rates, and have been associated with increased biomass of both planktonic and benthic microalgae, and benthic macroalgae in estuaries (Raffaelli *et al.*, 1989).

1. Planktonic Primary Production

In many macrotidal estuaries high turbidity limits the development of algal biomass, whereas in microtidal estuaries phytoplankton biomass appears to be higher (Monbet, 1992). Light limitation is particularly important in the low-salinity region of macrotidal estuaries where turbidity maxima (TMs) can occur (Dyer, 1997) and estuarine management can significantly affect the size and location of this TM. On moving seaward, suspended sediment concentrations usually decline allowing improved light conditions and consequently more primary productivity (Fisher *et al.*, 1988; Lebo and Sharp, 1992; Zwolsman, 1994a; Jickells, 1998). In general, cycling of DIN is dominated entirely by biological processes while phosphorus behaviour changes in different regions of the estuary. In the low-salinity turbid region, phosphorus is dominated by abiotic particle–water interactions, but on moving seaward the relative importance of these processes declines in comparison to cycling through the water column phytoplankton–bacteria community. Hence, in coastal waters in general, and entirely in oceanic waters, biological processes dominate the cycling of phosphorus.

Significant rates of primary production can occur, even in turbid estuaries, if the critical mixing ratio ($Z_{mix} : Z_{eu}$) is less than approximately 5–20 (Grobbelaar, 1990; Fichez *et al.*, 1992; Cloern, 1996; see also Underwood and Kromkamp, this volume). However, when available estimates of annual primary production are compared with annual nutrient loads it seems probable that planktonic primary production is unlikely to represent a major sink for the loads of DIN or DIP. Silicate, however, can be removed to minimal concentrations during the spring bloom period in temperate estuaries. Dubravko *et al.* (1995) have argued that removal of diatoms by increased silicate limitation in estuaries may have significant effects on coastal food webs. Particular parts of the DIN (e.g. nitrate or ammonium) may be used preferentially to one another during primary production. This preferential uptake of nitrate or ammonium is expressed as an f ratio; ratios < 0.5 indicate a preference for ammonium by the phytoplankton. Ammonium is taken up and assimilated preferentially to nitrate by algae, and even in the presence of much higher concentrations of nitrate algae will use much smaller concentrations of ammonium because of repression of *nitrate reductase* (Glibert *et al.*, 1982). Consequently, f ratios in estuaries may be low (< 0.5) and ammonium loads may be proportionately more affected by primary production than the nitrate load.

However, remineralization in such environments is also extremely rapid and the net impact of biological production on phosphorus budgets appeared to be small in the Great Ouse system (Rendell *et al.*, 1997; Sanders *et al.*, 1997b). In a very comprehensive study of the Delaware estuary, Lebo (1990) and Lebo and Sharp (1992) identified a marked contrast in phosphorus cycling between the turbid low-salinity region and the high-salinity estuarine region. The latter region involves a large bay-like structure, a geomorphology common on the US east coast. This structure will allow extensive estuarine sedimentation and the development of low turbidity conditions in the bay suitable for phytoplankton growth. In the turbid estuarine region of the Delaware, these workers reported biological DIP uptake primarily by bacteria rather than phytoplankton, a process that appears to be limited by carbon supply. DIP uptake by abiological processes also occurred in this region and in addition there may be regeneration from decomposition of the bacterial biomass. At the seaward end of the estuary, DIP uptake was primarily by phytoplankton rather than bacteria with rates of 0.8–26 nmol l^{-1} hr^{-1}, similar to the rates seen in other US east coast estuaries (Lebo and Sharp, 1992 and references therein), and a rate substantially lower than that seen in the low-salinity region. The uptake rates in the high-salinity region were proportional to DIP concentrations. Phosphorus regeneration rates were high in this region too, occurring predominantly in the water column rather than in the sediments and keeping pace with DIP demand throughout much of the year except during the spring bloom. Overall there was some net conversion of DIP to particulate phosphorus, but this particulate phosphorus was stored only temporarily within the estuary during the summer months before export during the autumn. Thus the picture that emerges from several estuarine studies is of intense biological phosphorus cycling but little net retention. Using a very different methodology, Kaul and Froelich (1984) reached a similar conclusion for the Ochlocknee estuary in Florida. In the plume of the Humber estuary in summer, Morris *et al.* (1995) estimated that primary production was a net sink for estuarine and coastal nutrients, although again this is likely to be a seasonal rather than a permanent storage mechanism.

2. Benthic Primary Production

The significance of benthic primary production as a sink for nutrients within an estuary is poorly understood. Primary production by microphytobenthos can contribute significantly to primary production in some estuaries (see Underwood and Kromkamp, this volume). Joint (1978) estimated for the Lynher estuary in the UK that 63% of total estuarine primary production could be accounted for by benthic production, whilst a smaller contribution of about 17% has been reported in the Westerscheldt (de Jong and de Jonge, 1995; Kromkamp *et al.*, 1995). During benthic primary production both N and P will be assimilated, with the nutrient demand being met either from the underlying sediment pore water or from overlying water. Indeed, benthic primary production can considerably moderate the sediment–water fluxes of nutrients (Sundbäck *et al.*, 1991; Feuillet-Girard *et al.*, 1997). However, data on microphytobenthic production are usually spatially restricted, and often the nutrient load to any particular estuary may be unknown, making extrapolation to the overall importance of benthic production difficult.

Where both benthic primary production data and nutrient loads are available, the amount of nitrogen required to support the estimated annual production (assuming Redfield ratios) has been expressed as a percentage of the DIN load entering the estuary. Where primary production rates are not available, published high and low values for benthic primary production have been used to estimate the limits of nutrient uptake by benthic production (Table 3). For the five estuaries for which DIN load data were available, the mean percentage uptake of nitrogen by benthic primary production was only about 5% of the DIN load. This may be an underestimate as no measure of turnover of N within the algal biomass is available. On the other hand, there may also be input loads of organic N to the estuary which we have not included and which would reduce the percentage uptake. Furthermore, as most benthic production will ultimately decay *in situ,* benthic primary biomass will be only a temporary sink for biologically important elements, which will be recycled by benthic mineralization.

We concluded that neither pelagic nor benthic primary production is likely to be a major sink for a significant part of either the N or P loads in these estuaries. However, in oligotrophic estuaries with lower nutrient loads it is possible that benthic primary production will account for a higher proportion of the nutrient input. Further work is required to elucidate this point.

E. Sources and Sinks in Estuaries

1. Particulate Load

The amount of suspended particulate material (SPM) in the water column along an estuary varies enormously (Figure 7), depending on the local load of both organic and mineral particles, and the availability of sedimentation areas or sediment accommodation space. Reclamation of areas such as salt marshes can, therefore, alter the turbidity of an estuary. Generally, SPM decreases seawards as deposition occurs within the estuary. Settlement is greatest where current velocities are low, and areas such as saltmarshes may be particularly important as the presence of plant stems decreases water velocities and enhances deposition of SPM. However, at some point along the estuarine salinity gradient, at the interface between fresh and sea waters, an increase in the turbidity of the water indicates the position of the TM (Uncles and Stephens, 1993), although, if the fluvial load of suspended particulate mineral material is high, the position may be obscured. The TM originates from both trapping and resuspension of fine

Table 3

Annual estimates of benthic primary production for intertidal areas (distinction between mud and sand where available), and the degree of attenuation for a given dissolved inorganic N load, assuming Redfield ratios

Estuary	Intertidal area (km^2)		Primary production ($g\ C\ m^{-2}\ year^{-1}$)		N demand ($mol\ N\ year^{-1}$)		DIN load ($mol\ N\ year^{-1}$)	Attenuation (%)
	Mud	Sand	Mud	Sand	Mud	Sand		
Western Scheldt[a]	63.3		136		1×10^8		3.2×10^9	3.1
Thames[b]	17	43	276	62	5.6×10^7	3.2×10^7	2.6×10^9	3.4
Colne[c]	0.94	0.54	276	62	3.1×10^6	4×10^5	3.5×10^7	10
Norsminde fjord[d]	0.35		276		1.1×10^6		1.5×10^7	7.3
Norsminde fjord[d]	0.35		62		2.6×10^5		1.5×10^7	1.7

[a] Area and production data from de Jong and de Jonge (1995), DIN load data from Alongi (1998).
[b] High and low production data from Colijn and de Jonge (1984), DIN load and area data (M. Trimmer, unpublished data).
[c] Production data as from Colijn and de Jonge (1995), area and DIN load data from Robinson et al. (1998).
[d] Production data as from Colijn and de Jonge (1995), area and DIN load data from Jørgensen and Sørensen (1985).

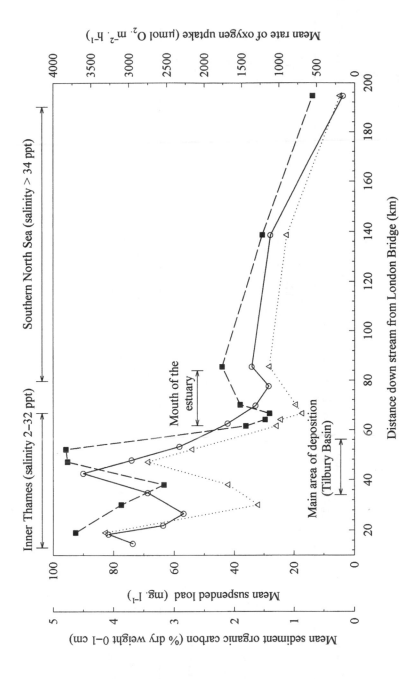

Fig. 7. Profiles along the Thames estuary of suspended particulate material (○), benthic oxygen uptake rates (■) and sedimentary organic carbon content (△). All points are the mean of data from six surveys from all sites between July 1996 to March 1998, except for the suspended particulate material data from 0 to 80 km, which is a 5 year mean from unpublished data (Environment Agency).

particulate material at this interface, and settlement of particles to the bottom in this region of the estuary is particularly intense.

The fate of mineral particulate matter in estuaries is uncertain (Milliman and Syvitski, 1992), although it seems likely that most fluvial sediment is retained in estuaries along the US east coast and around the North Sea (Nixon *et al.*, 1996). For example, in the Scheldt (Wollast and Peters, 1978) and Gironde (Allen *et al.*, 1976) estuaries, more than 75% of the riverine POM load is trapped within the estuary and does not reach the sea, while 60% of the particulate load of the Thames deposits within the estuary (Zwolsman, 1994a). Indeed there may well be net import of sediment from offshore (Nixon *et al.*, 1996; Jickells, 1998). However, sediment retained in estuaries may subsequently be exported offshore during extreme flood events (Milliman, 1991). Also, some large estuarine systems will export material directly on to the continental shelf or into the oceans (Milliman and Syvitski, 1992; Nixon *et al.*, 1996). The fate of particulate phosphorus in estuaries is poorly understood due to estuarine particle–water exchange reactions (see below). In addition, even if the majority of particulate matter is retained within an estuary, the finest fraction may still escape into coastal waters. This fraction may be of little importance in terms of mass budgets for sediment, but the finest fraction of the suspended sediment may carry a disproportionate amount of the particulate phosphorus, as has been shown for freshwaters (House *et al.*, 1998).

Coagulation of DOM at the salt–fresh water interface results in at least some of the DOM also flocculating to POM in the TM zone (Gardner and Menzel, 1974; Forsgren *et al.*, 1996) and settling out. The presence of the TM, therefore, enhances the tendency of an estuary to act as a trap for both particulate and dissolved organic loads, as well as mineral particles. As a consequence of this settlement, the organic matter content of the sediments in the region of the estuarine TM tends to be high (Trimmer *et al.*, 1998), and the sediments in this stretch of the estuary are particularly active sites of organic matter mineralization. Figure 7 shows the rates of benthic respiration along the Thames estuary, with sediment organic matter contents, and average SPM load. The presence of a TM in the Tilbury Basin is apparent, and in this region the sediments are particularly organically rich, and benthic respiration rates (organic matter mineralization) are highest. Suspended particulate material, benthic organic matter content and benthic oxygen uptake rates then decrease seaward in the estuarine plume, extending into the southern North Sea, as the riverine loads are deposited and diluted by coastal sea water.

Biological activity within the water column of the TM region may also be high relative to the rest of the estuary because of attachment of bacteria to suspended particles retained in the TM zone. A chlorophyll maximum also may be associated with the TM (Fisher *et al.*, 1995), presumably due to greater availability of nutrients in this region of rapid organic matter mineralization. What is effectively a fluidized bed reactor in the TM zone (Owens, 1986) may tend to enhance relatively slow biological processes such as nitrification in the water column by increasing the effective bacterial biomass retained on suspended particles in this region of estuary (Plummer *et al.*, 1978; Law *et al.*, 1992; Pakulski *et al.*, 1995). In other turbid estuaries there has been no detectable nitrification (Ogilvie *et al.*, 1997). Even when present, biological processes in the water column are unlikely to equal the magnitude of the same processes in the bottom sediments.

2. Particle–water Interactions

Phosphorus is a highly particle-reactive element in estuaries (Froelich, 1988), particularly in comparison with nitrogen (House *et al.*, 1998). The nature of these interactions has been ascribed to adsorption–desorption, co-precipitation–dissolution and ion exchange reactions with calcium (both as calcium carbonate and as apatite), iron and manganese, organic matter and clays (House *et al.*, 1998). In some cases colloidal material carried by the river system will be destabilized by the steep chemical gradients that occur early in estuarine mixing (House *et al.*, 1998).

The relative importance of these interactions probably varies in different systems. For instance, in the Amazon at low salinity, desorption of phosphorus from riverine particulate matter effectively doubles the flux of dissolved phosphorus from the Amazon into the Atlantic, compared with estimates based on the riverine dissolved phosphorus flux alone (Froelich, 1988). Global phosphorus budgets (e.g. Filipelli and Delaney, 1996), which are dominated by the largest and generally least contaminated rivers, also seem to require such large-scale release. In contrast, in the contaminated Humber estuary in central England, 80% of incoming dissolved phosphorus is removed within the estuary to particles. This removal occurs mainly at low salinities early in the estuarine mixing process and the ultimate fate of the particulate phosphorus formed is uncertain, as discussed above (Sanders *et al.*, 1997a). Different estuaries span the full range between these two extremes, as is evident from the compilation by Prastka *et al.* (1998).

The wide range in observed behaviour of phosphorus greatly complicates attempts to calculate the effects of any particular river inputs on adjacent coastal waters and prediction of how changes in such inputs will affect coastal ecosystems. Nixon *et al.* (1996) suggested that estuarine residence time may control the extent of estuarine phosphate removal, although they noted that the relationship is not particularly strong and it also cannot explain the release of DIP seen in some estuaries. More recently Prastka *et al.* (1998) have presented a simple model of phosphorus in estuaries using a K_d distribution coefficient to describe the relationship between the dissolved and particulate phases. The model is conceptually similar to the "phosphorus buffer mechanism" discussed by Froelich (1988). The model suggests that K_d values are relatively constant over a wide range of estuaries, although only 25–50% of the total particulate P appears to be involved in these exchange reactions. The model can explain successfully the different behaviour of phosphorus seen in various rivers (Figure 8), and suggests that the different behaviours are in part dependent on the riverine P concentrations. Increases in fluvial DIP inputs to some industrialized estuaries have raised DIP concentrations from < 1 μmol l^{-1} in many pristine systems (e.g. Froelich, 1988) to > 5 μmol l^{-1} in many industrialized estuaries (e.g. Justic et al., 1995; de Jonge, 1997; Prastka et al., 1998), shifting these estuaries from sources to sinks for dissolved phosphorus. The model also predicts that if the phosphorus-rich particles formed at low salinity are exported seaward, any adsorbed phosphorus will desorb. This prediction may be consistent with results from the Humber plume (Sanders et al., 1997a), but in some plume regions phytoplankton uptake may mask the process, or particulate levels in the outer estuary may not be high enough for such desorption reactions to be significant.

The Prastka et al. (1998) model will clearly require further evaluation, but it does highlight the critical role of particulate matter in estuarine processes. The concentration of suspended matter in estuaries is being extensively modified by many processes around the world, through the damming of rivers and changes in catchment land-use practices. These have a marked effect on fluvial

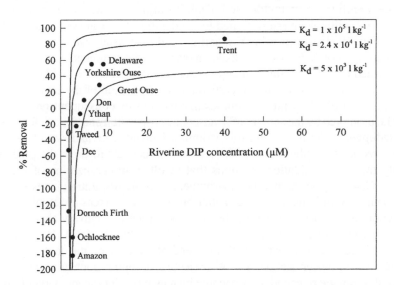

Fig. 8. Percentage removal (positive) or release (negative) of phosphate calculated according to the simple equilibrium model of Prastka et al. (1998) as a function of riverine DIP concentration. The continuous lines represent removal calculated using the annotated values of K_d; the labelled solid points represent literature-derived estimates of removal from various estuaries.

sediment fluxes (Milliman and Syvitski, 1992) as well as the fluxes of dissolved constituents (e.g. Humborg *et al.*, 1997). Furthermore, within many estuaries reclamation work and dredging has altered the suspended sediment load and removed the sites for sedimentation, thereby potentially modifying the capacity of the estuary to act as a source or sink for phosphorus (Jickells, 1998; Prastka *et al.*, 1998). Thus the nature of phosphorus cycling within estuaries may have changed fundamentally as a result of human interference. Decisions about the management of estuaries, particularly in the light of likely future climate change (Jickells, 1998), need to consider their effect on the flux of nutrients through the estuary as well as the direct effect on the estuarine habitat.

3. The Role of Sediments

The water column is less important than the bottom sediments in the budgets of biologically significant elements. Carbon and nitrogen assimilation during benthic primary production may exceed that in pelagic primary production (see Underwood and Kromkamp, this volume), and the bottom sediments exceed the water column as sites for the processing of nitrogen and phosphorus during organic matter degradation. Where extensive areas of saltmarsh or mudbanks increase benthic area in an estuary, the importance of benthic processing may be even greater.

The significance of bottom sediments to organic matter degradation and nutrient recycling is inversely related to the depth of the water column (Wollast and Billen, 1981; Jørgensen 1983), and in shallow coastal waters and estuaries the majority of organic matter breakdown occurs in the sediments, either because it is formed there by benthic primary production, or because of rapid deposition of POM from the water column, particularly in the region of the estuarine TM. Incorporation of POM into bottom sediments, and its consequent breakdown, may be promoted by the activity of invertebrate suspension and deposit feeders in the sediments (Herman *et al.*, this volume). However, breakdown of organic matter requires supplies of electron acceptors necessary to drive the predominantly microbial oxidation reactions that result in mineralization of organic matter. Generally, in the marine environment, aerobic respiration is responsible for about 50% of benthic organic matter breakdown, whereas anaerobic sulfate reduction largely accounts for the balance, as sulfate is relatively abundant (20 mmol l^{-1}) in seawater (Sørensen *et al.*, 1979; Nedwell, 1984). However, as salinity decreases towards the freshwater end of the estuary, the contribution of sulfate from seawater declines and other electron acceptors such as nitrate, introduced through the river end, may become increasingly significant, particularly in hypernutrified estuaries with high nitrate loads. For example, along the Great Ouse estuary nitrate respiration, as a percentage of total mineralization (indicated by benthic oxygen uptake), increased from 8% at the mouth to 18% in the upper estuary (Nedwell and Trimmer, 1996; Trimmer *et al.*, 1998).

4. Exchange at the Sediment–water Interface

In most estuaries sediments are sinks for nitrate, but are usually sources of ammonium and silicate to the water. Generally, exchange fluxes of phosphate seem to be small in most estuaries unless the sediment surface becomes deoxygenated (see below). Nitrate concentrations in sediment pore water are lower than in the water column because nitrate respiration within the sediment removes nitrate continually. (The exception to this is in sandy sediments, often near the mouth of estuaries, where nitrification occurs in the surface layers and nitrate can be exported to the water column (e.g. Ogilvie *et al.*, 1997).) The rate of transport of nitrate into the sediment will, therefore, be influenced by the nitrate concentration in the overlying water and saturation kinetics are commonly observed for uptake. Figure 9 shows the effect of nitrate concentration on nitrate flux into the sediments of the Great Ouse estuary. Increase of nitrate concentration in the water

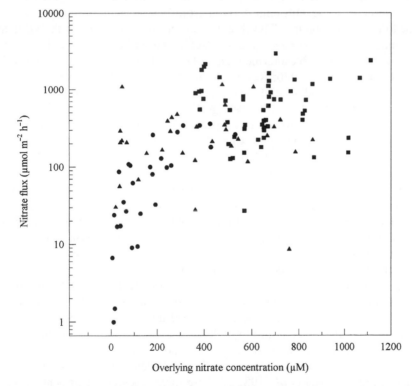

Fig. 9. Scatterplot of log nitrate flux into sediment against overlying estuarine water column nitrate concentrations in the Great Ouse. ■, Data for sites 1, 3 and 4 in the upper estuary (Nedwell and Trimmer, 1996); ▲, data for sites 4–9; ●, data for sites 10–12 in the lower estuary (Trimmer *et al.*, 1998). Each point is the mean of triplicate determinations.

up to about 400 μmol l^{-1} resulted in faster uptake of nitrate by the sediment, but at > 400 μmol l^{-1} there was no further increase as transport into the sediment became nitrate saturated (Trimmer *et al.*, 1998). Similar values for nitrate saturation of uptake rates have also been measured in other estuaries. As the maximum nitrate concentrations in the water near the head of the estuary were often > 400 μmol l^{-1}, the sediments in this region may have been taking up nitrate at their maximum capacity, but as the water column nitrate concentrations declined down the estuary due to dilution the sediments would have a spare capacity to take up nitrate. In effect, surge loads of nitrate into the estuary could be met by the spare capacity of downstream sediments to take up nitrate. In contrast, in oligotrophic low-nutrient estuaries, the capacity of the sediments to take up nitrate may never be exceeded by the water column concentration.

The sediment therefore acts as a sink for those solutes (O_2, SO_4^{2-}, NO_3^-) used as electron acceptors by organisms in the bottom sediments, with net fluxes into the sediments from the water column. Alternatively sediments may be a source of those nutrients derived from benthic mineralization fluxing back to the water column. The direction of net flux depends on the vertical concentration gradient of each solute. Usually the surface oxic layer is <1 mm deep in organically rich estuarine muds, to possibly a few millimetres in sands (e.g. Revsbech *et al.*, 1980; Revsbech and Jørgensen, 1986). The oxic layer depth may exhibit both spatial changes along an estuary in relation to organic matter availability, and seasonal changes in response to temperature (Ogilvie *et al.*, 1997), and its presence has a profound influence on the exchange of nutrients between sediment and water. Phosphate occurs in oxic sediments as the essentially insoluble Fe(III) species, but in anoxic environments iron is reduced biologically and phosphate may be solubilized. A surface oxic layer therefore acts as a solubility barrier, and soluble phosphate present in the anoxic layer cannot pass through the oxic layer to the water column. Any phosphate released within the sediments by the mineralization of deposited organic matter tends to be trapped until mobilized either by disappearance of the surface oxic layer, when soluble Fe(II) can move into the water, or by physical resuspension of sediment particles (see below).

Ammonium concentrations in sediment pore waters usually decrease towards the sediment surface because ammonification generates ammonium during mineralization of organic matter, and ammonium therefore diffuses upwards along the concentration gradient. Whether or not ammonium passes from the sediment to the water depends upon the extent to which it is nitrified to nitrate in the surface oxic layer of the sediment (Billen and Vanderborght, 1978; Billen, 1982; Nedwell *et al.*, 1982). Where there is a significant benthic oxic layer, it acts not only as a redox barrier for phosphate and Fe, but also as a potential nitrification barrier for ammonium. At the base of the oxic layer there are overlapping concentration gradients of ammonium diffusing upwards from the suboxic layer and oxygen diffusing downwards, where nitrifying bacteria are active. The greater the depth of the oxic layer, the greater is the probability that all of the ammonium will be intercepted and converted to nitrate (Billen and Vanderborght, 1978; Billen, 1982; Nedwell et al., 1982; Blackburn and Blackburn, 1993a,b). Where seasonal deposits of organic matter occur on the surface of the sediment, faster rates of oxygen removal deplete the depth of the oxic layer and facilitate the emission of ammonium (Nedwell et al., 1982). It has been shown (Blackburn and Blackburn, 1993a; Sloth et al., 1995) with numerical models of sediment–water exchange that in marine dominated sediments, with high rates of sulfate reduction, the degree of benthic nitrification may be strongly influenced by vertical migration of sulfide into the oxic layer where it will be reoxidized, competing with ammonium for the available oxygen, and reducing the importance of nitrification.

Urea, which can be assimilated as an N source by algae, can be a significant component of nitrogen exported from sediment to water, particularly where dense populations of sediment invertebrates are present and presumably arising from their excretory products (e.g. Boucher and Boucher-Rodini, 1988; Lomstein et al., 1989). However, bacteria such as Thiosphaera also can form urea during the degradation of ornithine (Pedersen et al., 1993a,b), and part of the export of urea from sediments may originate from this source. It appears that, in the absence of invertebrates, urea is exported from sediments in significant quantities only when there is an abundance of available organic matter (e.g. Lomstein et al., 1989; Pedersen et al., 1993a; Sloth et al., 1995) and urea export quickly declines back to negligible rates as the input of available organic substrates in the sediment is metabolized.

Particularly in sandy sediments with a significant oxic layer, a subsurface peak of nitrate concentration is often seen at the base of the oxic layer due to nitrification (e.g. Billen, 1982). This subsurface peak will result in bidirectional diffusion gradients of nitrate, both upwards out of the sediment into the water and downwards back into the suboxic layer where the nitrate may be denitrified. If nitrification is sufficiently intense, there may be no sediment–water export of ammonium, but only export of nitrate (e.g. Billen, 1982; Nedwell et al., 1982; Ogilvie et al., 1997). In temperate estuaries the depth of the benthic oxic layer varies seasonally, decreasing during summer as benthic respiration rates, increased by higher temperatures, mop up oxygen rapidly, but increasing during winter when the reduced rate of benthic respiration allows oxygen to penetrate more deeply. After winter, benthic nitrification then tends to increase during spring with initial seasonal temperature increase and the presence of an appreciable oxic layer depth but then declines in early summer as the temperature rises and the depth of the oxic layer diminishes and inhibits nitrification (Billen, 1982). There may be corresponding changes in sediment–water export of nitrogen, from nitrate during spring when nitrification is intense, to ammonium during summer when nitrification declines. In estuaries where either high temperature or high organic matter content permanently imposes a negligible oxic layer depth, nitrification may be essentially absent and export of nitrogen from the sediment is as ammonium.

5. Tidal Effects

Estuaries are dynamic environments whose intertidal sediments are inundated and exposed with the tidal ebb and flow. Sediment–water exchange occurs only when the sediment is water covered, and the duration of tidal cover decreases upshore. Sites in the high intertidal may differ greatly from those in the low intertidal which are almost permanently water covered. For example, drainage tends to increase up the intertidal gradient and tidal influence also declines towards the freshwater end of an estuary. As sediment–water export or import occurs only during the immersed period there may be intermittent inputs of solutes such as nitrate or sulfate, or export of products of mineralization such as ammonium or phosphate. Changes in the ionic strength of the water can also influence solute exchange. For example, NH_4^+ ions can be desorped from sediment particles by Na^+ in seawater. Gardner *et al.* (1991) have argued that mobilization of NH_4^+ from intertidal sediments by Na^+ in tidal water increases export of NH_4^+ and thereby diminishes benthic ammonium concentrations and nitrification in intertidal sediments, compared with freshwater sediments where ion pairing does not occur. Caetano *et al.* (1997) reported a pulsed release of NH_4^+ upon tidal inundation of intertidal sediments. Freshwater sediments generally have higher ammonium concentrations than estuarine or marine sediments, and a larger proportion (80–100%) of the ammonium flux is nitrified than in marine systems (40–60%) (Seitzinger, 1990).

6. Effect of Benthic Invertebrates on Sediment–water Exchange

Transport of solutes such as nitrate and oxygen into the sediment from the overlying water may be the result solely of molecular diffusion, although this is a relatively slow process. The irrigation of sediments by the respiratory streams of benthic invertebrates, or by their physical mixing activity, also stimulates transport of solutes into the sediment, effectively increasing the solute's vertical transport coefficient (e.g. Pelegri and Blackburn, 1996; see also Herman *et al.*, this volume). The burrows of benthic invertebrates also increase the available surface area across which solute transport can take place, stimulating sediment–water exchange rates (e.g. Fry, 1982). Irrigation of sediment by invertebrate activity may increase the oxygen input to the extent that the depth of the oxic layer is greatly increased (Henriksen *et al.*, 1980), potentially stimulating benthic nitrification, and hence nitrification–denitrification by the consequent greater availability of nitrate

(Pelegri and Blackburn, 1996). Gilbert *et al.* (1998) reported an approximate doubling of denitrification rates by polychaete bioturbation in muddy sand, compared with non-bioturbated sediment, whereas Rysgaard *et al.* (1995) showed, in a small Danish estuary, that benthic oxygen consumption, denitrification and coupled nitrification–denitrification were all linearly correlated with the density of the amphipod *Corophium*. Note, however, that Herman *et al.* (this volume) argue that bioturbation increases the proportion of organic matter mineralized anaerobically because it is buried deeper into the sediment where availability of oxygen is less.

7. Effect of Benthic Biofilms on Sediment–water Exchange

The presence of benthic algal biofilms can also have a major impact on sediment–water exchanges. Clearly, there can be no sediment–water exchange during sediment exposure at low tide (although drainage of pore water may occur), and the extent and duration to which algal biofilms will be illuminated when tidally covered will depend on their position in an estuary and the transparency of the estuarine water (see Underwood and Kromkamp, this volume). However, when photosynthesis is active the microphytobenthos may influence sediment–water fluxes in two ways. First, the algae effectively scavenge upwardly transported nitrogen and phosphorus, reducing under illumination the effluxes of ammonium and phosphate to the water column (e.g. Risgaard-Petersen *et al.*, 1994; Rysgaard *et al.*, 1995). Second, the generation of oxygen may extend the oxic layer depth and influence nitrification, denitrification and the exchange of DIP. Rysgaard *et al.* (1995) showed in the relatively oligotrophic Kertinge Nor estuary, Denmark, that during winter and spring, when denitrification of nitrate from the water column occurred, illumination decreased denitrification because it stimulated photosynthesis and caused a deeper benthic oxic layer, with a longer diffusion pathway for nitrate from the overlying water into the anoxic sediment. However, illumination increased coupled nitrification–denitrification, because the deeper benthic oxic layer stimulated the supply of nitrate from nitrification. In oligotrophic estuaries, where there are only low sedimentary concentrations of ammonium, benthic microalgae may compete with nitrifying bacteria for ammonium (Sundback and Granéli, 1988; Nielsen and Sloth, 1994), thereby reducing nitrification–denitrification.

F. Benthic Denitrification

1. Fate of Nitrate in Sediments

Nitrate acts as the terminal electron acceptor for anaerobic oxidation of organic matter by facultatively anaerobic, nitrate-respiring bacteria in the suboxic and anoxic regions of the sediment. The rate of nitrate-respiration is a function of both the concentration of nitrate and the concentration of available organic electron donors (organic substrates) within the sediment. The amount of available organic matter tends to be only a small fraction (< 1%) of the total organic matter present (e.g. Nedwell, 1987), the remainder being refractory. Experiments with anaerobic slurries of estuarine sediments have illustrated some of the factors that influence sedimentary nitrate reduction rates. First, with sediments from the hypernutrified Colne estuary it was apparent that pore water concentrations of nitrate were suboptimal for nitrate reduction, as addition of nitrate up to 1 mmol l^{-1} to slurries of sediment stimulated the rate of nitrate removal (Figure 10). This concentration of nitrate was seldom achieved even in the water column at the upper end of the estuary, and was very much lower within the sediment. Therefore, nitrate-reducing bacterial communities within the sediment must be always working at suboptimal concentrations of nitrate, and the nitrate-respiring potential of the benthic microbial community is never realized because the availability of nitrate within the sediment is limited by its transport into the sediment.

Furthermore, the end-products of microbial nitrate reduction vary. Nitrate can either be completely reduced to ammonium by bacteria with an essentially fermentative type of metabolism (a process termed *nitrate ammonification*

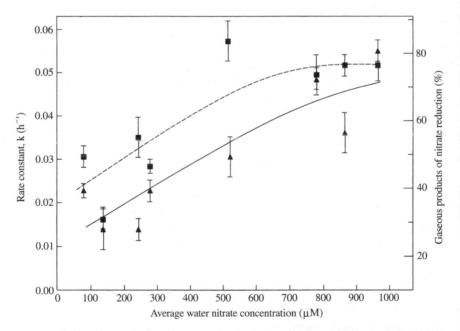

Fig. 10. Relationship between water nitrate concentration, the rate constant for nitrate removal (—▲—) and the percentage of gaseous end-products of nitrate reduction (--■--) in sediment from the River Colne estuary. Bars indicate standard error (*n* = 3). From King and Nedwell (1985).

(Cole and Brown, 1980)), or only partially reduced to nitrite (NO_2^-), or to gases such as N_2 or N_2O by bacteria with an oxidative or respiratory type of metabolism (a process termed *denitrification*) (Figure 11). Which of these processes predominates is ecologically very important as complete reduction of nitrate to ammonium conserves nitrogen within the aquatic environment, while denitrification results in the loss of nitrogen to the atmosphere. (Denitrification may be beneficial in ameliorating the effects of increased nitrogen loads within estuaries, but it may have other, less beneficial effects in the atmosphere; (see below).)

The rate of nitrate removal in slurries of Colne estuary sediments increased with nitrate concentration (King and Nedwell, 1985), and there was also a change in the end-products of nitrate reduction (Figure 10). The proportion of reduced nitrate being denitrified increased with nitrate concentration, to > 70% at > 1 mmol NO_3^- l^{-1}, while the proportion ammonified correspondingly decreased. Below 50 μmol l^{-1}, reduction of nitrate was almost entirely to ammonium. Clearly, the denitrifying bacterial community appeared to be competitively more successful for nitrate only at high nitrate concentrations, while nitrate ammonifiers were more successful at low nitrate concentrations. Subsequent work (Ogilvie *et al.*, 1997) has suggested that this is because nitrate ammonifiers have a higher affinity for nitrate at low nitrate concentrations, whereas the denitrifiers have a higher affinity for nitrate at high nitrate concentrations, although the reasons for this remain unclear. It is probable that the changes in end-products reflect not just changes in the ambient nitrate

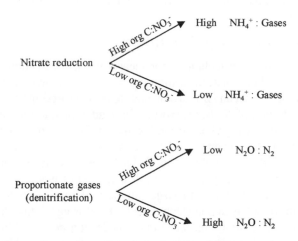

Fig. 11. End-products of nitrate reduction under conditions of either high ratios of organic carbon to nitrate (i.e. low nitrate concentrations) or low ratios of organic carbon to nitrate (i.e. high nitrate concentrations). The proportionate amounts of either N_2 or N_2O as gaseous end-products of denitrification for each scenario are also shown.

concentrations, but changes in the ratio of available organic matter to nitrate ($OM_{av} : NO_3^-$). At high $OM_{av} : NO_3^-$ nitrate-respiring bacteria are likely to be limited by the available electron sink, and complete reduction of NO_3^- to NH_4^+ optimizes electron disposal (Figure 11). At low $OM_{av} : NO_3^-$ ratios substrate availability, not the size of the electron sink, is the critical factor, and partial reduction of NO_3^- confers no disadvantage. As it is not possible to measure directly the amount of microbially available OM, the actual values of these ratios remain obscure, and their effects become apparent only in response to changes in the ambient NO_3^- concentrations affecting the ratios.

These observations imply that at the magnitude of nitrate concentrations typical of pristine estuaries (< 100 μmol l^{-1}), and certainly in the outer reaches of most estuaries where nitrate concentrations are reduced by dilution with coastal seawater, most nitrate reduced in bottom sediments will be converted to ammonium, conserving nitrogen within the system (Figure 11). In contrast, high nitrate concentrations, typical of hypernutrified upper estuaries, will result in an increase in the amount of nitrate denitrified, indicating the presence of a natural buffering mechanism with respect to nitrate concentrations in estuarine sediments. High nitrate concentrations in the water column stimulate rates of nitrate transport into the sediment, where increased pore-water nitrate concentrations will enhance rates of bacterial respiration of nitrate. At higher nitrate concentrations, as well as faster rates of nitrate reduction, greater proportions of the nitrate are removed by denitrification, together providing a larger nitrogen sink. The net effect of such benthic denitrification on the nitrogen load to the estuary will depend on the time that the nitrate resides in the estuary, in contact with the bottom sediments.

2. Methods of Measuring Denitrification

Probably the most direct technique for measuring denitrification would be to determine the benthic emission of dinitrogen gas (N_2). However, the difficulty with this approach is that any flux would have to be measured against a high background concentration of N_2. Some workers have attempted to measure N_2 fluxes after first removing the high background N_2 by purging sediment cores with inert gas such as argon (Seitzinger *et al.*, 1993; LaMontagne and Valiela, 1995), but it may take several days to reduce the N_2 background, during which time the sediment characteristics may change. While a number of workers have used this approach, it is not feasible with large numbers of replicated field samples. Probably the most widely applied technique for measuring benthic denitrification rates is the acetylene inhibition technique (Balderston *et al.*, 1976). Acetylene (10% v/v) blocks the enzyme *nitrous oxide reductase* so that during denitrification N_2O is not reduced further to N_2, and accumulates. It is possible to measure N_2O sensitively by gas chromatography, and so derive a denitrification rate from the rate of N_2O accumulation in the presence of acetylene. However, the method has a number of limitations, the first of which is that at even lower concentrations than those used to measure denitrification (10% v/v) acetylene also inhibits nitrification (Sloth *et al.*, 1992). Therefore, the acetylene block technique is likely to underestimate total benthic denitrification because sedimentary nitrate pools dependent upon nitrification will run down in the presence of acetylene. This is probably of more importance in relatively pristine sediments where overlying nitrate concentrations in the water column are low and nitrification is therefore of significance to nitrate supply, or in highly organic sediments where ammonification is intense, but of lesser importance in nutrified estuaries where denitrification is dominated by the supply of nitrate from the water column. Investigation of the relative importance of each of these sources of nitrate for benthic denitrification has been dramatically improved by the advent of the isotope pairing technique (Nielsen, 1992) which, with the addition of $^{15}NO_3^-$ to the water overlying a sediment core, permits the differentiation of N_2 originating from denitrification from nitrate from the overlying water (D_w) or from nitrate derived from coupled nitrification–denitrification within the sediment (D_n). Figure 12 shows the effect of changes in the concentration of nitrate in the overlying water on the rates of D_w and D_n measured in a sediment from the Colne estuary. It can be seen that

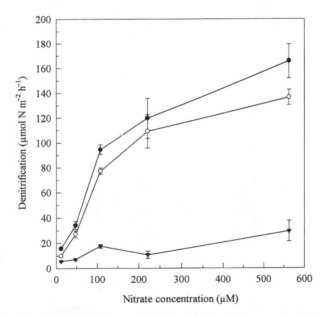

Fig. 12. Benthic denitrification at a site in the Colne estuary measured by the nitrogen isotope pairing technique (Nielsen, 1992), as a function of nitrate concentration in the overlying water. ●, Total denitrification; ○, denitrification of nitrate from the overlying water, D_w; ▼, denitrification coupled to nitrification, D_n. Measurements were made using artificial seawater at 15‰ with 100 μmol l^{-1} $^{15}NO_3^-$. Bars indicate standard error ($n = 4$).

increased water column nitrate indeed raised D_w, which depends upon nitrate from the overlying water column, but had no effect on D_n, which depends upon internal generation of nitrate from nitrification.

While the database from the isotope pairing technique currently remains rather limited, a number of features of estuarine denitrification have become apparent from applying this new technique. First, it is perhaps not surprising that the importance of D_w increases in nutrified areas of estuaries. Figure 13 shows the rates of benthic denitrification along the Colne estuary in the light and dark, and also the ratio of $D_w : D_n$ in the light and dark. Both the total rate of denitrification and the relative importance of D_w declined down the estuary to the sea, and D_n became more significant. What is also clear is the importance of internal nitrification–denitrification when nitrate concentrations in the water are low, or where there are high organic matter contents (and hence high ammonium levels) in sediments. There may also be seasonal changes in the significance of the two sources of denitrification, as D_w may vary with seasonal changes of nitrate load to the estuary, whereas D_n is dependent on ammonium generated internally by ammonification of organic matter and may be relatively constant (Rysgaard *et al.*, 1995). Figure 13 also shows that

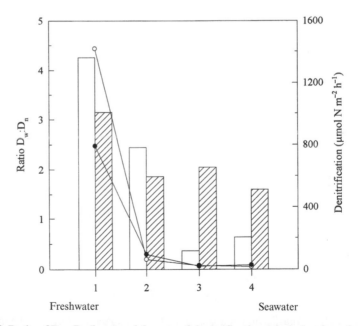

Fig. 13. Ratio of $D_w : D_n$ (bars) and the rate of denitrification (circles) at four sites in the River Colne estuary, UK, along a gradient of nitrate concentration decreasing towards the sea. Hatched bars and filled circles indicate measurements in the dark; open bars and circles indicate illuminated samples. Data are the annual means of monthly values.

benthic photosynthesis influences the balance of these processes. At sites 1 and 2 at the freshwater end of the estuary, illumination apparently increased the significance of D_w compared with the dark incubation, whereas in the two sites lower down the estuary illumination increased the significance of D_n.

3. Anammox Reaction

In addition to classical denitrification, which is driven by organic substrates, a process recently described by Mülder *et al.* (1995) may be significant in estuaries hypernutrified by STW effluents, and as a result of which contain high concentrations of both ammonium and nitrate. Originally isolated to treat industrial effluents with high ammonium levels, a consortium of bacteria from estuarine sediments was able to denitrify nitrate with ammonium as the electron donor.

$$5NH_4^+ + 3NO_3^- \Rightarrow 4N_2 + 9H_2O + 2H^+$$

This was termed the anammox process (Mülder *et al.*, 1995), and it can lead to large amounts of denitrification as both electron donor and electron acceptor are converted to N_2. While its role in the natural environment is largely unknown, its presence in Colne estuary sediments has been inferred at a site near a STW effluent outfall (Ogilvie *et al.*, 1997), and in the Thames estuary, near the main London STW works, where extremely high rates of coupled denitrification (D_n) have been detected which cannot be accounted for by nitrification coupled to denitrification (M. Trimmer and D. B. Nedwell, unpublished data). Anammox has the potential to be a major process of denitrification in such grossly polluted estuarine sediments, although its quantitative importance remains to be determined.

4. Significance of Attenuation of N Load in Estuaries

Subsequent to the first reports of significant denitrification in a nutrified estuary (Nedwell, 1974, 1975), it has become clear that there can be significant attenuation of N loads into estuaries, decreasing their impact on coastal seas. Seitzinger (1988) suggested that about 50% of the DIN load to estuaries was lost by denitrification, but the net attenuation is a function both of the rates of benthic processes removing nitrate from the water (see above) and of the retention time within the estuary of the freshwater bearing the high nutrient concentrations. Balls (1994) demonstrated a reciprocal relationship between the freshwater residence time and removal of nitrogen within a range of Scottish estuaries. Ogilvie *et al.* (1997) estimated that half the TOxN load to the Colne estuary was removed by denitrification. Nixon *et al.* (1996) emphasized the importance of the retention of the nutrient load within an estuary, and proposed a relationship between the proportion of the nitrogen load removed and the log of the freshwater residence time (Figure 14). Limitations in the precision of estimates of both estuarine flushing times and attenuation of N loads probably mean that the precise nature of the relationship between the two currently remains open to question, but it is clear that such estuarine attenuation has the potential to influence greatly the impact of terrestrially derived N loads on the coastal environment, and needs to be considered when the management of coastal ecosystems is being undertaken. For example, the North Quality Status Report (North Sea Task Force, 1993) does not consider these processes, but assumes that the N load monitored at gauging points at the river–estuary junction is the load that will impact the coastal sea. Similarly, the DIP cycling in estuaries discussed above is not always considered in assessing loads to coastal waters.

5. Estuarine Sources of N_2O

Denitrification attenuates the estuarine N load and removes N from the aquatic environment, much of it being converted to N_2. However, a small proportion of the nitrate is converted not to N_2 but to other gases such as NO and N_2O. Nitrous

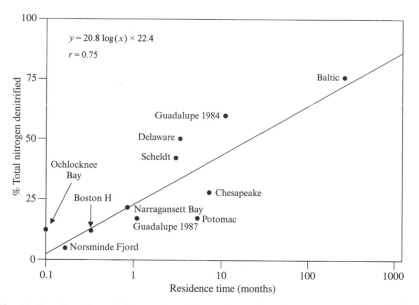

Fig. 14. Fraction of total nitrogen input from land and atmosphere that is denitrified in various estuarine systems as a function of mean water residence time. With kind permission from Kluwer Academic Publishers.

oxide (N_2O) particularly is a radiatively active trace gas which contributes to the greenhouse effect, having a global warming potential per mole some 270 times that of CO_2 over a 20-year period (Houghton et al., 1992). Nitrous oxide is increasing at about 0.3% per year in the atmosphere, and this increase is not matched by known N_2O sources (Houghton et al., 1992). The implication is that there are unknown global sources of N_2O. Recently it has become appreciated that estuaries are active sources of emission of N_2O (Law et al., 1992; Robinson et al., 1998) with N_2O saturation in the water many times that expected from air equilibration, and active emission to the atmosphere. The strong correlation of N_2O concentrations in Colne estuary water with nitrate concentrations suggested that it originated from benthic denitrification, rather than nitrification, and this would suggest in turn that the estuarine N_2O source strength would be greater in hypernutrified estuaries. An important feature is that N_2O emission from estuarine sediments seems to be maximum when sediments are immersed and the sedimentary nitrate pool is being constantly recharged from the water column. When tidally exposed, N_2O emissions from the sediments in the Colne decreased rapidly as the benthic nitrate pool ran down, but re-established immediately with the addition of nitrate (Robinson et al., 1998). This suggests that maximum N_2O emission will occur via the water–air interface, not the sediment–air interface, and may explain why in some studies where sediment–air emissions were measured directly (e.g. Middelburgh et al., 1995) the estuaries were apparently not important sources of N_2O.

It is difficult to decide whether estuaries are the missing source of N_2O, given the current poor database. Both Law et al. (1992) and Robinson et al. (1998) scaled up the N_2O emissions from the Tamar and Colne estuaries respectively, to estimate global estuarine N_2O emissions and concluded that about 0.15–0.45 Tg N year^{-1} N_2O-N were produced, about 15% of the "missing source". However, these estimates are dependent on the proportion of the denitrification that is converted to N_2O rather than N_2. Usually N_2O appears to be < 2% of denitrification, but the proportion is susceptible to environmental factors such as the organic carbon : nitrate ratio in sediments, and can be as high as 100% of denitrification in some sediments (García-Ruiz et al., 1998). An increase to about 13% of the total estuarine denitrification would be sufficient to account for the "missing source". At the time of writing (1999), estuarine sources of N_2O are not included in the global N_2O inventories.

G. Phosphorus in Sediments

Given the large reservoirs of phosphorus bound within sediments, the long-term fate of this material is important (Liss et al., 1991; Lebo and Sharp, 1992; de Jonge et al., 1993). For instance, Grant and Middleton (1990) have shown that modern sediments in the Humber estuary are enriched in phosphorus 5-fold over historical sediments of similar size and bulk chemistry. The phosphorus within sediments can be subdivided into the proportions associated with organic matter, iron and manganese oxyhydroxides, apatite, calcium phosphates, etc., using chemical leaching procedures (e.g. Ruttenberg, 1992; Slomp *et al.*, 1998). De Jonge *et al.* (1993) have used some novel experiments to compare these chemically available fractions with the biologically available component of the sediment-bound phosphorus. They suggest that 10–20% of the sediment phosphorus is directly available to bacteria. This implies that, in addition to adsorbed iron and manganese oxyhydroxide-bound phosphorus, some of the organic phosphorus or apatite or clay-bound phosphorus is available biologically.

As noted above, a significant fraction of phosphorus in estuarine sediments may be associated with oxidized Fe(III) (Lebo and Sharp, 1992; Berner and Rao, 1994; Prastka and Malcolm, 1994; Zwolsman, 1994b; Gunnars and Blomqvist, 1997; Slomp *et al.*, 1998) and possibly with manganese (Mn(IV)) oxyhydroxides (de Jonge *et al.*, 1993). The pattern seen in sediments from the Great Ouse estuary illustrates this well (Figure 15). Organic matter degradation in sediments releases phosphorus bound to the organic matter and also results in reducing conditions and the sequential use of a series of electron acceptors, including Fe(III), once free oxygen is exhausted (Jickells and Rae, 1997). The reduction of Fe(III) forms much more soluble Fe(II) and thus adsorption of phosphate by anoxic sediments can be much less than that by oxic sediments. Laboratory experiments conducted by Krom and Berner (1980) indicated that

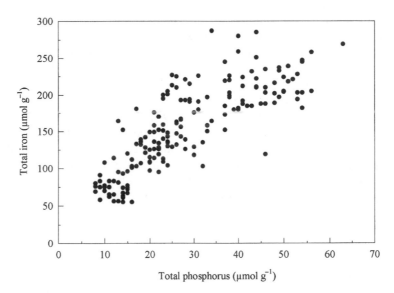

Fig. 15. Total sediment phosphorus (μmol g^{-1}) versus total sediment iron (μmol g^{-1}) from samples in the Great Ouse estuary, Norfolk, UK, illustrating the strong association between the two parameters ($p < 0.05$). Data from Prastka (1996).

the difference in the adsorption coefficient (a measure of the extent of adsorption) between oxic and anoxic sediments was of the order of 50-fold. De Jonge *et al.* (1993) reported 7-fold higher DIP fluxes from sediments under anoxic conditions compared with oxic conditions in passive flux experiments. Such redox processes are at least partly responsible for the pore water DIP concentrations increasing markedly with depth in many sediments, in addition to the release of phosphorus from the degradation of organic matter (Sundby *et al.*, 1992). Thus sediments that become anoxic at the sediment–water interface will release large amounts of adsorbed phosphorus, as is seen for example in the Scheldt River close to the estuary (Zwolsman, 1994b). This process can occur via passive diffusion, bioturbation (Sundby *et al.*, 1992) or tidal pumping (Shum and Sundby, 1996), although such extreme anoxia is not common in estuaries and even a shallow surface oxic layer may allow retention of phosphorus in sediments (Sundby *et al.*, 1992; van Raaphorst and Kloosterhuis, 1994; Prastka and Jickells, 1995). However, Slomp *et al.* (1998) have shown that when the oxic layer shrinks to less than 1 mm, bioturbation and the degradation of organic matter at the sediment surface can combine to increase the DIP flux from the sediments considerably over sediments with a deeper oxic layer. Gunnars and Blomqvist (1997) suggested that the dissolved iron : DIP ratio is an important determinant of DIP scavenging to ferric oxyhydroxides and that such scavenging may be more efficient in freshwater than marine systems because the ratio is higher.

IV. CONCLUSION

The major biological nutrients, N, P and Si, are all cycled in estuaries by different processes. The key processes influencing nitrogen loads are all biological, although of different significance. Primary production (both pelagic and benthic) would appear to be of limited importance even as a temporary sink for N, in relation to the total nitrogen loads to most estuaries, although it may be more important in low-nutrient estuaries. Benthic denitrification would appear to be of greater potential impact on the estuarine nitrogen load, and denitrification increases in importance in nutrified estuaries. Denitrification is a permanent sink for removal of nitrogen as gases. The retention time of freshwater (freshwater flushing time) in an estuary is a key factor influencing the degree to which the nitrogen load is attenuated, with longer retention times increasing the impact of benthic denitrification. Further work will be required if we are to define the role of estuaries as global sources of N_2O, particularly the environmental factors that govern the predominance of N_2O production over N_2. Silicate cycling also is dominated by biological processes, with silicate being almost entirely removed from water after the spring bloom, particularly in estuaries subjected to increased loads of N and P.

In contrast, phosphorus cycling is dominated by particle water interactions and sedimentation of particulate phosphorus. Sedimentation may be only a temporary sink for phosphorus. Its fate depends critically upon the sediment redox conditions (and hence organic matter deposition and sediment type) since recycling of phosphorus back to the water column occurs primarily under reduced conditions. Therefore, the extent of phosphorus removal in an estuary depends in part on the suspended sediment regime within an estuary, whereas nitrogen removal is related to the area of sediment. Both of these are subject to extensive modification as a result of human activity, and the overall impact of such changes on nutrient fluxes to coastal waters needs to be considered. Whilst DIP uptake on to particles can be demonstrated to occur in many estuaries and quantitatively explained in terms of the riverine phosphate concentration, we have relatively little information on the fate of the adsorbed phosphorus, although it seems to depend on the subsequent behaviour of the suspended sediment within an estuary. Thus, studies that couple the behaviour of dissolved and particulate phosphate are required.

We are beginning to understand in a general way the processes regulating nutrient transport within estuaries. However, we are still some way short of a quantitative and predictive capability, both to reduce requirement for major sampling programmes in all estuaries and also to allow for prediction of the impact of climatic change and managed realignment of estuaries on nutrient fluxes.

REFERENCES

Addiscott, T.M., Whitmore, A.P. and Powlson, D.S. (1991). *Farming, Fertilisers and the Nitrate Problem.* CAB International, Wallingford.

Allen, G.P., Savzay, G. and Castaing, J.H. (1976). Transport and deposition of suspended sediment in the Gironde estuary, France. In: *Estuarine Processes* (Ed. by M. Wiley), pp. 63–81. Academic Press, New York.

Allen, J.R., Slinn, D.J., Shammon, T.M., Hartnoll, R.G. and Hawkins, S.J. (1998). Evidence for eutrophication of the Irish Sea over four decades. *Limnol. Oceanogr.* **43**, 1970–1974.

Alongi, D.M. (1998). *Coastal and Ecosystem Processes.* Marine Science Series, CRC Press, London.

Amon, R.M.W. and Benner, R. (1994). Rapid cycling of high molecular weight dissolved organic matter in the ocean. *Nature* **369**, 549–552.

Azni, A.A.S. and Nedwell, D.B. (1986). The nitrogen cycle of an East Coast, UK, salt-marsh: nitrogen fixation, nitrification, denitrification, tidal exchange. *Estuar. Coastal Shelf Sci.* **22**, 689–704.

Balderston, W.L., Sherr, B. and Payne, W.J. (1976). Blockage by acetylene of nitrous oxide reduction in *Pseudomonas perfectomarinus. Appl. Environ. Microbiol.* **31**, 504–508.

Balls, P.W. (1994). Nutrient inputs to estuaries from nine Scottish east coast rivers: influence of estuarine processes on inputs to the North Sea. *Estuar. Coastal Shelf Sci.* **39**, 329–352.

Balls, P.W., Macdonald, A., Pugh, K. and Edwards, A. (1995). Long term nutrient enrichment of an estuarine system: Ythan Scotland (1958–1993). *Environ. Pollut.* **90**, 311–321.

Berner, R.A. and Rao, J. (1994). Phosphorus in sediments of the Amazon river and estuary: implications for the global flux of phosphorus to the sea. *Geochim. Cosmochim. Acta* **58**, 2333–2339.

Billen, G. (1982). Modelling the processes of organic matter degradation and nutrients recycling in sedimentary systems. In: *Sediment microbiology* (Ed. by D.B. Nedwell and C.M. Brown), pp. 15–52. Academic Press, London.

Billen, G. and Vanderborght, J.P. (1978). Evaluation of the exchange fluxes of materials between sediments and overlying waters from the direct measurements of bacterial activity and mathematical analysis of vertical concentration profiles in interstitial waters. In: *Biogeochemistry of Estuarine Sediments.* Proceedings of UNESCO/SCOR workshop, pp. 154–165. UNESCO, Paris.

Billen, G., Somville, M., De Becker, E. and Servais, P. (1985). A nitrogen budget of the Schellt hydrographical basin. *Neth. J. Sea Res.* **19**, 223–230.

Blackburn, T.H. and Blackburn, N.D. (1993a). Coupling of cycles and global significance of sediment diagenesis. *Marine Geol.* **113**, 101–110.

Blackburn, T.H. and Blackburn, N.D. (1993b). Models of nitrification and denitrification in marine sediments. *FEMS Microbiol. Lett.* **100**, 517–522.

Boucher, G. and Boucher-Rodini, R. (1988). *In situ* measurements of respiratory metabolism and nitrogen fluxes at the interface of oyster beds. *Mar. Ecol. Prog. Ser.* **44**, 229–238.

Boyle, E., Collier, R., Dengler, A.T., Edmond, J.M., Ng, A.C. and Stallard, R.F. (1974). On the chemical mass-balance in estuaries. *Geochim. Cosmochim. Acta* **38**, 1719–1728.

Caetano, M., Falcao, M., Vale, C. and Bebianna, M. (1997). Tidal flushing of ammonium, iron and manganese from intertidal sediment pore waters. *Mar. Chem.* **58**, 203–211.

Capone, D.G. (1983). Benthic nitrogen fixation. In: *Nitrogen in the Marine Environment* (Ed. by E.J. Carpenter and D.G. Capone), pp. 105–137. Academic Press, London.

Cloern, J.E. (1996). Phytoplankton bloom dynamics in coastal ecosystems: a review with some general lessons from sustained investigation of San Francisco Bay, California. *Rev. Geophys.* **34**, 127–168.

Cole, J.A. and Brown, C.M. (1980). Nitrite reduction to ammonia by fermentative bacteria: a short circuit in the biological nitrogen cycle. *FEMS Microbiol. Lett.* **7**, 65–72.

Colijn, F. and de Jonge, V. (1984). Primary production of microphytobenthos in the Ems-Dollard estuary. *Mar. Ecol. Prog. Ser.* **14**, 185–196.

Davidson, N.C., Laffoley, D.de'A, Doody, J.P., Way, L.S., Gordon, J., Key, R., Drake, C.M., Pienkowski, M.W., Mitchell, R. and Duff, K.L. (1991). *Nature Conservation and Estuaries in Great Britain.* Nature Conservancy Council, Peterborough.

Devol, A.H., Richey, J.E. and Forsberg, B.R. (1991). Phosphorus in the Amazon River mainstream: concentrations, forms and transport to the ocean. *Proceedings of Regional Workshop 3: South and Central America: Phosphorus Cycles in Terrestrial and Aquatic Ecosystems* (Ed. by H. Tiessen, D. Lopez-Hernandez and I.H. Salcedo), pp. 9–23. Saskatchewan Institute of Paedeology, University of Saskatchewan, Saskatoon.

DeVries, A.C. and Klavers, H.C. (1994). Riverine fluxes of pollutants: monitoring strategy first, calculation methods second. *Eur. Water Pollut. Cont.* **4**, 12–17.

Dicker, H.J. and Smith, D.W. (1980). Physiological ecology of acetylene reduction (nitrogen fixation) in a Delaware salt marsh. *Microb. Ecol.* **6**, 161–171.

DOE/CDEP (1986). *Nitrate in Water.* Pollution paper 26. HMSO, London.

Dubravko, J., Rabalais, N.N., Turner, R.E. and Dortch, Q. (1995). Changes in nutrient structure of river-dominated coastal waters: stoichiometric nutrient balance and its consequences. *Estuar. Coastal Shelf Sci.* **40**, 339–356.

Dyer, K.R. (1973). *Estuaries: A Physical Introduction.* Wiley, New York.

Dyer, K.R (1997). *Estuaries: A Physical Introduction.* 2nd Ed. Wiley, Chichester.

Feuillet-Girard, M., Gouleau, D., Blanchard, G. and Joassard, L. (1997). Nutrient fluxes on an intertidal mudflat in Marennes-Oléron Bay, and influence of the emersion period. *Aquat. Liv. Res.* **10**, 49–58.

Fichez, R., Jickells, T.D., and Edmunds, H.M. (1992). Algal blooms in high turbidity: a result of the conflicting consequences of turbulence on nutrient cycling in a shallow water estuary. *Estuar. Coastal Shelf Sci.* **35**, 577–592.

Filipelli, G.M. and Delaney, M.L. (1996). Phosphorus geochemistry of equatorial sediments. *Geochim. Cosmochim. Acta* **60**, 1479–1495.

Fisher, T.R., Harding, L.W., Stanley, D.W. and Ward, L.G. (1988). Phytoplankton, nutrients and turbidity in the Chesapeake, Delaware and Hudson estuaries. *Estuar. Coastal Shelf Sci.* **27**, 61–93.

Fisher, T.R., Melack, J.M., Grobbelaar, J.U. and Howarth, R.W. (1995). Nutrient limitation of phytoplankton and eutrophication of inland, estuarine, and marine waters. In: *Phosphorus in the Global Environment* (Ed. by H. Tiessen), pp. 301–322. SCOPE – Wiley, New York.

Froelich, P.N. (1988). Kinetic control of dissolved phosphate in natural rivers and estuaries: a primer on the phosphate buffer mechanism. *Limnol. Oceanogr.* **33**, 649–668.

Forsgren, G., Jansson, M. and Nilsson, P. (1996). Aggregation and sedimentation of iron, phosphorus and organic carbon in experimental mixtures of freshwater and estuarine water. *Estuar. Coastal Shelf Sci.* **43**, 259–268.

Fry, J.C. (1982). Interactions between bacteria and benthic invertebrates. In: *Sediment Microbiology* (Ed. by D.B. Nedwell and C.M. Brown), pp. 171–201. Academic Press, London.

Galloway, J.N., Howarth, R.W., Michaels, A.F., Nixon, S.W., Prospero, J.M. and Dentener, F.J. (1996). Nitrogen and phosphorus budgets of the North Atlantic Ocean and its watershed. *Biogeochemistry* **35**, 3–25.

García-Ruiz, R., Pattinson, S.N. and Whitton, B.A. (1998). Denitrification and nitrous oxide production in sediments of the Wiske, a lowland eutrophic river. *Sci. Tot. Environ.* **210/211**, 307–320.

Gardner, W.S. and Menzel, D.W. (1974). Phenolic aldehydes as indicator of terrestrially derived organic matter in the sea. *Geochim. Cosmochim. Acta* **38**, 813–822.

Gardner, W.S., Seitzinger, S.P. and Malczyk, J.M. (1991). The effects of sea salts on the forms of nitrogen released from estuarine and freshwater sediments: does ion pairing affect ammonium flux? *Estuaries* **14**, 157–166.

Gieskes, W.W.C. and Kraay, G.W. (1977). Continuous plankton records: changes in the plankton of the North Sea and its eutrophic Southern Bight from 1949 to 1975. *Neth. J. Sea Res.* **11**, 334–364.

Gilbert, F., Stora, G. and Bonin, P. (1998). Influence of bioturbation on denitrification activity in Mediterranean coastal sediments: an *in situ* experimental approach. *Mar. Ecol. Prog. Ser.* **163**, 99–107.

Glibert, P.M., Goldmann, J.C. and Carpenter, E.J. (1982). Seasonal variation in the utilization of ammonium and nitrate by phytoplankton in the Vineyard Sound, Massachussetts, USA. *Mar. Biol.* **70**, 237–249.

Gordon, D.C., Boudreau, P.R., Mann, K.H., Ong, J.-E., Silvert, W.L., Smith, S.V., Wattayakorn, G., Wulff, F. and Yanagi, T. (1995). *LOICZ Biogeochemical Modelling Guidelines. LOICZ/R&S/95–5.* LOICZ, Texel, The Netherlands.

Grant, A. and Middleton, R. (1990). An assessment of metal contamination of sediments in the Humber estuary. *Estuar. Coastal Shelf Sci.* **31**, 71–85.

Grobbelaar, J.U. (1990) Modelling phytoplankton productivity in turbid waters with small euphotic to mixing depth ratios. *J. Plankton Res.* **12**, 923–931.

Gunnars, A. and Blomqvist, S. (1997) Phosphate exchange across the sediment–water interface when shifting from anoxic to oxic conditions—an experimental comparison of freshwater and brackish marine systems. *Biogeochemistry* **37**, 203–226.

Henriksen, K., Hansen, J.I. and Blackburn, T.H. (1980). The influence of benthic infauna on exchange rates of inorganic nitrogen between sediment and water. *Ophelia Suppl.* **1**, 249–256.

Hickel, W., Mangelsdorf, P. and Berg, J. (1993). The human impact in the German Bight: eutrophication through three decades. *Helgolander Meeresuntersuchungen* **47**, 243–263.

Houghton, J.T., Callander, B.A. and Varney, S.K. (1992). *Climate change 1992. The Supplementary Report to the IPCC Scientific Assessment. Intergovernmental Panel on Climate Change.* Cambridge University Press, Cambridge.

House, W.A., Jickells, T.D., Edwards, A.C., Prastka, K.E. and Denison, F.H. (1998). Reactions of phosphorus with sediments in fresh and marine waters. *Soil Use Manag.* **14**, 139–146.

Howarth, R.W. (1988). Nutrient limitation of net primary production in marine ecosystems. *Annu. Rev. Ecol.* **19**, 89–110.

Howarth, R.W., Jensen, H.S., Marino, R. and Postma, H. (1995). Transport to and processing of P in near-shore and oceanic waters. In: *Phosphorus in the Global Environment* (Ed. by H. Tiessen), Scientific Committee on Problems in the Environment (SCOPE) **54**, pp. 323–356. Wiley, New York.

Howarth, R.W., Billen, G., Swaney, D., Townsend, A., Jaworski, N., Lajtha, K., Downing, J.A., Elmgren, R., Caraco, N., Jordan, T., Berendse, F., Freney, J., Kudeyerov, V., Murdoch, P. and Zhaoliang, Z. (1996). Regional nitrogen budgets and riverine N and O fluxes for the drainage to the North Atlantic Ocean: natural and human influences. *Biogeochemistry* **35**: 75–139.

Humborg, C., Ittekott, V., Cociasu, A. and v. Bodungen, B. (1997). Effects of Danube river dam on Black Sea biogeochemistry and ecosystem structure. *Nature* **386**, 385–388.

Jickells, T.D. (1998). Nutrient biogeochemistry of the coastal zone. *Science* **281**, 217- 222.

Jickells, T.D. and Rae, J.E. (1997). *Biogeochemistry of Intertidal Sediments.* Cambridge University Press, Cambridge.

Joint, I.R. (1978). Microbial production of an estuarine mudflat. *Estuar. Coast. Mar. Sci.* **7**, 185–95.

de Jonge, V.N. (1997). High remaining productivity in the Dutch Wadden Sea despite decreasing nutrient inputs from riverine sources. *Mar. Pollut. Bull.* **34**, 427–436.

de Jong, D.J. and de Jonge, V.N. (1995). Dynamics of microphytobenthos chlorophyll-a in the Scheldt estuary (SW Netherlands). *Hydrobiologia* **311**, 21–30.

de Jonge, V.N. and Essink, K. (1991). Long term changes in nutrient loads and primary and secondary production in the Dutch Wadden sea. In: *Spatial and Temporal Intercomparisons*, pp. 307–316. Olsen and Olsen, Denmark.

de Jonge, V.N., Engelkes, M.M. and Bakker, J.F. (1993). Bioavailability of phosphorus in sediments of the western Dutch Wadden sea. *Hydrobiologia* **253**, 151–163.

Jørgensen, B.B. (1983). Processes at the sediment–water interface. In: *The Major Biogeochemical Cycles and their Interactions* (Ed. by B. Bolin and R.B. Cook), pp. 477–509. SCOPE 21. Wiley, Chichester.

Jørgensen, B.B. and Sørensen, J. (1985). Seasonal cycles of O_2, NO_3^- and SO_4^{2-} reduction in estuarine sediments: the significance of a NO_3^- reduction maximum in the spring. *Mar. Ecol. Prog. Ser.* **48**, 147–154.

Justic, N. Rabalais, N.N., Turner, R.E. and Dortsch, Q. (1995). Changes in nutrient structure of river-dominated coastal waters: stoichiometric nutrient balance and its consequences. *Estuar. Coastal Shelf Sci.* **40**, 339–356.

Kaul, L.W. and Froelich, P.N. (1984). Modelling estuarine nutrient geochemistry in a simple system. *Geochim. Cosmochim. Acta* **48**, 1417–1433.

King, D. and Nedwell, D.B. (1985). The influence of nitrate concentration upon the end-products of nitrate dissimilation by bacteria in anaerobic salt marsh sediment. *FEMS Microbiol. Ecol.* **31**, 23–28.

Klein, A.W.O. and van Buuren, J.T. (1992). *Eutrophication of the North Sea in the Dutch Coastal Zone 1976–1990.* Ministry of Transport, Public Works and Water Management, Tidal Waters Division, The Hague.

Krom, M.D. and Berner, R.A. (1980). Adsorption of phosphate in anoxic marine sediments. *Limnol. Oceanogr.* **25**, 797–808.

Kromkamp, J.J., Peene, J., van Rijswijk, P., Sandee, A. and Goosen, N. (1995). Nutrients, light and primary production by phytoplankton and microphytobenthos in the eutrophic, turbid Westerschelde estuary (The Netherlands). *Hydrobiologia* **311** (*Dev. Hydrobiol.* **110**), 9–19.

LaMontagne, M.G. and Valiela, I. (1995). Denitrification measured by a direct N_2 flux method in sediments of Waquoit Bay, MA. *Biogeochemistry* **31**, 63–83.

Law, C.S., Rees, A.P. and Owens, N.J.P. (1992). Nitrous oxide: estuarine sources and atmospheric fluxes. *Estuar. Coastal Shelf Sci.* **35**, 301–314.

Lebo, M.E. (1990). Phosphate uptake along a coastal plain estuary. *Limnol. Oceanogr.* **35**, 1279–1289.

Lebo, M.E. and Sharp, J.H. (1992). Modeling phosphorus cycling in a well-mixed coastal plain estuary. *Estuar. Coastal Shelf Sci.* **35**, 235–252.

Lebo, M.E., Sharp, J.H. and Cifuentes, L.A. (1994). Contribution of river phosphate variations to apparent reactivity estimated from property salinity diagrams. *Estuar. Coastal Shelf Sci.* **39**, 583–594.

Liss, P.S. (1976). Conservative and non-conservative behaviour of dissolved constituents during estuarine mixing. In: *Estuarine Chemistry* (Ed. by J.D. Burton and S.P. Liss), pp. 93–100. Academic Press, London.

Liss, P.S., Billen, G., Duce, R.A., Gordeev, V.V., Martin, J.-M., McCave, I.N., Meincke, J., Milliman, J.D., Sicre, M.A., Spitzy, A. and Windom, H.L. (1991). What regulates boundary fluxes at ocean margins? In: *Ocean Margin Processes in Global Change* (Ed. by R.F.C. Mantoura, J.-M. Martin and R. Wollast), pp. 111–126. Wiley, New York.

Lomstein, B.A, Blackburn, T.H. and Henriksen, K. (1989). Aspects of carbon and nitrogen cycling in the northern Bering Shelf sediment. I: The significance of urea turnover in the mineralization of NH_4^+. *Mar. Ecol. Prog. Ser.* **57**, 237–247.

Middelburgh, J.J., Klaver, G., Nieuwenhuize, J., Markusse, R.M., Vlug, T. and van der Nat, F.J.W.A. (1995). Nitrous oxide emissions from estuarine intertidal sediments. *Hydrobiologia* **311**, 43–55.

Milliman, J.D. (1991). Flux and fate of fluvial sediment and water in coastal seas. In: *Ocean Margin Processes in Global Change* (Ed. by R.F.C. Mantoura, J.-M. Martin and R.J. Wollast), pp. 69–90. Wiley, New York.

Milliman, J.D. and Syvitski, J.P.M. (1992). Geomorphic/tectonic control of sediment discharge to the ocean: the importance of small mountain rivers. *J. Geol.* **100**, 525–544.

Monbet, Y. (1992). Control of phytoplankton biomass in estuaries: a comparative analysis of microtidal and macrotidal estuaries. *Estuaries* **15**, 563–571.

Moore, W.S. (1996). Large groundwater inputs to coastal waters revealed by 226-Ra enrichments. *Nature* **380**, 612–614.

Morris, A.W., Allen, J.I., Howland, R.J.M. and Wood, R.G. (1995). The estuary plume zone: source or sink for land derived nutrient discharges. *Estuar. Coastal Shelf Sci.* **40**, 387–402.

Morris, A. (1986). Removal of trace metals in the very low salinity zone of the Tamar estuary England. *Sci. Total Environ.* **49**, 297–304.

Mülder, A., van der Graaf, A.A., Robertson, L.A. and Kuenen, J.G. (1995). Anaerobic ammonium oxidation discovered in a denitrifying fluidized bed reactor. *FEMS Microbiol. Ecol.* **16**, 177–184.

Nedwell, D.B. (1974). Sewage treatment and discharge into tropical coastal waters. *Search* **5**, 187–190.

Nedwell, D.B. (1975). Inorganic nitrogen metabolism in a eutrophicated tropical mangrove estuary. *Water Res.* **9**, 221–231.

Nedwell, D.B. (1984). The input and mineralization of organic carbon in anaerobic aquatic sediments. *Adv. Microb. Ecol.* **7**, 93–131.

Nedwell, D.B. (1987). Distribution and pool sizes of microbially available carbon in sediment measured by a microbiological assay. *FEMS Microbiol. Ecol.* **45**, 47–52.

Nedwell, D.B. and Azni, A.A.S. (1980). Heterotrophic nitrogen fixation in an intertidal saltmarsh sediment. *Estuar. Coastal Marine Sci.* **10**, 699–702.

Nedwell, D.B. and Trimmer, M. (1996). Nitrogen fluxes through the upper estuary of the Great Ouse, England: the role of the bottom sediments. *Mar. Ecol. Prog. Ser.* **142**, 273–286.

Nedwell, D.B., Hall, S.E., Andersson, A., Hagstrøm, Å.F. and Lindstrøm, E.B. (1982). Seasonal changes in the distribution and exchange of inorganic nitrogen between sediment and water in the northern Baltic (Gulf of Bothnia). *Estuar. Coastal Shelf Sci.* **17**, 169–179.

Nielsen, L.P. (1992) Dentrification in sediment determined from nitrogen isotope pairing. *FEMS Microbiol. Ecol.* **86**, 357–362.

Nielsen, L.P. and Sloth, N.P. (1994). Denitrification, nitrification and nitrogen assimilation in photosynthetic microbial mats. In: *Microbial Mats, Structure Development and Environmental Significance* (Ed. by L.J. Stal and P. Caumette), pp. 319–324. NATO ISI Series G, Ecological Sciences. Springer, Berlin.

Nielsen, K., Nielsen, L.P. and Rasmussen, P. (1995). Estuarine nitrogen retention independently estimated by the denitrification rate and mass balance methods: a study of Norsminde Fjord, Denmark. *Mar. Ecol. Prog. Ser.* **119**, 275–283.

Nixon, S.W., Ammerma, J.W., Atkinson, L.P., Berounsky, V.M., Billen, G., Biocourt, W.C., Boynton, W.R., Church, T.M., DiToro, D.M., Elmgren, R., Garber, J.H., Giblin, A.E., Jahnke, R.A., Owens, N.J.P., Pilson, M.E.Q. and Seitzinger, S.P. (1996). The fate of nitrogen and phosphorus at the land–sea margin of the North Atlantic Ocean. *Biogeochemistry* **35**,141–180.

North Sea Task Force (1993). *North Sea Quality Status Report 1993. Oslo and Paris Commissions, London.* Olsen and Olsen, Fredensborg.

Officer, C.B. (1979). Discussion of the behaviour of nonconservative dissolved constituents in estuaries. *Estuar. Coastal Shelf Sci.* **9**, 91–94.

Ogilvie, B.G., Nedwell, D.B., Harrison, R.M., Robinson, A.D. and Sage, A.S. (1997). High nitrate muddy estuaries as nitrogen sinks: the nitrogen budget of the River Colne estuary. *Mar. Ecol. Prog. Ser.* **150**, 217–228.

Owens, N.J.P. (1986). Estuarine nitrification: a naturally occurring fluidized bed reactor? *Estuar. Coastal Shelf Sci.* **22**, 31–44.

Pacés, T. (1982). Natural and anthropogenic flux of major elements from central Europe. *Ambio* **11**, 206–208.

Pakulski, J.D., Benner, R., Amon, R., Eadie, B. and Whitledge, T. (1995). Community metabolism and nutrient cycling in the Mississippi River plume: evidence for intense nitrification at intermediate salinities. *Mar. Ecol. Prog. Ser.* **117**, 207–218.

Parr, W. and Wheeler, M.A. (1996). *Trends in Nutrient Enrichment of Sensitive Marine Areas in England.* English Nature, Peterborough.

Pedersen, H., Lomstein, B.A. and Blackburn, T.H. (1993a). Evidence for bacterial urea production in marine sediments. *FEMS Microbiol Ecol.* **12**, 51–59.

Pedersen H., Lomstein, L.A., Isaksen, M.F., Blackburn, T.H. (1993b). Urea production by *Thiosphaera pantotropha* and by anaerobic enrichment cultures from marine sediments. *FEMS Microbiol Ecol.* **13**, 31–36.

Peierls, B., Caraco, N., Pace, M. and Cole, J. (1991). Human influence on river nitrogen. *Nature* **350**, 386–387.

Pelegri, S.P. and Blackburn, T.H. (1996). Nitrogen cycling in lake sediments bioturbated by *Chironomus plumosus* larvae, under different degrees of oxygenation. *Hydrobiologia.* **325**, 231–238.

Plummer, D.H., Owens, N.J.P. and Herbert, R.A. (1978). Bacteria–particle interactions in turbid estuarine environments. *Continental Shelf Res.* **7**, 1429–1433.

Prastka, K.E. (1996). Phosphorus cycling in intertidal sediments. *PhD Thesis,* University of East Anglia, Norwich.

Prastka, K.E. and Jickells, T. (1995). Sediment geochemistry of phosphorus at two intertidal sites on the Great Ouse estuary south east England. *Neth. J. Aq. Ecol.* **29**, 245–255.

Prastka, K.E. and Malcolm, S.J. (1994). Particulate phosphorus in the Humber estuary. *Neth. J. Aq. Ecol.* **28**, 397–403.

Prastka, K., Sanders, R. and Jickells, T. (1998). Has the role of estuaries as sources or sinks of dissolved inorganic phosphorus changed over time? Results of a K_d study. *Mar. Pollut. Bull.* **36**, 718–728.

Raffaelli, D., Hull, S. and Milne, H. (1989). Long term changes in nutrients, weed mats and shorebirds in an estuarine system. *Can. Mar. Biol.* **30**, 259–270.

Rendell, A.R., Horrobin, T.M., Jickells, T.D., Edmunds, H.M., Brown, J. and Malcolm, S.J. (1997). Nutrient cycling in the Great Ouse estuary and its impact on nutrient fluxes to the Wash, England. *Estuar. Coastal Shelf Sci.* **45**, 653–668.

Risgaard-Petersen, N., Rysgaard, S., Nielsen, L.P. and Revsbech, N.P. (1994). Diurnal variation of denitrification and nitrification in sediments colonized by benthic microphytes. *Limnol. Oceanogr.* **39**, 573–579.

Revsbech, N.P. and Jørgensen, B.B. (1986). Microelectrodes: their use in microbial ecology. *Adv. Microb. Ecol.* **9**, 293–352.

Revsbech, N.P., Sørensen, J., Blackburn, T.H. and Lomholt, J.P. (1980). Distribution of oxygen in marine sediments measured with microelectrodes. *Limnol. Oceanogr.* **25**, 403–411.

Richey, J.E., Victoria, R.L., Salati, E. and Forsberg, B.R. (1991). The biogeochemistry of a major river system: the Amazon case study. In: *Biogeochemistry of Major World Rivers* (Ed. by E.T. Degens, S. Kempe and J.E. Richey), pp. 57–73. Wiley, New York.

Robinson, A.D., Nedwell, D.B., Harrison, R.M. and Ogilvie, B.G. (1998). Hypernutrified estuaries as sources of N_2O emission to the atmosphere: the estuary of the River Colne, Essex, UK. *Mar. Ecol. Prog. Ser.* **164**, 59–71.

Ruttenberg, K.C. (1992). Development of a sequential technique for different forms of phosphorus in marine sediments. *Limnol. Oceanogr.* **37**, 1460–1468.

Rysgaard, S., Christensen, P.B. and Nielsen, L.P. (1995). Seasonal variation in nitrification and denitrification in estuarine sediment colonized by benthic microalgae and bioturbating infauna. *Mar. Ecol. Prog. Ser.* **126**, 111–121.

Sanders, R. Jickells, T., Malcolm, S., Brown, J., Kirkwood, D., Reeve, A., Taylor, J., Horrobin, T. and Ashcroft, C. (1997a). Nutrient fluxes through the Humber estuary. *J. Sea Res.* **37**, 3–23.

Sanders, R., Klein, C. and Jickells, T.D. (1997b). Biogeochemical cycling in the Great Ouse estuary, Norfolk, UK. *Estuar. Coastal Shelf Sci.* **44**, 543–555.

Scudlark, J.R. and Church, T.M. (1993). Atmospheric input of inorganic nitrogen to Delaware Bay. *Estuaries* **16**, 747–759.

Seitzinger, S.P. (1988). Denitrification in freshwater and coastal marine ecosystems: ecological and geochemical significance. *Limnol. Oceanogr.* **33**, 702–724.

Seitzinger, S.P. (1990). Denitrification in aquatic systems. In: *Denitrification in Soil and Sediment* (Ed. by N.P. Revsbech and J. Sørensen), pp. 301–322. Plenum Press, New York.

Seitzinger, S.P. and Sanders, R.W. (1997). Contribution of dissolved organic nitrogen from rivers to estuarine eutrophication. *Mar. Ecol. Prog. Ser.* **159**, 1–12.

Seitzinger, S.P., Nielsen, L.P., Caffrey, J. and Christensen, P.B. (1993). Denitrification measurements in aquatic sediments: a comparison of three methods. *Biogeochemistry* **23**, 147–167.

Shiller, A.M. (1996). The effect of recycling traps and upwelling on estuarine chemical flux estimators. *Geochim. Cosmochim. Acta* **60**, 3177–3185.

Shum, K.T. and Sundby, B. (1996). Organic matter processing in continental shelf sediments—the subtidal pump revisited. *Mar. Chem.* **53**, 81–87.

Simpson, J.H., Gong, W.K. and Ong, J.E. (1997). The determination of the net fluxes from a mangrove estuary system. *Estuaries* **20**, 103–109.

Slomp, C.P., Malschaert, J.F. and van Raaphorst, W.V. (1998). The role of adsorption in sediment water exchange of phosphate in the North Sea continental margin sediments. *Limnol. Oceanogr.* **43**, 832–846.

Sloth, N.P., Nielsen, L.P. and Blackburn, T.H. (1992). Nitrification in sediment cores measured with acetylene inhibition. *Limnol. Oceanogr.* **37**, 1108–1112.

Sloth, N.P., Blackburn, T.H., Hansen, L.S., Risgaard-Petersen, N. and Lomstein, B.A. (1995). Nitrogen cycling in sediments with different organic loading. *Mar. Ecol. Prog. Ser.* **116**, 163–170.

Soetart, K. and Herman, P.M. (1995). Nitrogen dynamics in the Westerschelde estuary (SW Netherlands) estimated by means of the ecosystem model MOSES. *Hydrobiologia.* **311**, 225–246.

Sørensen, J., Jørgensen, B.B. and Revsbech, N.P. (1979). A comparison of oxygen, nitrate and sulfate respiration in coastal marine sediments. *Microb. Ecol.* **5**, 105–115.

Sundbäck, K. and Granéli, W. (1988). Influence of microphytobenthos on the nutrient flux between sediment and water: a laboratory study. *Mar. Ecol. Prog. Ser.* **43**, 63–69.

Sundbäck, K., Enoksson, V., Granéli, W. and Pettersson, K. (1991). Influence of sublittoral microphytobenthos on the oxygen and nutrient flux between sediment and water: a laboratory continuous-flow study. *Mar. Ecol. Prog. Ser.* **74**, 263–279.

Sundby, B., Gobeil, C., Silverberg, N. and Mucci, A. (1992). The phosphorus cycle in coastal marine-sediments. *Limnol. Oceanogr.* **37**, 1129–1145.

Trimmer, M., Nedwell, D.B., Sivyer, D.B. and Malcolm, S.J. (1998). Nitrogen fluxes through the lower estuary of the river Great Ouse, England: the role of the bottom sediments. *Mar. Ecol. Prog. Ser.* **163**, 109–124.

Turner, R.E. and Rabalais, N.N. (1991). Changes in Mississippi River water quality this century: implications for coastal food webs. *BioScience* **41**, 140–147.

Uncles, R.J. and Stephens, J.A. (1993). Nature of the turbidity maximum in the Tamar estuary, UK. *Estuar. Coastal Shelf Sci.* **36**, 413–431.

van Raaphorst, W. and Kloosterhuis, H.T. (1994). Phosphate sorption in superficial sediments. *Mar. Chem.* **48**, 1–16.

Vannekom, A.J. and Salomons, W. (1980). Pathways of nutrients and organic matter from land to ocean through rivers. In: *River Inputs to Ocean Systems*, (Ed. by J.-M. Martin, J.D. Burton and D. Eisma), pp. 33–51. UNESCO, Paris.

Vitousek, P.M. and Howarth, R.W. (1991). Nitrogen limitation on land and sea: how can it occur? *Biogeochemistry* **13**, 87–115.

Walling, D.E. and Webb, B.W. (1985). Estimating the discharge of contaminants to coastal waters by rivers: some cautionary comments. *Mar. Pollut. Bull.* **16**, 488–492.

Wollast, R. (1983). Interactions in estuaries and coastal waters. In: *The Major Biogeochemical Cycles and their Interactions* (Ed. by B. Bolin and R.B. Cook), pp. 385–407. SCOPE–Wiley, New York.

Wollast, R. and Billen, G. (1981). The fate of terrestrial organic carbon. In: *The Flux of Carbon from the Rivers to the Oceans*, US Department of Energy, CONF-8009/40/UC. USDOE, Washington, DC.

Wollast, R. and Peters, J.J. (1978). Biogeochemical properties of an estuarine system: the River Scheldt. In: *Biochemistry of Estuarine Sediments* (Ed. by G.D. Goldberg), pp. 279–293. UNESCO, Paris.

Zwolsman, J.J.G. (1994a). *North Sea Estuaries as Filters for Contaminants*. Report T1233. Delft Hydraulics, The Netherlands.

Zwolsman, J.J.G. (1994b). Seasonal variability and biogeochemistry of phosphorus in the Scheldt estuary, southwest Netherlands. *Estuar. Coastal Shelf Sci.* **39**, 227–248.

Primary Production by Phytoplankton and Microphytobenthos in Estuaries

G.J.C. UNDERWOOD AND J. KROMKAMP

I. SUMMARY

Phytoplankton and microphytobenthos are important components of estuarine ecosystems. In this review we discuss the factors known to influence the photosynthesis, primary production and biomass of these two groups of primary producers, and the application of new techniques.

There is a large interplay of variables influencing the rate of phytoplankton photosynthesis (nutrient or light limitation, osmotic stress) and factors influencing biomass like grazing, washout, resuspension and deposition. Phytoplankton in estuaries may experience rapid changes in the type of limitation (nutrients, light) and different physical environments (mixing, salinity), and these changes may influence species composition. Interannual variability in primary production can to a large extent be explained by changing weather

ADVANCES IN ECOLOGICAL RESEARCH VOL. 29
ISBN 0–12–013929–4

conditions and changing land use, as the watershed and rainfall determine the nutrient and sediment input into estuaries. These inputs can both stimulate primary production when the system is nutrient limited or when the light conditions improve due to vertical density stratification, or it can decrease primary production as sediment-laden water can decrease the light availability or flush out populations. There is a wealth of evidence that, due to increased land use and the concomitant increasing nutrient load, many estuaries have undergone eutrophication. The effect may to a large extent be dampened out when grazing by suspension feeders is important, as might have been the case in the Eastern Scheldt estuary, but other estuaries have shown signs that eutrophication might lead to the development of harmful algal blooms. Even in this case, total primary production will not necessarily change (in the case of turbid systems), but the changes in concentration and nutrient ratios might influence species composition, and it is important to realize that changes in the dominant species can have ecological implications. Thus, in order to understand ecosystem functioning better, we need to be able to understand what determines phytoplankton species composition and succession during blooms, what happens when blooms decay, how contaminants influence bloom dynamics and what is the interplay between nutrient enrichment and harmful algal blooms. These questions should preferably be addressed in integrated research programmes.

The relationship between irradiance and photosynthesis of microphytobenthos is fairly well described, although more research is needed into the effects of microspatial distribution within the vertical profile, the degree of coupling between photosynthesis and population growth, and the different facets of nutrient limitation (in its broadest sense) on cell physiology. Realization that the species composition of a biofilm may be important, especially given that new techniques measure at a scale comparable with individual cells, requires that more than the (all too common) general description of a "diatom biofilm" should be attempted. Many questions remain about what factors control microphytobenthic biomass on muddy shores (resuspension, nutrients, grazing, exposure, desiccation, etc.), and whether predictive models can be generated. Such data are needed to permit accurate estimation of whole estuarine microphytobenthic primary production. Indeed, while only a few estimates of the contribution of microphytobenthos production to total estuarine production are available, statements about the importance of microphytobenthic activity in such systems are common. There is little evidence that microphytobenthic assemblages in cohesive sediments are nutrient limited, although carbon dioxide limitation of photosynthesis has been suggested. However, nutrient limitation can refer to either biomass accumulation or rates of processes, and within an estuary nutrient concentrations can co-vary with other factors that also influence microphytobenthos. Microphytobenthic biofilms may play an important role (or barrier) in the exchange of nutrients between the sediments and the water phase, and in effecting bacterial processes. In low-nutrient systems, release of nutrients to the water phase from the sediment might be decreased by uptake of microphytobenthos, and this effect might be more pronounced when light is available when the sediments are submerged (i.e. in microtidal estuaries). Perhaps one of the obvious gaps in microphytobenthic data sets is the general scarcity of long-term data sets. With the increase in research on cohesive sediment microphytobenthos over the past 5 years, it can be hoped that longer-term data sets will become available. Such data may contribute to our ability to predict biomass and in determining the long-term response of microphytobenthic assemblages to eutrophication or improving water quality.

II. INTRODUCTION

Estuaries are at the interface between the terrestrial and marine environments. Rivers concentrate inputs from the catchment which are transported over a short distance (i.e. an estuary or delta) into coastal seas. Estuaries are valuable environments, both ecologically and economically. Large proportions of the world's human population live close to estuaries, and their influence causes problems in terms of environmental degradation and management (Viles and Spencer, 1995). Projections for growth and industrial and urban development for the twenty-first century indicate that these issues will be of increasing concern. As such, there are pressing reasons to develop our scientific under-standing of how estuaries function. The functioning of an estuary is dependent on a wide range of factors (e.g. geomorphology, tidal and freshwater flushing time, nature and extent of intertidal area, inputs, climatic conditions, tidal range), which results in there being many different types of estuarine systems. Of these, temperate macrotidal estuaries are probably the most intensively studied, and other types of estuarine systems may function quite differently. This review examines the role of autochthonous primary production within estuaries, focusing on macrotidal temperature estuaries.

Estuaries are among the most productive marine ecosystems of the world, with primary production estimates for phytoplankton ranging between 7 and 875 g C m^{-2} year^{-1} (Boynton et al. 1982; Table 1), and those for microphyto-benthos between 29 and 234 g C m^{-2} year^{-1} (Heip et al., 1995; MacIntyre et al., 1996; Table 2). There is a large degree of uncertainty about some of these estimates. The high degree of spatial and temporal heterogeneity of phyto-plankton and especially of microphytobenthos within estuaries (see Sections IV.C and D) can cause significant problems with regard to scaling up measurements (Shaffer and Onuf, 1985), while the various methods and approaches used (see below) generate estimates of gross photosynthesis, net photosynthesis and net community production that may not be directly comparable. Although differences in estimates of annual primary production

Table 1

Phytoplanktonic primary production in different estuaries

Estuary	References	Station (code) or time period	Annual production (gC m⁻²)	Mean chlorophyll a (mg m⁻³)	Max. daily production (gC m⁻²)	Max chlorophyll a (mg m⁻³)
Europe						
Oosterschelde	Wetsteyn and Kromkamp (1994)	Outer (P5)	223–540	3.0–7.4	4–10	14–43
		Central (P3)	242–406	3.0–6.6	4–8.5	16–36
		Inner (O21)	230–460	3.6–9.1	4–10	10–50
Westerschelde	van Spaendonk et al. (1993)	Outer	212–230	7.7–8.7	2.4–3.6	21–24
	Kromkamp et al. (1995)	Central	184–197	8.5–11.1	1.2–3.7	26–33
		Inner	75–100	7.7–20.1	0.9–1.7	21–41
		Freshwater	500–875	32.2–50.9	2.5–2.8	72–120
Bristol Channel	Joint and Pomroy (1981)	Outer	165			1–2
		Central	49			2–10
		Inner	7–8			1–3
Ems-Dollard	Colijn (1983)	Outer	166–408	4.2–12.0	1.7–6.1	16–44
		Central	26–154	4.6–9.9	1.6–4.5	13–45
		Inner	70	3.9–11.9	0.6–0.9	11–45
Marsdiep	Cadée (1986)	1955–1975	140–200	3–6	1.5–3.5	
	Cadée and Hegeman (1993)	1975–1992	250–390	6–15	5–6.5	38–42
United States						
San Francisco Bay	Cole and Cloern (1984, 1987)	South Bay	150		2.2–3.3	10–70
	Cloern (1996)	San Pablo	110–130		0.9–1.2	6–14
		Suisun Bay	93–98	1.1–14.6	<0.55	20–50
Tomales Bay	Cole (1989)	Outer (1–4)	440–520	7–8	2–5	6–12
		Central (6–12)	360–810	5–11	3–9	15–83
		Inner (14–18)	60–210	2.8–4	0.1–0.5	3–9
Neuse River Estuary	Paerl et al. (1998) Pinckney et al. (1998).	Integrated	360–531		2.5–3.4	40 to >50

Location	Reference	Zone				
Chesapeake Bay	Harding (1994)	Outer (I)		0.8–8.9		
	Boynton et al. 1982)	Central (III–IV)	337–782	4.0–10.5		15
	Harding (1994)					
	Harding (1994)	Inner (VI)		2.1–16.1		
Hudson River	Cole et al. (1992)	Freshwater	70–220	18–22	1.2	30–55
Bash Harbor Marsh	Kinney and Roman (1998)	Lower marsh	11	2.6	12	11
Delaware estuary	Malone (1977)	Outer	344			
		Central	296			40
		Inner	200			
Peconic Bay	Bruno et al. (1980, 1983)	Outer (8–9)	162–191	3.0–3.5	2.1–4.6	14
		Central (4–7)	213	2.9–3.2	1.4	13
		Inner (1–3)		3.5–10.3		
Asia						
Pearl River estuary	Lee (1990)	Pond 18	20	3–5		7–16

When data for several years were available, the range is given.

Table 2

Microphytobenthic primary production from intertidal sediments in different estuaries. Adapted from Heip *et al.* (1995) and MacIntyre *et al.* (1996) with additions

Estuary	Method	Annual production (g C m^{-2})	Mean chlorophyll *a* (mg m^{-2})	Daily production (mg C m^{-2})	References
Europe					
Ythan estuary, Scotland	^{14}C	116	25–34 μg g^{-1} sediment	9–226	Leach (1970)
Wadden Sea, Netherlands	^{14}C	60–140	40–400	15–1120	Cadée and Hegeman (1974) and Cadée and Hegeman (1977)
Lynher estuary, UK	^{14}C	29–188	3–13 μg g^{-1}	<50–920	Joint (1978)
Ems-Dollard, Netherlands	^{14}C	143	30–80 μg g^{-1}	5–115 (h^{-1})	van Es (1982)
	O$_2$	69–314	—	0–1900	Colijn and de Jonge (1984)
Oosterschelde, Netherlands	^{14}C	62–276	33–184	600–1370	Nienhuis *et al.* (1985)
Pre 1985	^{14}C	105–210	99–212		de Jong *et al.* (1994)
Post 1985	^{14}C	150	115		
	^{14}C	242	195		
Westerschelde, Netherlands	^{14}C	136	113	—	de Jong and de Jonge (1995)
Ria de Arosa, Spain	^{14}C	54			Varela and Penas (1985)
Tagus estuary, Portugal	O$_2$	47–178		5–32 (h^{-1})	Brotas and Catarino (1995)
Africa					
Langebaan Lagoon, South Africa	^{14}C	63 (sand) / 253 (mud)		17–69	Fielding *et al.* (1988)
North America					
False Bay, Washington	O$_2$	143–226	30–70 μg g^{-1}		Pamatmat (1968)
Falmouth Bay, Massachusetts	^{14}C	106	—	5–85 (h^{-1})	van Raalte *et al.* (1976)
Bay of Fundy, Canada	O$_2$	47–83	10–500		Hargrave *et al.* (1983)
Graveline Bay, Mississippi	^{14}C			5–56 (h^{-1})	Sullivan and Montcreiff (1988)
North Inlet, South Carolina	O$_2$ micro	93	60–120	19–180	Pinckney and Zingmark (1993)
	O$_2$ micro	56–234			Pinckney (1994)
Weeks Bay, Alabama	O$_2$ micro	90.1	0.2–30.7	10–750	Schreiber and Pennock (1995)

are due partly to differences in methodology, primary production is also the outcome of interplay between several bottom-up and top-down processes. This review will discuss several of these, how they interact, the impact new techniques have had on our understanding, and suggest some areas where further investigations are required.

III. COMPARISON: METHODOLOGY

The term primary production is used widely, but there is no generally accepted definition of its meaning. Generally it is viewed as the assimilation of inorganic carbon and nutrients into organic matter by autotrophs, and this is also the definition used in dictionaries of biological terms (Abercrombie *et al.*, 1973). Hence, primary production is a rate. Although this definition also includes production by chemoautotrophs, this is normally not measured, because most primary production measurements on phytoplankton are made with the ^{14}C method, and with this method the dark-bottle values are usually subtracted from light-bottle values in order to obtain a true phytoplankton measurement (Banse, 1993).

There is more confusion about the term gross primary production. This is partly caused by methodological differences. Platt *et al.* (1984) defined gross primary production as the rate of photosynthetic energy conversion of light energy into chemical energy, and this reflects the view held by theoretical ecologists and plant biophysicists/physiologists who view primary production as energy conversion (i.e. as the first steps in photosynthesis). The term gross photosynthesis is perhaps more correctly used to describe the rate of electron flow from water to terminal electron acceptors in the absence of respiratory losses (Falkowski and Raven, 1997), which can be measured as gross oxygen evolution. Williams (1993a) defined gross primary production as organic carbon production by the reduction of CO_2 as a consequence of the photosynthetic process over some specific period. This is more the viewpoint of the community ecologist and biogeochemist, who interpret primary production in terms of nutrient flow (Heip *et al.*, 1995). For a review on production terms, the reader is referred to Williams (1993a) and Falkowski and Raven (1997).

Net primary production is normally defined as gross primary production minus autotrophic respiration (Platt *et al.*, 1984; Williams, 1993a). Although this definition is quite straightforward, its measurement is not. Phytoplankton primary production is usually measured by the ^{14}C method (incorporation of radioactive dissolved inorganic carbon into either total or particulate organic C). Because of the uncertainties associated with the ^{14}C method, (i.e. does it measure gross or net photosynthesis or something in between?) comparison between studies can be difficult. Some authors have used 24-h incubations to arrive at net production; others have used short-term incubation times to obtain an estimate of gross production. Williams (1993b) argued on the basis of modelling that most C-fixation measurements were probably closer to gross photosynthesis, especially with short incubation times and low growth rates. Williams (1993b) recognized that a weakness in the models was the possibility of intracellular recycling of fixed C. In a later paper Williams and Lefèvre (1996) argued convincingly that the ^{14}C method most likely measures rates close to net photosynthesis, even with short incubation times. If we assume this to be true, most of the ^{14}C measurements determine what is sometimes called net photic zone (NPZ) production. This facilitates the comparison between phytoplankton primary production in estuaries. In order to calculate net water column production from NPZ production, the respiratory losses in the aphotic zone and during the night have to be estimated. We will demonstrate later in this review that this is no easy task, and that the evidence in the literature is conflicting.

The ^{14}C method is normally used to measure phytoplankton primary production; a few studies use oxygen exchange techniques (Randall and Day, 1987; Lee, 1990; Kinney and Roman, 1998). In microphytobenthos production studies, different techniques are widely used: (1) bell jars or benthic chambers, (2) ^{14}C uptake in slurries (e.g. Blanchard and Cariou-Le Gall, 1994; MacIntyre and Cullen, 1995; Blanchard et al., 1996; Barranguet et al., 1998), and (3) the oxygen microelectrode technique (Revsbech and Jørgensen, 1983).

(1) The bell jar (or benthic chamber) technique measures the change in oxygen concentration (Pomeroy, 1959; Pamatmat, 1968; Lindeboom and de Bree, 1982), or ^{14}C uptake from the overlying water over a specified time period (Rasmussen et al., 1983; Colijn and de Jonge, 1984). In the case of the oxygen exchange method, it measures net community production, and when the dark values are added to the light values an estimate of the gross production rate is obtained (with the assumption that respiration in the light and dark does not differ). When the technique is used with ^{14}C uptake, the results tend to underestimate carbon fixation rates because it is not possible easily to measure the specific activity of the DIC pool within the thin photosynthetically active layer (Heip et al., 1995; Smith and Underwood, 1998; Vadeboncoeur and Lodge, 1998). Modification of the technique, percolating seawater of known specific ^{14}C activity, resulted in higher estimates of carbon fixation (Jönsson 1991) and can be used successfully in porous non-cohesive sediments (Sundbäck et al., 1996). However, the percolation technique cannot be used with cohesive sediments. It is possible with in situ chambers and ^{14}C incubation to obtain good relationships between carbon fixation and illumination (Sundbäck and Jönsson, 1988; Smith and Underwood, 1998), and if carbon pathways within sediments are being studied, then the technique is useful. However, the problem of underestimation of real rates means that the ^{14}C technique is not very suitable for estimating photosynthetic parameters in microphytobenthos.

(2) The slurry technique is a valuable tool for measuring photosynthetic parameters, as it is easy to generate photosynthesis light curves in a photosynthetron (Blanchard and Cariou-Le Gall, 1994; MacIntyre and Cullen, 1995; Blanchard *et al.*, 1996). However, with slurries the existing microgradients in the sediment are destroyed, and this technique therefore measures potential production. Only when nutrients (CO_2 included) are not limiting might it be expected that production estimates based on slurry and oxygen microelectrode techniques give similar results (Barranguet *et al.*, 1998).

(3) Revsbech and Jørgensen's (1983) paper introduced the use of oxygen microelectrodes, and the technique is now used widely (Glud *et al.*, 1992; Pinckney and Zingmark, 1993; Yallop *et al.*, 1994; Heip *et al.*, 1995; Kromkamp and Peene, 1995a; Kühl *et al.*, 1996). Oxygen microelectrode measurements give gross primary production rates. However, construction of a complete photosynthesis profile curve is time consuming; the time taken to generate enough replicate production profiles can be greater than some of the temporal properties of the biofilm (e.g. endogenous vertical migration). To avoid this problem, production rates are sometimes calculated from the oxygen depth profile under a fixed irradiance, assuming diffusion and porosity coefficients (Epping and Jørgensen, 1996; Wiltshire *et al.*, 1996). Net oxygen production can also be calculated from the slope in oxygen concentrations out of the sediment, but with exposed sediments this can be problematic. Significant amounts of variation in oxygen production profiles can also be generated by the patchiness often observed in the distribution of microphytobenthic biomass. Oxygen microelectrodes are an important tool for measuring the spatial distribution of photosynthesis within sediments and response of photosynthesis to environmental variation, but scaling up these measurements to larger areal rates is contentious.

IV. FACTORS INFLUENCING PRIMARY PRODUCTION

Primary production will be viewed here as the photosynthetic reduction of CO_2 and uptake of nutrients, leading to the production of new algal biomass. Hence, it encompasses both the photosynthetic process and growth, but not factors affecting biomass. Therefore, algal bloom formation can be the result of two, often simultaneously operating, processes: (1) bottom-up control of photosynthesis and (2) top-down control of biomass. The interplay of these factors will determine whether there is an increase in biomass. Mathematically the growth of phytoplankton can be expressed as

$$\Delta B/\Delta t = (P - R - E)B - G_P - G_B \pm B_E \pm B_X \qquad (1)$$

where $\Delta B/\Delta t$ is the biomass increase with time, P is the rate of gross primary production, which is a function of light availability and nutrients, R is the rate of respiration, E is excretion or cell lysis, G_P and G_B are losses due to pelagic and benthic grazing respiration. B_E stands for the exchange of biomass between the bottom sediments and the overlying water column. In some estuaries this exchange can be high, so that it is difficult to distinguish between phytoplankton and microphytobenthos (Delgado et al., 1991; de Jonge, 1992; de Jonge and van Beusekom, 1992). B_X is the biomass exchange due to advection at the freshwater and marine boundaries. $P-R-E$ is the net growth rate and $(P-R-E)B$ is the gross rate of biomass production. It is clear that when R is larger than P, the growth rate is negative and the population will decrease, unless compensated for by import.

This formula is nearly identical to equation (1) in Cloern (1996) except that we replaced his growth rate μ for P and added excretion, which is part of Cloern's (gross) μ.

A. Factors Influencing the Rate of Photosynthesis

1. Light and Phytoplankton

When nutrients are abundant, phytoplankton primary productivity (P in eqn. 1) is directly proportional to light availability in the euphotic zone of the water column (Wofsy, 1983; Cole and Cloern, 1984; Pennock, 1985; Cole et al., 1986). When measured over 24 h, it is sometimes referred to as net photic zone production (NPZ). Combining data from Puget Sound, New York Bight, and North and South San Francisco Bay, Cole and Cloern (1987) found that NPZ (mg C m^{-2} day^{-1}) could be described as a linear function of biomass B (mg chlorophyll m^{-3}), photic zone depth Z_p and daily incident irradiance (I_0, mol photons m^{-2} day^{-1})

$$\text{NPZ} = 150 + 0.73BI_0Z_p \qquad (r^2 = 0.82, n = 211) \qquad (2)$$

This relationship, although with varying coefficients, could also be used to describe primary production at the more marine, downstream stations in the Western Scheldt estuary (Kromkamp et al., 1995) and in some stations in the Eastern Scheldt estuary (Heip et al., 1995). It is difficult to see why the relationship does not hold for all stations in the Western Scheldt, as nutrients were in surplus at all stations. As argued by Cole and Cloern (1987), deviations from the estimates might be caused by nutrient limitation (lower estimates), toxins or measuring errors. Cole and Cloern (1987) did not find a significant difference between the regression slope and the regression intercept (= 0) between the four estuaries. However, the standard error in the estimate was 410 mg C m^{-2} day^{-1}. For the individual estuaries the regression coefficients could vary substantially. To explain this, Heip et al. (1995) developed a simple analytical model in which photosynthesis per unit chlorophyll (P^B) was described as a hyperbolic function of incident irradiance I_0:

$$P^B = P^B_{max} \frac{I_0}{K_p + I_0} \qquad (3)$$

where K_p is the irradiance when $P^B = 0.5P^B_{max}$. The irradiance at depth z is calculated from the attenuation coefficient K_d using the Lambert–Beer extinction model. With these parameters, the depth-integrated photic zone primary production P^BZ can be calculated

$$P^BZ = \int_{z=0}^{z=\infty} \left[P^B_{max} \frac{I_0 e^{-k\,dz}}{K_p + I_0\, e^{-k_d z}} \right] dz \qquad (4)$$

P^BZ is a linear function of I_0/K_d when I_0 is constant and only K_d varies. This implies that during a short period in the season with little variation in I_0, a good linear fit of the depth-integrated production would be obtained, with an intercept close to zero. If K_d is kept constant, P^BZ varies hyperbolically with I_0. This situation would be representative of a system with a relatively constant K_d but large seasonal or weather-driven variation in surface irradiance. Fitting data sets with a linear function will result in a poor fit and a regression intercept greater than zero, especially in systems with a relatively high transparency, such as the Eastern Scheldt estuary (Heip et al., 1995). More details about simple models and empirical models can be found in this reference.

2. Light and Microphytobenthos

There are major spatial and temporal gradients in light availability in microphytobenthic habitats. Steep gradients of irradiance occur within the sediment, and intertidal sites are subject to varying patterns of diel illumination periods mediated by periods of tidal immersion. The shifting pattern of tidal exposure (approximately 55 min day^{-1}) within diel light curves, and the fortnightly cycle of spring and neap tides can result in periods when microphytobenthos are exposed to very high irradiance during low tide at solar noon (exceeding 2000 µmol m^{-2} s^{-1}), and at other times (up to periods of days for some regions of the intertidal) when little or no light reaches the sediment surface (Brotas and Catarino, 1995; Guarini et al., 1997). Thus benthic microalgae have to be able to adapt to a widely fluctuating light climate.

Characterization of the light climate within both non-cohesive (sandy) and cohesive sediments has been carried out using fibreoptic microprobes (Lassen et al., 1992ab; Ploug et al., 1993; Kühl et al., 1994). Such systems can measure at a scale of approximately 100 µm, and can determine both directional (usually perpendicular to the sediment surface) and scattered (scalar) irradiance (Lassen et al., 1992a; Ploug et al., 1993). Attenuation of light is rapid (attenuation coefficients (k mm^{-1}) between 1 and 3.5 for sandy and cohesive sediments respectively, Kühl et al., 1994; D.M. Paterson et al., unpublished data) with much scattering, leading to a dominance of scalar irradiance within a sediment (Lassen *et al.*, 1992b). Such attenuation results in euphotic zone depths (1% of incident light) usually less than 2 mm (Paterson *et al.*, 1998). Light intensity at the sediment surface can actually be higher than the incident light, particularly at wavelengths > 700 nm, due to back-scatter effects within the sediment (Ploug *et al.*, 1993). In sandy sediments this can result in increases in light intensity of 200% at the surface. However, in cohesive sediments this effect is much reduced (D.M. Paterson *et al.*, unpublished data; Paterson *et al.*, 1998). There are also changes in the spectral quality of light within sediments (Ploug *et al.*, 1993; D.M. Paterson *et al.*, unpublished data), and spectral quality and irradiance is further modified by increased light attenuation due to the presence of microalgal biofilms in the surface layers of sediments (Ploug *et al.*, 1993; Kühl *et al.*, 1994; D.M. Paterson *et al.*, unpublished data).

Because of the shallow depth of the euphotic zone within sediments, and the periodic deposition of fresh sediment material, microphytobenthos need to be able to reposition themselves within the euphotic zone to photosynthesize. The majority of epipelic microphytobenthos are motile; epipelic diatoms moving through the extrusion of extracellular polymeric substances from the raphe slit in the silica frustule (Edgar and Pickett-Heaps, 1984; Smith and Underwood, 1998), cyanobacterial filaments by gliding (Stal, 1994) and non-flagellated euglenids by amoeboid movement. In many intertidal habitats the microphytobenthos exhibit rhythmic vertical migration, linked to both the diel and tidal cycles (Round, 1981; Admiraal, 1984; Serôdio *et al.*, 1997; Kromkamp *et al.*, 1998; Paterson *et al.*, 1998; Smith and Underwood, 1998). These endogenous rhythms of migration, which can be maintained in the laboratory in the absence of external stimuli for up to 11 days (Palmer and Round, 1967; Paterson, 1986; Serôdio *et al.*, 1997), result in the vertical migration of cells remaining in synchrony with the the daily shift in the tidal cycle within the diel light frame, and the fortnightly sequence of spring and neap tides. The overlying water in many estuaries is turbid, preventing light penetration to the sediment during periods of tidal immersion. In these situations, migratory rhythms result in accumulation of microalgae on the sediment surface during daylight tidal emersion periods, resulting in high cell densities (up to $10^4 - 10^6$ cells cm^{-2}; Paterson, 1989; Underwood *et al.*, 1998) and high chlorophyll *a* concentrations within the top few microns during the period of low tide (Wiltshire *et al.*, 1997). However, in other situations (e.g. subtidal mats of *Gyrosigma balticum*) cells accumulate at the surface during dawn and dusk, and move down into the sediment during the high light, midday period (Jönnson *et al.*, 1994; Underwood *et al.*, 1999), or demonstrate no major patterns of vertical migration.

Measurements of gross photosynthesis using oxygen microelectrodes have demonstrated that within cohesive sediments the majority of photosynthesis occurs within the top 200–400 μm of the sediment (Yallop *et al.*, 1994; Figure 1). In sandy sediments, where light penetration is greater, gross photosynthesis can occur deeper than this, and may even show a bimodal distribution due to distinct

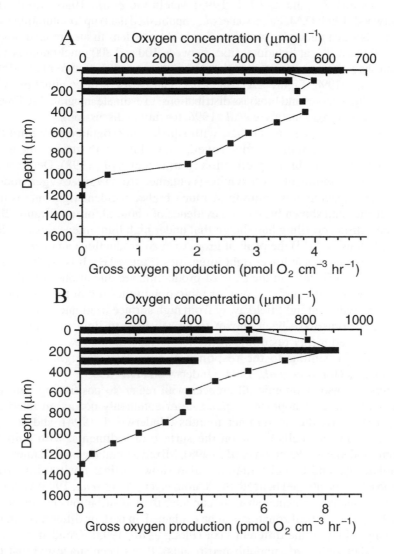

Fig. 1. Microprofiles of oxygen concentration (■) and gross oxygen production (bars) in a cohesive intertidal sediment supporting a biofilm of epipelic diatoms. (A) Irradiance 600 μmol m^{-2} s^{-1}; (B) irradiance 2100 μmol m^{-2} s^{-1}. Note the shift of the peak of gross oxygen production to a depth of 200 μm in the sediment. From Paterson *et al.* (1997).

vertical separation of diatoms and cyanobacterial layers (Yallop et al., 1994). Maximum integrated rates of photosynthesis have been found to occur when low tide corresponds to the solar maximum (Pinckney and Zingmark, 1991), and integrated photosynthesis is usually closely correlated with light intensity (Pinckney and Zingmark, 1991, 1993; MacIntyre et al., 1996; Smith and Underwood, 1998; D.M. Patterson et al., unpublished data) up to saturating irradiances. Isolated microphytobenthos (i.e. in slurries, lens tissue preparations or cultures) reach P^B_{max} at light intensities between 100 and 800 μmol photons m^{-2} s^{-1} and show decreases in P^B at higher light intensities (MacIntyre et al., 1996; Hartig et al., 1998; Figure 2A). Such data can be used, in conjunction with sediment light curves and biomass distributions, to estimate integrated sediment photosynthesis (see MacIntyre et al., 1996, for further discussion).

It should be kept in mind that, with slurries, microgradients within the chemical environment are destroyed, and slurries therefore measure potential rather than in situ production (Barranguet et al., 1998). Depth-integrated rates of sediment photosynthesis obtained from in situ oxygen microelectrode measurements saturate at much higher incident light intensities than slurries and shown little or no evidence of photoinhibition (Figure 2B). Microelectrode profiling has shown that under high light intensities (> 1200 μmol photons m^{-2} s^{-1}) the peak of gross oxygen production can occur lower in the sediment than at lower light intensities (Figure 1B). It is not known to what extent such shifts in the peak of production are due to photoinhibition of microalgae in the surface layer, or migration of the bulk of the population within the sediment light field away from high surface irradiances. Intertidal epipelic diatoms have been shown to be sensitive to light intensity, with shading of sediments resulting in higher surface biomasses (unpublished observations), similar to the response of subtidal Gyrosigma balticum assemblages (Jönnson et al., 1994; Underwood et al., 1999).

There are also taxonomic differences with regard to positioning with the light field. The euglenophyte Euglena deses commonly occurs on intertidal flats, particularly during summer months (Underwood, 1994), and at high incident irradiance cells occur on the surface of sediments, with epipelic diatoms underneath (Paterson et al., 1998). Mixed assemblages of filamentous cyanobacteria and epipelic diatoms also show vertical positioning, with cyanobacteria positioned beneath the diatom layer (Ploug et al., 1993; Yallop et al., 1994) and experimental shading can alter these distributions (Wiltshire et al., 1997). This vertical zonation can be explained, in part, by differences in the photophysiology of the different taxa (Ploug et al., 1993; Kromkamp et al., 1998; Oxborough et al., unpublished results). It has been suggested that the high productivity of intertidal epipelic biofilms and the lack of any measurable photoinhibition may be due to the ability of cells to migrate out of the high light zone and be replaced by cells from deeper layers, thus maintaining the overall productivity of the biofilm (Kromkamp et al., 1998). For some of the larger

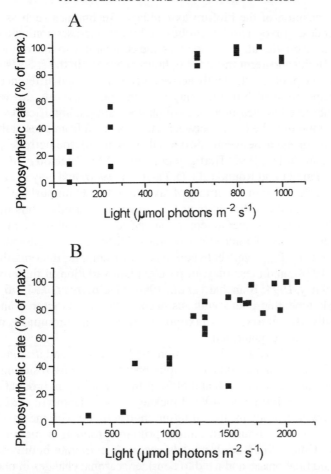

Fig. 2. Photosynthesis versus irradiance curves for the epipelic diatom *Navicula phyllepta* (Kütz) in (A) a sediment-free suspension and (B) *in situ* within a biofilm on cohesive sediment. From Paterson *et al.* (1997).

epipelic diatom taxa (e.g. *Gyrosigma* and *Pleurosigma* species, 50–200 μm long; *Nitzschia epithemioides*, 50 μm long), cells need to move only one or two cell lengths to place themselves at significantly lower light intensities, and with movement rates in sediment of 0.2 μm s^{-1} (Hay *et al.*, 1993) such cycling could easily occur. The development of techniques of microsectioning (Wiltshire *et al.*, 1997) and image analysis of *in situ* biofilms (section VI.B) may resolve this issue.

A criticism of some studies of microphytobenthic photosynthesis and primary production is the lack of identification and quantification of the species composition of the biofilm assemblage. Techniques such as oxygen microelectrodes, fibreoptic microprobes, sediment microsectioning and fluorescence characteristics all measure at a scale comparable to that of individual algal cells. Both short-term (minutes to hours) changes in the species composition within the photosynthetically active layer and seasonal differences in the species composition of biofilms may all contribute towards the variation frequently observed in measured rates of photosynthesis. Significant variation in P_{max}^B can sometimes be found between samples taken from sites at the same time of year, or even between individual oxygen microelectrode profiles (MacIntyre and Cullen, 1995; Hartig *et al.*, 1998; D.M. Paterson *et al.*, unpublished data). Palmer and Round (1967), Paterson (1986) and Hay *et al.* (1993) have shown differences in the timing of taxa migrating to the surface and their positioning within the sediment over tidal emersion periods. Thus measurements of photosynthesis over an emersion period may actually be measuring different populations of microphytobenthos. Differences in photosynthetic parameters (α, I_o, P_{max}^B, etc.) between species of epipelic diatoms do occur (Admiraal, 1984), and these may, in part, explain variation in field measurements (Pinckney, 1994; Blanchard *et al.*, 1996; MacIntyre *et al.*, 1996). Given that microphytobenthic photosynthesis occurs within a structured and taxonomically diverse matrix, then taxonomic effects on community primary production need to be quantified.

In general, there is a fairly clear relationship between P^B and irradiance (and therefore tidal phase) in intertidal microphytobenthic systems, resulting in irradiance explaining between 30% and 60% of the variability in production, and biomass explaining further 30–40% (Pinckney, 1994; MacIntyre *et al.*, 1996). There are, however, problems in scaling-up short-term measurements of photosynthesis (seconds to hours) to annual primary production measurements (Shaffer and Onuf, 1985). Such models need to incorporate both changes in irradiance and tidal phase, and need to consider seasonal changes in photosynthetic parameters due to adaptation of the algae, effects of desiccation and temperature during emersion periods (Pinckney, 1994; Guarini *et al.*, 1997) as well as seasonal changes in species composition (Underwood, 1994; Blanchard *et al.*, 1996). There is also significant spatial and temporal heterogeneity in microphytobenthic biomass, which means estimation of annual estuarine (or even mudflat) microphytobenthic primary production should be done with extreme caution (see section IV.C.2).

3. Nutrients and Phytoplankton

Nutrient limitation will affect both the photosynthesis (P in eqn. 1; Kolber and Falkowski, 1993), i.e. Blackman type of limitation (Falkowski *et al.*, 1992) and biomass (B in eqn. 1; Liebig type of limitation). It is generally assumed that most marine systems are nitrogen limited, whereas most freshwater systems are probably phosphorus limited. However, as Hecky and Kilham (1988) argue, this conclusion is based mostly on nutrient concentrations and nutrient ratios in comparison to the nutrient composition of algal cells, for which the Redfield ratio (C : N : Si : P = 106 : 16 : 16 : 1) is used. Nevertheless, even with a low nutrient concentration, a high nutrient flux may be possible, in which case limitation might not occur. Until recently, it was not possible to measure nutrient limitation directly, but the development of the variable fluorescence technique (see section VI.B) seems to make this possible. Additional evidence for nutrient limitation can be obtained from bioassay experiments involving enrichment studies or from physiological indicators like the maximal uptake rate V_{max} (Zevenboom et al., 1982; Riegman and Mur, 1984). A bioassay experiment is prone to artefacts and should be interpreted with caution.

Additional evidence for nutrient limitation in the system can be obtained by investigating the relationship between nutrient load or nutrient concentration and primary production. This analysis is normally based on yearly averages. This method may overlook short periods of limitation by a particular nutrient and this may have a large effect on species succession and thus on the functioning of the ecosystem. Short periods of P limitation have been observed in the Eastern Scheldt estuary (SW Netherlands; Kromkamp and Peene, 1997), Delaware estuary (Pennock and Sharp, 1994) and Chesapeake Bay (Fisher et al., 1988). Often a positive relationship between the dissolved inorganic nitrogen (DIN) load or DIN concentration and annual primary production is found. Boynton et al. (1982) described a linear relationship between the annual DIN load and areal annual primary production, but the relationship explained less than 50% of the variability, ($r^2 = 0.38$). Expressing the load on an areal basis improved the relationship ($r^2 = 0.60$). Monbet (1992) studied the relationship between chlorophyll a content and DIN. He found that for a given mean DIN input, the mean annual chlorophyll a concentrations were higher in microtidal estuaries, but the relationship between chlorophyll a concentration and DIN loading was not very good. However, when the chlorophyll a concentration was plotted against the mean annual DIN concentration, the relationship was very positive (Figure 2 in Monbet, 1992). Microtidal estuaries were more sensitive to DIN concentrations and showed a higher biomass concentration at a given DIN concentration. Although nutrient loadings and concentrations are related, the morphology of the basin influences the relationship between loading and concentration, and it therefore seems better to use concentrations as indicators of nutrient loading.

Cloern (in press) developed a simple index for coastal ecosystems to estimate the relative importance of light and nutrient limitation of phytoplankton growth (μ). μ is described as a function of P^B, the daily carbon assimilation rate per unit chlorophyll, the chlorophyll : carbon ratio (chl : C) and the specific respiration rate r (taken as 0.015 day^{-1}; Cloern et al., 1995):

$$\mu = 0.85 P^{B} \frac{\text{chl}}{C} - 0.015 \tag{5a}$$

$$\frac{\text{chl}}{C} = 0.03 + 0.0154[e^{0.05T}][e^{-0.059I}]\left[\frac{N}{K_{N} + N}\right] \tag{5b}$$

where T is the temperature, I the mean daily irradiance averaged over depth, N the concentration of the limiting nutrient and K_{N} the half-saturation constant for algal growth. A ratio (R) is calculated, based on the increment in growth rate with respect to either I or N. When $R > 1$, the system is more sensitive to light, and when $R < 1$, the system is more sensitive to changes in nutrients. Based on eqn. 5, a resource map is made in which R is plotted as a function of DIN and light availability. This resource map can easily function as a management tool, as it gives an indication of the susceptibility of the system to changes in nutrients. The Eastern Scheldt (Netherlands) and Tomales Bay (California) are classified as systems potentially sensitive to changes in nutrient conditions, whereas the very turbid North San Francisco Bay and the Westerschelde are not sensitive to changes in nutrient concentrations. However, as argued by Cloern (1999), the model is based on potential changes in growth rate, and this does not necessarily imply that phytoplankton biomass will change, as this is also influenced by physical conditions (residence times) as well as by grazing. Large-scale works performed in the Eastern Scheldt demonstrated that grazing can stabilize ecosystems (Herman and Scholten, 1990), as primary production did not change in the Eastern Scheldt despite large changes in light and nutrient conditions (Wetsteyn and Kromkamp, 1994; see also section V on long-term time series).

Most analyses between nutrient loading or concentration and phytoplankton biomass or primary production are based on inorganic dissolved nutrients. However, due to regeneration of nutrients, dissolved organic matter can be an important nutrient source as well. It is now well recognized that phytoplankton is an integral part of the microbial loop, and in some estuaries the microbial loop is stimulated by the import of organic material (van Spaendonk et al., 1993; Goosen et al., 1997). Heip et al. (1995) found a very significant relationship between annual system primary production and allochthonous organic import: annual primary production = 2.4 × net organic imput + 177 ($r^2=0.80$). Below the constant 177 mg C m^{-2} year^{-1}, primary production is assumed to be light limited, and this value is close to the 160 mg C m^{-2} year^{-1} reported by Smith and Hollibaugh (1993) to be representative for coastal (not estuarine) production. Heip et al. (1995) observed that their relationship between primary production and organic import was better than the relationship between DIN import and primary production given by Boynton et al. (1982). This supports the model of Smith et al. (1989) in which it is the organic loading, rather than the DIN loading, that controls primary production in semi-closed systems. This is because the biomass is limited by P, but production by N (Heip *et al.*, 1995).

4. Nutrients and Microphytobenthos

The importance of nutrients to microphytobenthos depends in part on the sediment type concerned. Fine cohesive estuarine sediments usually have a high organic matter content, with high rates of bacterial mineralization and high porewater concentrations of dissolved nutrients, while sandflats are more oligotrophic (Admiraal, 1984; Heip *et al.*, 1995). There is thus an increased possibility that microphytobenthos inhabiting sediments of a larger grain size will be nutrient limited. The spatial distribution of sediments within estuaries is also pertinent to whether nutrient limitation will occur, in that many estuaries exhibit significant nutrient gradients along their length (e.g. Ogilvie *et al.*, 1997; Trimmer *et al.*, 1998) and areas of extensive mudflats supporting microphytobenthos may coincide with regions of high nutrient concentration.

Published data on spatial distribution of microphytobenthos in estuaries often shows correlation between nutrient gradients and biomass (Underwood *et al.*, 1998). However, there are few experimental data on nutrient effects on intertidal microphytobenthos independent of other co-varying factors which also affect primary production and biomass (shelter, salinity, etc.). Where nutrient enrichment experiments have been carried out on mudflats (with mixed cyanobacterial–diatom assemblages), no consistent pattern of increased photosynthesis has been observed (Paerl *et al.*, 1993). In contrast, enrichment experiments using nitrate and phosphate additions with subtidal microphytobenthic assemblages inhabiting non-cohesive (sandy) substrata, and cyanobacterial mat assemblages in nutrient poor habitats, have showed varying levels of stimulation of both microphytobenthic photo-synthesis and biomass development (Nilsson *et al.*, 1991; Nilsson and Sundbäck, 1991; Paerl *et al.*, 1993; Pinckney *et al.*, 1995). However, these studies are generally concerned with low nutrient environments (nitrate concentrations $< 20 \ \mu mol \ l^{-1}$). In more eutrophic, cohesive estuarine sediments, short-term experiments with nutrient addition have shown little effect on epipelic algal biomass (Underwood *et al.*, 1998), although long-term reductions (over 16 years) in nutrient inputs have been shown to result in declines in biomass (Peletier, 1996).

It is therefore generally considered that epipelic (i.e. cohesive sediment inhabiting) microalgae are not nutrient limited (Admiraal *et al.*, 1982; Underwood *et al.*, 1998), and that they obtain nutrients both from within the sediment, particularly during migration during tidal immersion, and from the overlying water. However the term "nutrient limitation" includes both

Liebig-type limitation (on final biomass) and short-term effects on rates of photosynthesis and growth. Short-term experiments have shown significant porewater depletion of ammonium during periods of tidal emersion and photosynthesis (Thornton *et al.*, 1999), resulting in either uptake into, or reduced ammonium efflux from, sediments during subsequent immersion periods. Microphytobenthos have been found to influence nutrient fluxes and bacterial processes in sediments (Nedwell *et al.*, this volume). To what extent these short-term nutrient dynamics influence the rate of photosynthesis and growth of microphytobenthos is not known. Until there is a clearer picture of the degree of coupling–uncoupling between rates of oxygen evolution, carbon fixation, nutrient uptake and luxury storage, migration patterns within the photic zone and periods of cell division, the effects of short-term "limitation" on long-term reduction in photosynthesis and growth rates is open to debate.

There is increasing evidence that, although nutrients may not limit primary production or biomass in muddy sediments, the nutrient environment is important in determining species composition. There is a fairly extensive literature on species distribution of epipelic diatoms in estuaries (see Underwood *et al.*, 1998; Sullivan, 1999) and certain taxa do appear to show preferences for particular levels of specific nutrients. However, all such field-based correlations incorporating seasonal variation in species occurrence are complicated by co-varying seasonal variation in nutrient concentrations, light climate, temperature and disturbance events. Experimental manipulation is one approach to address this issue. In subtidal microphytobenthic systems, increased nitrogen and phosphorus concentrations resulted in increased diatom abundance in previously nutrient-limited cyanobacterial mats (Pinckney *et al.*, 1995), while addition of nitrogen altered the taxonomic composition of both epipsammic and epipelic subtidal assemblages (Nilsson and Sundbäck, 1991; Nilsson *et al.*, 1991; Sundbäck and Snoeijs, 1991). Ammonium concentrations in benthic sediments appear to be an important factor influencing the distribution of benthic diatom species in both saltmarshes (Sullivan, 1978) and mudflats (Underwood *et al.*, 1998). Concentrations of ammonium between 500 and 1000 μmol l^{-1} are selective for some taxa of benthic algae, with the toxic effects of ammonia being enhanced in high pH conditions (Admiraal and Peletier, 1979, 1980; Admiraal, 1984; Underwood *et al.*, 1998). Increases in the relative abundance of *Navicula phyllepta, Navicula flanatica* and *Pleurosigma angulatum* in the Ems-Dollard estuary have been recorded coincident with decreasing ammonium concentrations due to the installation of waste treatment processes by industry (Peletier, 1996). There is also evidence that sediment organic content and tolerance to sulfide influence the species composition of microalgal biofilms (Admiraal, 1984; Peletier, 1996; Underwood, 1997; Sullivan, 1999). Given the steep gradients in pH, oxygen and sulfide within fine cohesive sediments, these are likely to be important selective factors determining the species diversity of microphytobenthic assemblages (Sullivan, 1999).

5. Stratification, Vertical Mixing and Salinity and Temperature

Freshwater input into estuaries can stimulate primary production by importing nutrients into the system. On the other hand, these inputs will also cause salinity stress, and marine phytoplankton will be replaced by more brackish and ultimately freshwater phytoplankton (e.g. Kromkamp and Peene, 1995a).

The stability of the water column influences light and nutrient availability. Variation in the stability of the water column results from the balance between buoyancy forces (solar heating, freshwater runoff) and mechanical energy inputs (from wind, tides and also freshwater runoff) (Mann and Lazier, 1991; Monbet, 1992; Ragueneau et al., 1996). Freshwater input may also cause vertical salinity stratification. This will improve the light conditions in the upper layer, and may stimulate primary production (Malone et al., 1988; Cloern, 1991). Hence, it is clear that the light and nutrient conditions are very much influenced by morphological and meteorological conditions. This may be an important reason why estimates of primary production between estuaries are so variable.

The spring bloom is often initiated when the critical depth (i.e. the depth at which daily production and (algal) respiration is balanced) is exceeded (Sverdrup, 1953, discussed in detail by Smetacek and Passow, 1990 and by Tett, 1990). Tidally driven resuspension and riverine sources of sediments might be important mechanisms in influencing suspended matter concentrations, and thus photic depth, in the water column (Cloern et al., 1989). Semi-diurnal cycles of sediment erosion and deposition as a result of the ebb and flood currents are important in macrotidal estuaries (estuaries with a mean tidal range > 4 m). The suspended sediment concentrations in macrotidal estuaries is 5–100 times higher than in microtidal estuaries (mean tidal range < 2 m) (Monbet, 1992). For San Francisco Bay (California, USA), it was concluded that when the mixing depth (Z_m) was less than six times the euphotic depth (Z_{eu}), the spring bloom was initiated (Cole and Cloern, 1984; Alpine and Cloern, 1988). Other authors (Grobbelaar, 1990; Tett, 1990) also observed a $Z_m : Z_{eu}$ ratio of 6. However, Kromkamp and Peene (1995a) argued that Z_m/Z_{eu} should be much higher to allow phytoplankton growth in the Westerschelde estuary. Smetacek and Passow (1990) had shown that the critical depth is dependent on the way in which the respiration in the aphotic zones is calculated. This is discussed in more detail in section IV.E.1.

Temperature is another factor influencing rates of primary production. Temperature positively influences phytoplankton as enzymes are involved in the photosynthetic process. Because photosynthesis at low light is limited by the rate of photon absorption, light-limited rates of photosynthesis are not expected to be influenced by temperature (Wilhelm, 1990). This has indeed been shown in a number of studies (Post et al., 1985; Harrison and Platt, 1986; Smith et al., 1994). For this reason, gross primary production and respiration will not necessarily show the same temperature dependency. We will not discuss temperature effects on phytoplankton any further here.

6. Temperature Effects and Potential CO_2 Limitation of Microphytobenthos

Temperature can have a major influence on rates of microphytobenthic photosynthesis. The temperature of a mudflat can change rapidly during a tidal emersion period, at up to 2–3°C h^{-1}, with daily ranges of 20°C and seasonal ranges between 0 and 35°C (Blanchard and Guarini, 1999). The effects of these temperature changes have been studied extensively by Blanchard and co-workers. They found that there is a clear relationship between P^B_{max} and temperature, with an optimal temperature for intertidal diatoms at their sites of 25°C, and values of P^B_{max} nearly doubling over a 10°C temperature range (Blanchard et al., 1996). Modelling the effects of temperature over tidal and seasonal cycles has shown that significant inhibition of microphytobenthic photosynthesis can occur on the upper intertidal during certain times of the year (Guarini et al., 1997; Blanchard and Guarini, 1999). Assuming this to be a common phenomenon, then models of benthic photosynthesis based only on light and exposure period may well overestimate production by up to 30% over 75% of the area of an intertidal flat during periods of high sediment temperature (Blanchard and Guarini, 1999).

The other environmental variable to consider with regard to benthic photosynthesis is the availability of CO_2. Various workers have commented on the possibility of CO_2 limitation within dense microphytobenthic biofilms (Admiraal et al., 1982; Ludden et al., 1985; Glud et al., 1992; Kromkamp et al., 1998), especially under conditions of high pH and oxygen supersaturation. One of the problems in studying potential CO_2 limitation has been the difficulty in measuring accurately porewater CO_2 concentrations at a scale relevant to a microphytobenthic biofilm. This is now possible using a rapid-response CO_2 microelectrode (de Beer et al., 1997). Using model systems (of Synechococcus and Commamonas), de Beer et al. showed that CO_2 depletion could occur within the top 1 mm of a photosynthetically active biofilm, and that the carbonate system was significantly out of equilibrium. However, the oxygen profile simultaneously measured could be explained only by the assumption that both CO_2 and HCO_3^- were used for photosynthesis. Cyanobacteria are known to be able to use HCO_3^- as an inorganic carbon source (Glud et al., 1992; de Beer et al., 1997), and there is increasing evidence of HCO_3^- usage, carbonic anhydrase activity and CO_2 concentrating mechanisms in many algal groups, including diatoms (see Raven, 1997, for a comprehensive review).

Determination of the relationship between CO_2 limitation, photosynthesis and photorespiration, and microalgal vertical migration (which may serve to alleviate CO_2 limitation; Kromkamp *et al.*, 1998) within biofilms is required. These are important issues which require collaboration between physiological and ecological disciplines, especially with regard to the importance of the structural nature of microbial biofilms and migratory behaviour of microphytobenthos.

7. Phytoplankton Respiration: Calculating Photic Zone Versus Water Column Production

Estimating net water column primary production is difficult. A major reason for this is that it is very difficult to measure the true dark respiration (R in eqn. 1). Respiration rates are often derived from measured chlorophyll-normalized photosynthesis irradiance (PI) curves, yielding an intercept equivalent to approximately 10% of P_{max}. This value is then used to calculate respiratory losses (Cole and Cloern, 1984; Smetacek and Passow, 1990; Cole *et al.*, 1992). Langdon (1993) studied the relationship between respiration and P_{max} for several algal classes. On average he found that the R/P_{max} value was 0.16 for diatoms, 0.13 for green algae, 0.10 for cyanobacteria, 0.35 for Dinophyceae and 0.16 for Prymnesiophyceae, but the spread in the data was large. Assuming an R/P_{max} ratio of 0.1, Cole and Cloern (1984) calculated net (NPW) and gross (PGW) annual water column production for several stations in San Francisco Bay (Table 3). The ratio between these two variables could be described by a linear function as NPW/GPW = $1 - 0.16\, Z_m/Z_{eu}$. This again indicates a critical depth of 6, but, as noted by the authors, this ratio is dependent on the rate of respiration. As can be seen in Table 3, the NPW is often negative. This led Cole *et al.* (1992) to the conclusion that the oberved dynamics in chlorophyll were due to advection from the intertidal areas. Kromkamp and Peene (1995a,b) observed the same for the Western Scheldt estuary. However, they argued (based on model calculation from Soetaert *et al.*, 1994) that resuspended microphytobenthos from the intertidal areas could not explain the observed chlorophyll dynamics and that they were the result of local phytoplankton production. To explain this Kromkamp and Peene (1995a) used a model by Langdon (1993) which described respiration as a function of a constant maintenance respiration rate and a variable rate dependent on the water column production. The model allowed the rate of respiration in the light to differ from that in the dark:

$$R_{w,t} = BZ_m\,[24DR_0 + (24 - D)R_D P^B_{Zm} + DR_D R_L P^B_{zm}]$$ (6)

where $R_{w,t}$ = the respiration in the mixed layer (or watercolumn), B the biomass (mg chl m^{-3}), Z_m = the depth of the mixed layer (or watercolumn), D the

Table 3

Comparison between annual net (NPZ) or gross (GPZ) photic zone production and annual net water column production (NPW) for different estuaries. Adapted from Heip et al. (1995)

Estuary	Reference	Station	NPZ or GPZ (g C m^{-2})	NPW (g C m^{-2})
San Francisco Bay[a]	Cole and Cloern (1984)			
South Bay		27	150	30
		162	150	101
San Pablo Bay		13	130	32
		318	110	48
Suisun Bay		6	98	−152
		418	93	39
Hudson River	Cole et al. (1992)			
Estuary[b]		Poughkeepsee	131	0
		Fort Montgomery	88	−263
Oosterschelde[c]	Kromkamp and Peene (1995b)	O1	238	149
		O2	348	98
		O3	385	57
		O4	379	6
Westerschelde[c]	Kromkamp and Peene (1995b)	3 psu	280	−60
		11 psu	70	−108
		21 psu	146	−91
		29 psu	265	−134

[a]NPZ based on 24 h ^{14}C incubations and $R/P_{max} = 0.1$.
[b]GPZ, based on 2–3 h ^{14}C incubations and $R/P_{max} = 0.05$.
[c]GPZ, based on 2 h ^{14}C incubations and $R/P_{max} = 0.05$.

daylength, R_o = the biomass specific maintenance respiration rate, R_D = photosynthesis rate-dependent linear coefficient of dark respiration, R_L = ratio of light to dark respiration, and P^B_{Zm} is the biomass-specific rate of primary production averaged for Z_m. This model is more realistic than the assumption that R is a fixed percentage of P_{max} and is based on physiological evidence (Laws and Bannister, 1980; Falkowski et al., 1985; Geider and Osborne, 1986). Using this model Kromkamp and Peene (1995a) showed that, using coefficients for diatoms, net water column primary production was possible, and that the results resembled those obtained with the ecosystem model made for the Western Scheldt (Soetaert et al., 1994). When coefficients were used for Prymnesiophytes, net production was possible only in the inner and central Western Scheldt (Kromkamp and Peene, 1995b).

The model of Langdon assumes that "short-term" ^{14}C incubations yield gross production rates. However, according to Williams and Lefèvre (1996), ^{14}C incubation results are more likely to be closer to net than to gross production. The effect of this was investigated for phytoplankton in the Western and Eastern Scheldt estuaries. Figure 3 shows the results of calculations assuming that in the light no respiratory correction has to be made. For the inner station Antwerp (salinity approximately 2), the difference between gross and net is clearly visible. When it is assumed that the ^{14}C incubation measures net incubation, the respiratory losses compared with the photic zone production are relatively small. However, when the same incubation is performed with coefficients for Prymnesiophytes, no net water column production is possible if the ^{14}C method measures gross production. These simple calculations show the importance of measuring respiration in calculations of net production and the effect of species composition on the estimate. Measuring phytoplankton respiration in the field is, however, very difficult and new developments are eagerly awaited. Possibly the use of membrane inlet mass spectrometry gas analysis systems will be a useful way forward (Grande *et al.*, 1989; Weger *et al.*, 1989; Kana, 1992; Kana *et al.*, 1994).

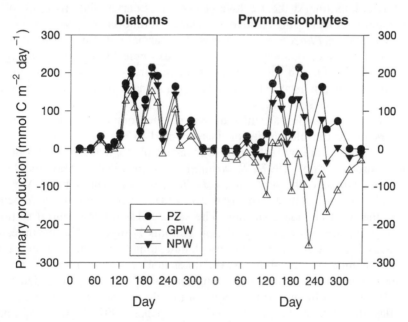

Fig 3. Calculation of net and gross production for diatoms and prymnesiophytes for station "Antwerp" in the Western Scheldt estuary, according to eqn 5. (PZ, gross C fixation in the water column during the light period (no correction for respiration); GPW, gross daily water column production (assuming C fixation measures gross photosynthesis; PZ corrected for respiration in the light and dark); NPW, net water column production (assuming C fixation measures net C fixation; PZ corrected for respiration in the dark only).

B. Factors Influencing Biomass of Phytoplankton

1. Phytoplankton Species Composition and Distribution

The estuarine habitat is complex, with large spatial and temporal changes in nutrients, light availability and salinity. Freshwater phytoplankton tend to dominate the upper limnic regions of estuaries, and are gradually replaced by marine species. Species richness can be high, with 263 and 354 species of phytoplankton identifed in the Delaware and Chesapeake estuaries respectively (Marshall and Alden, 1993). Species assemblages and seasonal development between these two estuaries were similar for the same salinity range. Cluster analyses revealed three assemblages: tidal freshwater, oligohaline and meso-polyhaline. River flow was important in determining the spatial and temporal variation of these three assemblages (Marshall and Alden, 1990).The fact that more species were observed in Chesapeake was attributed to a more favourable light climate and a more diverse nutrient regime compared with Delaware Bay (Marshall and Alden, 1993).

Salinity appears to be an important factor in determining the composition of the phytoplankton. At the freshwater end of estuaries diatoms tend to be dominant, with increasing relative abundance of dinoflagellates and prymnesiophytes (often *Phaeocystis*) towards the marine end. Euryhaline diatom species are very common in estuaries, for example *Skeletonema costatum* (Rijstenbil, 1991), and this species is an important component of spring blooms, although in many estuaries it can be abundant in summer, autumn and winter as well (Bakker *et al.*, 1994).

There are also seasonal shifts in assemblage composition. In the turbid Westerschelde the bloom period is restricted to the spring and summer months. Diatoms are by far the most abundant phytoplankton species at the inner stations (Tripos, 1997). *S. costatum* was abundant in May and June, whereas *Thalassiosira leptopa* (= *T. lineata*) was found from May to September. Dinoflagellate diversity was low, with *Heterocapsa rotunda* being by far the most abundant species. The same holds for the central station, although the species diversity of non-diatoms is higher. The prymnesiophyceaen *Phaeocystis* was also observed, indicating that it can occur in conditions of salinity of around 20. At the outer marine station a typical coastal spring bloom was observed, consisting mainly of diatoms, succeeded by a large *Phaeocystis* bloom. At the height of the bloom, the ratio of flagellate to non-flagellate cells was approximately 1 : 1 (Tripos, 1997). Despite the small distances between the stations (20–40 km), the species diversity was different.

The construction of a storm-surge barrier in the Eastern Scheldt (completed in 1986) changed the system from a light (and silicate) – limited estuary in the pre-barrier period to a Si- and N-limited lagoon in the post-barrier period (Wetsteyn and Kromkamp, 1994). In the pre-barrier period, the phytoplankton assemblage was diatom-dominated, and light availability was a steering factor in the seasonal succession (Bakker *et al.*, 1994). Cryptophycean flagellates were abundant and did not show a clear seasonality. Due to the improved light conditions in the post-barrier period, the growth season started earlier and lasted longer, and the seasonal trends decreased. New fragile, weakly silicified diatoms appeared (*Rhizoselenia fragilissima*, *Leptocylindrus minimus*, *Thalassiosira fallax*), and flagellates became abundant, especially in summer, possibly as a consequence of low DIN or silicate concentrations (Bakker *et al.*, 1994). Because of the disappearance of salinity changes, *S. costatum* decreased in abundance in the post-barrier period. The nuisance algae *Phaeocystis* was able to bloom also in late summer and autumn in the post-barrier period. This was attributed to an increase in the light availability. As demonstrated in section IV.A.7, Prymnesiophytes need more light in order to grow, owing to a higher respiration rate. As diatoms seem to be able to cope with low light conditions quite well, due to low respiration rates (Langdon, 1993), it is not surprising that they form the dominant part of the phyto-plankton assemblage in turbid estuaries throughout the year or in winter – early spring and late autumn.

2. Grazing: Zoobenthos and Zooplankton

Phytoplankton is grazed (*G* in eqn. 1) by mesozooplankton (200–2000 μmol l^{-1}), mainly copepods and microzooplankton (20–200 μmol l^{-1}) consisting of heterotrophic protozoans (flagellates, ciliates) and juvenile stages of copepods. Meiobenthos and macrobenthos also feed on algae. The latter group can be divided into two broad categories: the benthic suspension feeders and benthic deposit feeders. Because of their relatively shallow depth, estuarine ecosystems are often characterized by very intense benthic–pelagic coupling. However, in times of stratification the impact of the benthic suspension feeders will be more or less decoupled from phytoplankton primary production as there is less interaction between the water layers (Cloern, 1991).

Besides the reduction of phytoplankton biomass, *in situ* measurements have showed uptake of particulate organic matter (including phytoplankton) and release of dissolved inorganic nutrients by bivalve beds, and consequently enhanced regeneration of inorganic nutrients. This may result in a promotion of primary production as well as eutrophication control (Smaal and Prins, 1993). In many estuaries the total biomass (and potential grazing rate) of benthic organisms is higher than the biomass of copepods and other zooplankton (Cloern, 1982).

On theoretical grounds, picoplankton are better competitors for light (Raven, 1984) and nutrients (Thingstad and Sakshaug, 1990) and, because of their small size, sinking losses are minimal. However, owing to their small size they are vulnerable to microzooplankton grazing. Microzooplankton generation times are in the same order as those of phytoplankton and consequently they can grow as rapidly as phytoplankton, and are therefore able to impose a stringent control on phytoplankton biomass. This has been known for some time in oceanic studies which are dominated by picoplankton and in which regeneration of nutrients by the microbial loop is the main source of nutrients and hence production (Dugdale and Goering, 1967; Dugdale and Wilkerson, 1986; Dugdale *et al.*, 1990; Thingstad and Sakshaug, 1990). Similar regeneration systems occur in estuaries and coastal ecosystems. Gallegos and Jordan (1997) found for the Rhode River estuary that summer nutrient regeneration by microzooplankton grazing was not sufficient to account for phytoplankton production, but that in spring and autumn nutrient limitation was relieved by regeneration of nutrients by grazers. They used Landry and Hassett's (1982) dilution technique to determine microzooplankton grazing rates and found that grazing rate approximated to 73% of phytoplankton growth rate. For the Marsdiep tidal inlet to the Wadden Sea (The Netherlands) it was observed that small algae and grazers dominated the light-limited winter and early spring period (Riegman *et al.*, 1993). When light conditions improved, larger phytoplankton that could escape microzooplankton grazing control became dominant, before grazing by mesozooplankton and sedimentation in turn regulated their biomass. From this example it is clear that size-differential grazing control is important, and that the size of the phytoplankton determines largely whether the linear foodchain is dominant or the microbial foodweb is more important.

3. Cell Lysis and Viruses

Natural mortality was until recently hardly studied, but can be an important loss factor (E in eqn. 1). Information on the rates of cell lysis is needed. Cell lysis can occur due to environmental stress, causing cessation of blooms (Reynolds *et al.*, 1982; van Boekel *et al.*, 1992). Depending on their ability to cope with unfavourable conditions, such as nutrient depletion, ultraviolet radiation, photoinhibition and light limitation, algal cells can escape these situations by forming resting stages (French and Hargraves, 1980; McQuid and Hobson, 1995) or die by autolysis. Brussaard *et al.* (1995) demonstrated that cell lysis accounted for 75% of the decline of a coastal *Phaeocystis* bloom. This lysis stimulated bacterial growth through the production of DOM.

Viral infections can also cause cell lysis. Virus-like particles have been described for many eukaryotic algae (van Etten *et al.*, 1991; Reisser, 1993), cyanobacteria (Suttle *et al.*, 1993) and natural phytoplankton communities (Peduzzi and Weinbauer, 1993). Viruses have been held responsible for the collapse of *Emiliania huxleyi* blooms in mesocosms (Bratbak *et al.*, 1995) and in the North Sea (Brussaard *et al.*, 1996), and have been shown to induce lysis of *Chrysochromulina* (Suttle and Chan, 1993). Virus-induced lysis has been proposed as a mechanism to release dimethylsulfonium proprionate (Bratbak *et al.*, 1995). Viruses are also thought to be an important alternative food source for heterotrophic flagellates (Peduzzi and Weinbauer, 1993). Because viruses are sometimes strain-specific, they can increase genetic diversity (Nagasaki and Yamaguchi, 1997). Nevertheless, despite the increasing number of reports, their role in the ecology of the marine ecosystem is still far from understood.

4. Interrelationships

It is clear that several factors can act simultaneously to control the biomass of phytoplankton. In the absence of grazing, phytoplankton can reach high biomass, even when the growth rates are low due to bottom-up limitations. For example, high phytoplankton primary production due to a high (but not very active) biomass in the upper freshwater region of the Western Scheldt was due to the nearly anoxic conditions in the water column, making living conditions for grazers impossible (Kromkamp et al., 1995). As mentioned above, microtidal estuaries are more sensitive to changes in nutrients than macrotidal estuaries (Monbet, 1992).

Development of phytoplankton blooms is possible only when the net rate of biomass accumulation exceeds the losses, i.e. when the residence time is larger than the phytoplankton population doubling time (B_E in eqn. 1; Cloern et al., 1983). Small estuaries with short residence times are more prone to flushing than large estuaries. The upper region of the Shannon estuary (Ireland) has a flushing time of 1–3 tidal cycles, preventing development of the phytoplankton population (McMahon et al., 1992). Hurricanes and associated rainfall also can cause large freshets, responsible for flushing out populations (e.g. Paerl et al., 1998). However, as mentioned above, horizontal advection can also import phytoplankton population from rivers (de Madariaga et al., 1989, 1992) or adjacent coastal systems (Malone, 1977; Cloern, 1996).

Recruitment of cells (or resting stages) from the bottom can also be an important factor in bloom formation (B_X, in eqn. 1; Rengefors, in press). Exchange of algal cells between the benthos and water column can be very intense in some estuaries, making it difficult to make a distinction between phytoplankton and microphytobenthos (de Jonge and van Beusekom, 1995; Irigoien and Castel, 1997). Resuspension of benthic microalgae from the inter-tidal and shallow areas in the Hudson River estuary was thought to be the main reason why net primary productivity was calculated in this turbid estuary. Thus, advection processes play an important role in phytoplankton dynamics.

C. Factors Influencing Biomass of Microphytobenthos

1. Small-scale Heterogeneity in Microphytobenthos

Microphytobenthic assemblages exhibit a high degree of spatial heterogeneity in biomass and species composition. This patchiness occurs on a scale of micrometres to many tens of metres (Pinckney and Sandulli, 1990; Underwood and Paterson, 1993b; Saburova *et al.*, 1995; MacIntyre *et al.*, 1996; Oxborough *et al.*, unpublished result). There are also patterns of vertical distribution within sediments, with the bulk of the active biomass (determined as chlorophyll *a*) found within the top few of millimetres of cohesive sediments (Yallop *et al.*, 1994; MacIntyre and Cullen, 1995; Wiltshire *et al.*, 1997). However, viable cells and chlorophyll *a* can be isolated from deeper layers (Baillie and Welsh, 1980; Admiraal, 1984; MacIntyre and Cullen, 1995). This heterogeneity is one of the major problems associated with scaling up measured rates of microphytobenthic activity to larger areal rates, and in comparing different studies.

The accurate determination of the active biomass of microphytobenthos is important when attempting to characterize photosynthetic parameters. The depth of sediment sampled for chlorophyll *a* often ranges between 2 and 10 mm, dependent on sediment type and the nature of the investigation (see MacIntyre *et al.*, 1996). Given that the photic depth in such sediments can be an order of magnitude smaller than this, values of P^B are significantly underestimated by the inclusion of "non-active" chlorophyll *a* in the values used to normalize the rate of photosynthesis (Wiltshire *et al.*, 1997). This "dilution" by non-active photopigments may be more of a problem now that high-performance liquid chromatography (HPLC) is used to measure pigments in sediments. Because HPLC can measure a suite of photopigments, information on microspatial variation in pigment composition can be lost due to inclusion of photopigments from below the photic zone. This effect has been held responsible for problems in determining diadinoxanthin : diatoxanthin ratios in microphytobenthic assemblages exposed to increased ultraviolet B radiation (Sundbäck *et al.*, 1997; Underwood *et al.*, 1999). It is now possible to avoid these problems by using *in situ* freezing techniques followed by microsectioning with a cryomicrotome (Wiltshire *et al.*, 1997).

The importance of "buried" chlorophyll *a* and cells depends in part on the degree of mixing and resuspension with the habitat. Although biofilms of epipelic microphytobenthos can increase sediment stability and reduce resuspension (Underwood and Paterson, 1993a; Paterson and Black, this volume), some studies have found that between 30% and 50% of the epipelic microphytobenthic biomass can be resuspended into the estuarine water column (de Jonge and van Beusekom, 1995; Irigoien and Castel, 1997) with the degree of resuspension dependent partly on wind speed and fetch. Erosion of 3–4 mm day^{-1} of cohesive sediment can occur even during 'calm' conditions (Underwood and Paterson, 1993a), with storm events removing substantially greater depths (Underwood and Paterson, 1993b). Therefore, on an estuarine scale, there may be substantial proportions of the total microphytobenthic population within an estuary either resuspended within the water column or buried within the sediment. This proportion of the biomass may be coupled only indirectly to measured *in situ* rates (and annual estimates) of benthic primary production, but may be important to other trophic levels within the ecosystem.

2. Large-scale Heterogeneity

A major determining factor in the abundance of microphytobenthos is the nature of the substratum. Sandy silts and sands support significantly lower concentrations of microalgal biomass than sites with fine cohesive sediments (Cammen, 1982; Montagna et al., 1983; Cammen, 1991; de Jong and de Jonge, 1995; Underwood and Smith, 1998a). With increasing sediment grain size, the proportion of epipelic, motile taxa decreases (Sundbäck and Snoeijs, 1991; Yallop et al., 1994; Saburova et al., 1995) and microphytobenthic assemblages in intertidal sands consist predominantly of small, virtually immobile, araphid and monoraphid diatom taxa which produce an extracellular polymeric substance pad for attachment (Hoagland et al., 1993). Sands tend to be both lower in nutrients and more frequently resuspended than cohesive sediments, and these characteristics probably contribute towards lower microphytobenthic biomass. Nutrient enrichment studies in sandy substratum usually indicate nutrient limitation (section IV.A.4), while cultures of epipelic diatom species grown in agitated sand have higher rates of cell damage compared with attached epipsammic taxa (Delgado et al., 1991).

There are also patterns of biomass and species distribution across intertidal mudflats. Microphytobenthic biomass tends to be greater towards the upper shore (Colijn and Dijkema, 1981; Paterson et al., 1990; Underwood and Paterson, 1993b; Underwood, 1997; Brotas and Catarino, 1995; de Jong and de Jonge, 1995; Santos et al., 1997). Lower shore sediments tend to have higher water contents and are less stable than sediment at the middle and upper shore (Underwood and Paterson, 1993b), due partly to the energy of tidal flow and regular resuspension of sediments. Periods of illuminated emersion are also more limited on the low intertidal (Blanchard and Guarini, 1999) where light penetration is restricted by highly turbid estuarine waters. This suggests that low-shore microphytobenthos are probably light limited (in terms of available photoperiod per 24 h), whereas biomass accumulation is prevented by frequent disturbance. Many of the characteristic lower-shore diatom taxa (*Rhaphoneis, Cymatosira*) are non-motile species, and appear to be cast up on to the shore after periods of storms, rather than actively growing *in situ* (Underwood, 1994). At higher tidal heights on a mudflat, the pattern of illuminated emersion periods and reduced resuspension contribute to create conditions favourable for epipelic microphytobenthos. However, upper-shore stations are also subject to greater desiccation and temperature effects (Underwood, 1994; Guarini et al., 1997; Santos et al., 1997), which can be increased by long periods of exposure during neap tide periods (Underwood and Smith, 1998b). These factors usually result in a monomodal distribution of biomass across an intertidal flat, with the peak somewhere between mid tide level and mean high water neap tide level, and not necessarily at the highest bathymetric level (Guarini et al., 1998).

3. Temporal Variation

Microphytobenthic biomass shows a high level of variability both within and between years (see Table 3 in MacIntyre *et al.*, 1996). Chlorophyll *a* concentrations ranging from 1 to 560 mg m^{-2} and 0.1–460 µg g^{-1} sediment are reported (MacIntyre *et al.*, 1996; Underwood, 1997), although these values are derived from samples of varying depth. Many studies of intertidal habitats have found that there are increases in epipelic diatom biomass during the summer months (Colijn and Dijkema, 1981; Admiraal *et al.*, 1982; Montagna *et al.*, 1983; Colijn and de Jonge, 1984; Guarini *et al.*, 1998). MacIntyre *et al.* (1996) concluded that in the northern hemisphere (where there is a reasonably large data set) the biomass peak was of shorter duration and occurred later in the year with increasing latitude. However, peaks of biomass also occur frequently at other times of the year (Underwood and Paterson, 1993b; Brotas *et al.*, 1995; Peletier, 1996; Underwood, 1997; Guarini *et al.*, 1998), and it would appear that epipelic diatom assemblages are less seasonally influenced than phytoplankton communities (Cadée and Hegeman, 1974). High temporal variability in biomass is a common feature of intertidal epipelic diatom biofilms, with biomass dependent on local environmental changes such as erosion and deposition events, desiccation linked to tidal exposure and weather conditions, and periods of rapid growth during favourable conditions (Underwood and Paterson, 1993a,b; Yallop *et al.*, 1994; de Jonge and van Beusekom, 1995; Saburova *et al.*, 1995; Underwood *et al.*, 1998). Therefore, under suitable conditions, surface biofilms dominated by diatoms can rapidly develop on intertidal flats. Grazing by deposit feeding invertebrates and fish can also cause a significant reduction in biomass, particularly during summer months (Montagna, 1984; Morrisey, 1988; Smith *et al.*, 1996).

D. Relative Importance of Planktonic and Benthic Primary Production in Estuaries

Is it possible to estimate the percentage contribution of phytoplankton and microphytobenthos to whole estuarine primary production? Such calculations rely on good data sets and models which adequately describe the key processes in primary production and the variability in planktonic and benthic biomass. As has been discussed, this is not an easy task, particularly with estimates of total microphytobenthic biomass. However, a number of studies have generated estimates (Table 4), which show that microphytobenthic primary production can account for up to 50% of whole estuarine primary production.

The proportional contribution of microphytobenthic primary production to total autochthonous production depends on a number of variables. Phytoplanktonic primary production will be enhanced in estuaries with low flushing times, low suspended solid loads and high nutrient concentrations.

Table 4
Relative contribution of phytoplankton and microphytobenthos to total estuarine autochthonous primary production

Estuary	Phyto-plankton (%)	Micro-phytobenthos (%)	Other	Reference
Wadden Sea, Netherlands	50	50		Cadée and Hegeman (1974)
Lynher estuary, UK	36.5	63.5		Joint (1978)
	33	33	33% detritus	Reise (1985)
Langebaan Lagoon, South Africa	23	22	55% macrophytes	Fielding et al. (1988)
North Inlet, South Carolina, USA	18.4	36.3	13.7% macroalgae 31.7% *Spartina*	Pinckney (1994)
Westerschelde, Netherlands		17		de Jong and de Jonge (1995)
Oosterschelde, Netherlands				
Pre-1985		16		de Jong et al.
Post-1985		30		(1994)
Weeks Bay, Alabama, USA		20.6		Schreiber and Pennock (1995)

With greater light penetration, the potential for nutrient limitation increases. Thus, in turbid estuaries that are severely light limited, a monomodal peak of phytoplankton activity during the summer months related to light availability would be observed (van Spaendonk *et al.*, 1993). In less turbid situations, a more "classical" seasonal pattern of phytoplankton activity would be observed. Estimating microphytobenthic primary production relies mainly on data describing the response to irradiance and the distribution of biomass. Various models have been produced using photosynthetic parameters and diel, tidal and annual light curves to estimate biomass specific production (μg C (μg chlorophyll $a)^{-1}$). These models need to be linked to information on estuarine morphology and both the area and the sediment type of intertidal flats. Larger areas of cohesive sediments will support more microphytobenthic biomass than mobile sandflats, although the elevation of such mudflats is also important. Even if such data exist, the natural heterogeneity in microphyto-benthic biomass occurrence and distribution makes the scaling up of these values to whole estuarine estimates very problematic. Until we have a better understanding of the predictable and stochastic processes that influence microphytobenthic biomass, it will be difficult to model benthic production over large areas. Recent developments in remote sensing of mudflats may enable better estimates of microphytobenthic biomass distribution to be made (Paterson *et al.*, 1998, and this volume). With such information, a reasonable estimate of benthic production may be obtainable.

V. INTER-ANNUAL CHANGES IN PRIMARY PRODUCTION: LONG-TERM TIME SERIES AND ANTHROPOGENIC EFFECTS

Large inter-annual variability in phytoplankton productivity estimates have been observed (Cloern, 1991, 1996; Harding, 1994; Wetsteyn and Kromkamp, 1994; Paerl *et al.*, 1998; Table 5). From the preceeding discussion, it is obvious that there are several ways to explain these interannual differences (Cloern, 1996). Nutrient enrichment has stimulated primary production in coastal areas and estuaries, causing eutrophication (Boynton *et al.*, 1982; Cadée, 1986; Cadée and Hegeman, 1993; Smayda, 1990; Harding, 1994; Nixon, 1995) and harmful algal blooms of *Phaeocystis*, *Chrysochromulina* or *Phiesteria* and red tides. Harmful algal blooms are a concern for water management and public health (for more information visit the Harmful Algal Bloom webpage from NSF/NOAA at Woods Hole Oceanographic Institution: http: //www.redtide.whoi.edu/hab/ and the special issue of *Limnology and Oceanography* 1997, **42**(5)). The mechanisms behind species succession as a result of eutrophication are still largely unknown. Because of this, it is very difficult to predict whether attempted restoration of eutrophic ecosystems will restore the *status quo ante*. It is also very difficult to separate human influences from the naturally occurring variability in primary production. For this, long-term time series are necessary, and they are scarce.

The effect of nutrient enrichment (mainly from the river Rhine) on primary production in the Marsdiep (tidal inlet between the Dutch Wadden Sea and the North Sea) is well documented (Cadée, 1986; Cadée and Hegeman, 1993; de Jonge and van Raaphorst, 1995). Despite efforts to combat high P and N loadings, N loads have not decreased. Measures to decrease P loads were more successful, but as yet no proportional decline in primary production has been observed (Cadée and Hegeman, 1993; de Jonge, 1997). Due to the changes in nutrient input, the N : P and N : Si ratios have changed considerably, and it has been suggested that these shifts might play an important role in the occurrence of nuisance blooms (Hallegraeff, 1993). Besides changes in N : P ratio, the NH_4^+: NO_3^- ratio has also changed, and Riegman *et al.* (1992) and Stolte *et al.* (1994) demonstrated that this could influence species succession.

Rousseau and Lancelot (personal communication) analysed a time series data set for a Belgian coastal station and observed that the relative importance of diatoms and *Phaeocystis* in the spring bloom was very variable from year to year. Only one of the three "typical" winter diatom communities observed, a

Table 5

Overview of variation in annual phytoplankton primary production and possible mechanism responsible for variation

Estuary	Interval	Range of annual primary production (g C m^{-2})	Control mechanism	References
Marsdiep	1964–1992	140–390	P and N enrichment	Cadée and Hegeman (1993)
Oosterschelde estuary	1982–1990	176–450	Unknown, but alteration in	Wetsteyn and Kromkamp (1994)
Inner		240–406	light, Si, P and	
Central outer		223–540	N limitation	
Delaware estuary	1981–1987	190–400	Unknown, but stimulation by nutrients from river input, and periods with light limitation	Pennock and Sharp (1986, 1994)
Neuse River estuary	1988–1991	202–320	High watershed runoff and	Mallin et al. (1993)
	1994–1996	360–531	N-input stimulus eutrophication; washout due to high discharge (hurricanes)	Paerl et al. (1998)
Chesapeake Bay	1972–1977	337–782	High discharge (increased nutrient input)	Harding (1994)
San Francisco Bay	1975–1996		High discharge, grazing	Cloern (1996)

highly silicified late winter–early spring assemblage, was able to outcompete *Phaeocystis* colonies at temperatures below 9°C, providing Si was not limiting. In dry years, Si was limiting and *Phaeocystis* established large populations. It is very likely that this process will operate in some estuaries as well. *Phaeocystis* colonies are often poorly grazed, at least when diatoms are present (Davies *et al.*, 1992; Weisse *et al.*, 1994). Collapse of a *Phaeocystis* bloom is thus not caused by grazing, but by disintegration and lysis (Passow and Wassmann, 1994; Brussaard *et al.*, 1996) fuelling the microbial foodweb. A change from a more linear foodchain to a microbial foodweb trophic structure has also been observed in other estuaries (Riegman *et al.*, 1993; Lewitus *et al.*, 1998). These changes due to shifts in species composition may also affect fisheries, as the transfer of C and N to higher trophic levels will be dependent on the importance of the microbial foodweb in the overall foodweb. The effect of phycotoxins on trophic structures is hardly understood, and is complicated by the fact that some grazers are affected, whereas others are not (Smayda, 1997; Turner and Tester, 1997). This is furthermore complicated by the fact that higher up in the foodweb mortality can be induced by phyco-toxins, as was shown for *Chrysochromulina* (Estep and MacIntyre, 1989) and *Phiesteria* (Burkholder and Glasgow, 1997). It may also inhibit competing, co-occurring phytoplankton species (Freeberg *et al.*, 1979; Windust *et al.*, 1996). Algal species succession might also influence trophic transfer in a different way: some species contain more polyunsaturated fatty acids (PUFAs) than others, and also the type of PUFA can differ from species to species (Fraser *et al.*, 1989). As PUFAs are essential dietary factors for grazers and animals, a change in species composition can also influence the grazer community, and hence transfer efficiencies of C to higher trophic levels (Cloern, 1996).

Phytoplankton blooms in the Neuse River estuary followed periods of increased riverine N input, except when flushing rates were high during extremely high runoff periods (Mallin *et al.*, 1993). The blooms caused organic matter loads capable of inducing hypoxic or anoxic conditions, inducing wide-spread death of fish and shellfish. Small *et al.* (1990) observed an increase in biomass in the Columbia River estuary as a result of very high discharge of sediment-laden water caused by eruption of Mount St Helens. In the nutrient-rich Delaware estuary, primary production is generally determined by light availability (Pennock, 1985), and in the period from 1981 to 1985 production varied between 290 and 400 g C m^{-1} year^{-1}. In a later paper, Pennock and Sharp (1994) demonstrated by means of a variety of techniques that in late spring P-limited phytoplankton growth occurred in the middle estuary. Downstream, N limitation was observed after the period of P limitation, based on dissolved nutrient ratios and concentrations. This was not corroborated by particulate N : P ratios, and Pennock and Sharp (1994) concluded that sufficient N was supplied by advection and remineralization. As a result, grazing most likely controlled phytoplankton biomass. An alternation between factors limiting photosynthesis and biomass has also been observed in other estuaries (Fisher *et al.*, 1988; Kromkamp and Peene, 1997; Wetsteyn and Kromkamp, 1994).

Boynton *et al.* (1982) observed that the timing and frequency of blooms were quite similar between 1972 and 1977 in Chesapeake Bay, but the magnitude of the bloom varied from 337 to 782 g C m^{-2} year^{-1}. It was specu-lated that the variability was related to nutrient input from watershed sources, which varied with river discharge. However, analyses of a 40-year data set of Chesapeake Bay revealed large interannual variations in timing, position and magnitude of the winter–spring diatom bloom, which was attributed to vari-ability in the freshwater inflow from the Susquehanna River. The associated nutrients and suspended solids regulate the light and nutrient environment

(Harding, 1994, and references therein). Nutrients are especially important in the lower polyhaline region of the Bay. This study also revealed that the short-term effect of freshwater input is less than 4 months (most likely depending on the residence time), but that "multi-year" effects exist (in this case reduced nutrient loading during periods of low freshwater flow influenced phyto-plankton abundance over a longer time period).

The Eastern Scheldt estuary is probably one of the few estuaries that has undergone a system-wide experiment. The storm flood disaster of 1953 in SW Netherlands resulted in a large-scale engineering project in the "delta" area of the rivers Rhine, Meuse and Scheldt (Nienhuis and Smaal, 1994). As a result, a storm-surge barrier was constructed in the mouth of the estuary. This barrier allowed the tides to enter freely, which was necessary to keep the shellfish fisheries intact, but the cross-section was reduced by 75%. As a result, the mean tidal range changed from 3.75 to 3.25 m, and the reduced flow velocities increased the water transparency considerably. At the head of the estuary, compartment dams blocked the inflow of freshwater. Hence, the residence times increased substantially, and the freshwater input (with concomitant nutrients) decreased by 65%. It was concluded that the system changed from a light- and silicate-limited estuary in the pre-barrier period to a silicate- and nitrogen-limited lagoon in the post-barrier period (Wetsteyn and Kromkamp, 1994). Although phytoplankton biomass changed (partly due to longer residence times), primary production did not. Later analyses showed that nitrogen did not limit phytoplankton production (Kromkamp and Peene, 1997; see also Figure 5), but that, due to very high regeneration rates, the phytoplankton in spring was most likely Si and P limited. In contrast, in summer, due to high regeneration rates, phytoplankton grew rapidly and biomass was controlled by grazing. It is remarkable that despite these large changes the primary productivity did not change. However, the phyto-plankton composition changed considerably, and a species community previ-ously normal for the summer developed both early and later in the year, most likely as a response to the improved light conditions in the post-barrier period (Bakker et al., 1994). Also, small, weakly silicified diatoms became more abundant, and small flagellates increased in summer. The results obtained for the Eastern Scheldt show that estuarine systems can be rather robust. Model simulations corroborated that the Eastern Scheldt is relatively insensitive to changes in nutrient load, and this was attributed to suspension feeders, which are thought to be able functionally to stabilize estuarine ecosystems (Herman and Scholten, 1990).

Cloern et al. (1989) demonstrated for San Francisco Bay that the spring bloom was associated with density stratification of the water column. The magnitude of the spring bloom was correlated with the magnitude of the river flow during the wet season, demonstrating that climatic and hydrological forcing can be an important cause of inter-annual variation in primary production. Introduction of allochthonous species can also change phyto-plankton primary production and biomass significantly, as was demonstrated when North San Francisco Bay was colonized by large numbers of the clam *Potamocorbula amurensis* in 1987 (Cloern, 1996). As a result, the average chlorophyll content decreased dramatically, and the large blooms disappeared.

There have been very few studies on long-term changes in microphyto-benthos. Peletier (1996) revisited sites in the Ems-Dollard that had been actively investigated between 1977 and 1980. Significant reductions in organic inputs to the intertidal flats over this period were found to have resulted in changes in the species composition of microphytobenthos, and a change in the seasonal pattern of peaks in biomass. Conversely, increases in annual microphytobenthic production in the Marsdiep between 1968 and 1984 corresponded to increased nutrient loadings from the River Rhine and Lake IJssel (de Jonge *et al.*, 1996). Thus, in the longer term, even epipelic micro-phytobenthic communities, buffered by their close association with a nutrient-rich substratum, will respond to such environmental perturbations.

VI. ALTERNATIVE METHODS AND NEW DEVELOPMENTS

A. N-Uptake

Phytoplankton primary production is normally measured in the field using the ^{14}C fixation method. C fluxes can also be estimated from incorporation of ^{15}N (Dugdale and Goering, 1967) and the Redfield C : N ratio of 6.6. By using both $^{15}N-NO_3$ and $^{15}N-NH_4$, the preference of the algae for either substrate can measured (reviewed by Dortch, 1990), and the f ratio (ratio of new to total production based on nitrate uptake to total nitrogen uptake; Eppley and Peterson, 1979; Dugdale and Wilkerson, 1986) can be determined. In ocean water under steady-state conditions, the f ratio should equal the export production (Platt *et al.*, 1992), as the vertical flux into the euphotic zone of nitrate regenerated by bacterial nitrification in deep water is considered to fuel the new production. In estuaries the f ratio cannot be computed as easily as in the open ocean because of the absence of steady-state conditions: ammonia can be advected into the estu-aries by the rivers, leading to new production, and not to regenerated production. Oxidation of ammonia via nitrite to nitrate by nitrifiers could be an important source for regenerated nitrate. Euphotic zone nitrification varied between 47% and 142% of the nitrate assimilation rates at the oligotrophic North Pacific Times series station ALOHA (Dore and Karl, 1996). Considering the sometimes relatively high concentrations of nitrite in estuaries, chemoautotrophic bacterial nitrification can add substantially to the regeneration of nitrogen in the form of nitrate, and this should be taken into account when computing f ratios; for example, nitrification in the upstream region of the Western Scheldt estuary was estimated to be maximally 0.3 g.nitrate m^{-3} day^{-1}, leading to 3 mg N production m^{-3} day^{-1} by nitrifiers. This was equivalent to or even exceeding phytoplankton N production (Soetaert and Herman, 1995). It should also be taken into consideration that heterotrophic bacteria can take up DIN, and that the observed N uptake rates may overestimate phytoplankton production rates (Kirchman *et al.*, 1992; Boyer *et al.*, 1994; Dauchez *et al.*, 1995). Correction for this by filtration can improve results, although a significant, but variable, proportion of the bacteria is attached to particles (Laanbroek and Verplanke, 1986).

An advantage of the ^{15}N uptake method is that it measures net production, whereas it is still unknown whether the ^{14}C method measures net or gross production (or something in between; see Williams *et al.*, 1993a,b; Williams and Lefèvre, 1996). As with C uptake, nitrate uptake is light dependent, and uptake rates in the dark are very low. Ammonia uptake is less light dependent, with uptake rates in the dark varying between 30% and 95% of the maximal uptake rate in the light (Pennock, 1987; Boyer *et al.*, 1994).

Estuarine phytoplankton in general show a preference, as do most algae, for ammonia (Figure 4) (McCarthy *et al.*, 1977; Pennock, 1987; Boyer *et al.*, 1994). The preference for a particular nitrogen source is often expressed as the

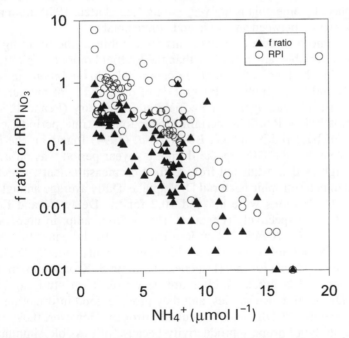

Fig. 4. Relative Preference Index (RPI) for nitrate and the f ratio as a function of ambient ammonia concentration.

Relative Preference Index (RPI, McCarthy *et al.*, 1977). For example the RPI for nitrate is

$$RPI_{NO_3^-} = \frac{\rho NO_3^-/\rho DIN}{[NO_3^-]/[DIN]} \tag{7}$$

i.e. the RPI for nitrate is the ratio of nitrate transport rate ($\mu mol\ NO_3^-\ \mu mol\ PN^{-1}\ h^{-1}$) to the total dissolved inorganic nitrogen (DIN) transport rate over the ratio of nitrate to DIN. When the $RPI_{NO_3^-} > 1$, nitrate uptake is preferred. As stated previously, the RPI for nitrate is generally less than 1. The use of the RPI has been criticized because this complex ratio is difficult to interpret (see Dortch, 1990). The precision of the RPI is low because of errors resulting from combining many variables (Collos and Slawyk, 1996). It cannot be calculated when nitrate is below the detection limit (when it might be preferred), and its numerical value can change in response to changing ambient concentrations without changes in uptake rates. Boyer *et al.* (1994) found an increase in the RPI with increasing salinity in the Neuse River estuary, which was attributed to the changing phytoplankton community, possibly adapted to low ammonia concentrations (Pennock, 1987). This pattern is, however, not observed in all estuaries (Middelburg *et al.*, in preparation). It is also clear from these papers and Figure 4 that, although there is a preference for ammonia, the inhibition of nitrate uptake by ammonia is not very strong (see Dortch, 1990, for a review on the interaction between ammonia and nitrate uptake).

When C and N uptake measurements are combined, the resulting C : N ratio is very variable, and often less than the Redfield ratio, especially when short-term incubations in the light are compared. This points to an uncoupling of N and C metabolism. Integration of dark and light measurements over a 24-h cycle can approach Redfield stoichiometry (Pennock, 1987). C : N ratios below Redfield, certainly over longer time periods, can be indicative of bacterial N uptake. Boyer *et al.* (1994) found for the Neuse River estuary that the DIN uptake during a 4-year period was on average twice as high as that estimated from C fixation measurements, which was attributed to heterotrophic bacterial DIN uptake. Daily average annual depth integrated C : N ratios were 8.5 and 10.2 for the Delaware and Eastern Scheldt estuary respectively (Pennock, 1987; Kromkamp, in preparation). Although a low C : N ratio below Redfield can also be caused by luxury uptake of DIN, possibly caused by too high enrichments with ^{15}N-DIN, there is hardly any evidence for this (Dauchez *et al.*, 1995; Kromkamp, in preparation). Hence, ^{15}N uptake studies are very useful for studying phytoplankton DIN preference studies, and they provide good information about the turnover rates of DIN and particulate nitrogen. However, they are less suited for studying primary productivity because of possible simultaneous bacterial uptake.

B. Variable Fluorescence

The primary production measurements given in Table 1 are based mainly on C fixation measurements performed on hourly to daily time scales. Primary production is a rate, hence it tells us the speed of the primary production. It can be argued that primary production is equivalent to the rate of the dark reactions of photosynthesis, a process with a time scale of several milliseconds (Platt and Sathyendranath, 1993). Because such a time scale is ecologically not very relevant, this rate is amplified to at least the time scale of the measurement. Variable fluorescent techniques have recently emerged as promising new tools to measure (gross) photosynthesis. These techniques measure the activity of the photosystem II (PSII) reaction centre, hence they measure the rate of production of electrons by the water-splitting system of PSII. Although these techniques measure rates with time scales of less than a second, they are based on optical techniques and they therefore have the capacity for *in situ* measurements, avoiding all bottle artefacts. They also open up the possibility of making measurements very frequently, with a more dense measuring grid or more frequent measurements, avoiding the problems associated with undersampling (Wiggert *et al.*, 1994; Yin *et al.*, 1997).

We will briefly describe the two methods used (the pump and probe method, developed by Falkowski and co-workers (e.g. Falkowski *et al.*, 1986; Kolber and Falkowski, 1993), and pulse amplitude modulated (PAM) fluorometry, developed by Schreiber *et al.* (1986). Both methods are based on the same principles, made possible by the fact that under *in vivo* conditions, fluorescence changes originate almost exclusively from chlorophyll *a* in PSII and the associated antenna pigments. In very dim light, with insignificant rates of photosynthesis and when the primary electron acceptor in PSII, Q_A, is completely oxidized, nearly all fluorescence emanates from the pigment bed, and is minimal (see Dau, 1994; Krause and Weis, 1991, for a reviews on the basics of chlorophyll fluorescence and photosynthesis). This level (operationally determined after a sufficient dark adaptation) is called the minimum fluorescence level, F_o. Upon application of a brief, very strong, pulse of light, the photochemical energy conversion of PSII (and PSI) becomes saturated and Q_A, the first stable electron acceptor in PSII, becomes completely reduced, causing the fluorescence to rise to a maximal value (F_m). From these two parameters, the maximum PSII quantum efficiency can be estimated as follows

$$\text{maximum PSII quantum efficiency} = \Phi_P^o = (F_m - F_o)/F_m = F_v/F_m = \Delta\phi_m \quad (8)$$

(Genty *et al.*, 1989). During the pulse, the quantum yields of fluorescence and of non-radiative energy dissipating processes will be increased transiently to maximal values. The three processes, photochemistry, fluorescence and thermal dissipation, are competitive, and depend on the redox state of Q_A. The sum of the probability of these processes is always 1. In the light, the fluorescence will rise to a steady state level, F. Analogous to eqn. 8, the effective PSII quantum efficiency can be calculated as (Genty *et al.*, 1989)

$$\Phi_P = (F'_m - F)/F'_m = \Delta F/F'_m = \Delta\phi \qquad (9)$$

where F'_m is the maximum fluorescence of a light-adapted alga. The photosynthetic electron transport rate (ETR) can be calculated from these parameters

$$ETR = \Delta F/F'_m \times E \times \sigma_{PSII} \qquad (10)$$

(Kolber and Falkowski, 1993; Hofstraat *et al.*, 1994), where E is the irradiance and σ_{PSII} the functional absorption cross section of PSII, i.e. the amount of light absorbed by a PSII to drive photochemistry. As this latter parameter is not always measured, the ETR is often approximated by the product of $\Delta F/F'_m$ and irradiance. The basic differences between the PAM fluorometry and the fast repetition rate fluorometer (FRRF) techniques is that the PAM fluorometer uses a multiple turnover flash, saturating both the Q_A, Q_B and PQ pools, whereas the FRRF uses a single turnover flash, which reduces only the Q_A pool. As a result, the F_m values will differ, as will the calculated quantum efficiencies (Schreiber *et al.*, 1995). Hence, although both techniques measure similar but not identical processes, it remains to be seen whether these differences are important for most ecological studies. It seems unlikely that this will be the case.

As stated, these active fluorescence methods measure the rate of photosynthetic electron transport. As oxygen formation is a byproduct of the water-splitting process, a good correlation might be expected between oxygen production and PSII electron transport. This has indeed been observed in a number of comparative studies in phytoplankton (Falkowski *et al.*, 1991; Kolber and Falkowski, 1993), higher plants (Genty *et al.*, 1989; Seaton and Walker, 1990; Krall and Edwards, 1991) and microphytobenthos (Hartig *et al.*, 1998). However, under non-optimal conditions (e.g. high light), the relationship between oxygen production and ETR often becomes non-linear, especially at saturating irradiances (Geel *et al.*, 1997; Flameling and Kromkamp, 1998), where oxygen production saturates whereas ETR continues to increase with higher irradiances. Non-linearity has also been observed at low light (Falkowski *et al.*, 1986; Kroon, 1994; Prasil *et al.*, 1996). The most probable causes are a decrease in the functional PSII cross-section (Kolber and Falkowski, 1993) and an increase in the pseudocyclic electron flow (Mehler reaction) (Kana, 1992; Geel *et al.*, 1997; Flameling and Kromkamp, 1998). Other possible causes, such as increased dark respiration, photorespiration, cyclic electron flow around PSII or PSII heterogeneity, are discussed in Flameling and Kromkamp (1998).

In microphytobenthos, the relationship between C fixation and PSII electron flow has hardly been studied. Hartig *et al.* (1998) found a reasonable linearity in sediment slurries between C fixation rates and ETR multiplied by the cross-sectional chlorophyll absorptivity. Close examination revealed that the relationship at high light was variable. In contrast to Hartig *et al.* (1998), Barranguet *et al.* (1998) found a large deviation from linearity at high irradiances. Clearly more results are needed. It should be stressed that the PAM technique allows rapid assessment of photosynthesis, which is needed for estimations of spatial and temporal distribution patterns with high resolution.

Apart from measurements of the light–response curve, the fluorescence measurements give an indication of the physiological state of the phytoplankton. Nutrient limitation has been shown to decrease F_v/F_m (Kolber *et al.*, 1988, 1994; Falkowski *et al.*, 1992; Flameling, 1998) and alter the PSII cross-section (Kolber *et al.*, 1988). This latter parameter is, however, also influenced by the irradiance (Falkowski *et al.*, 1986; Kolber *et al.*, 1988). With some caution (e.g. under non-photoinhibitory conditions) F_v/F_m can be used to assess the nutrient status of natural populations, e.g. in Delaware Bay (Falkowski *et al.*, 1992). Kromkamp and Peene (1997) followed the nutrient status of phytoplankton in the Eastern Scheldt estuary (The Netherlands) throughout the year. It was found that F_v/F_m values decreased only in spring, following a decrease in phosphate and silicate to potentially limiting concentrations (Figure 5). An interesting observation was that limiting nutrient concentrations were not followed immediately by decreasing F_v/F_m ratios. Apparently the conditions were not limiting at the start of the period with low nutrient concentrations, or the cells continued to grow on internal nutrient concentrations. They might also have taken some time to develop a nutrient-limited physiology. This technique has also been used to demonstrate Fe limitation in the equatorial Pacific Ocean (Kolber *et al.*, 1994). Decreases in F_v/F_m attributed to nutrient limitation have also been measured in subtidal *Gyrosigma balticum* mats, the effect being reversed by nutrient addition (Underwood *et al.*, 1999).

As the minimal fluorescence level, F_0, originates from the pigment bed, it is also a good indicator for biomass. Serôdio *et al.* (1997) and Kromkamp *et al.* (1998) used this to measure non-destructive vertical migration of microphytobenthos. In the latter study, photosynthetic electron transport (light–response curves) were made, and the seasonal changes in photosynthetic parameters of microphytobenthos were studied on an intertidal mudflat in the Western Scheldt estuary. Changes in the maximum rate of ETR suggested photoacclimation. To investigate the effect of high irradiance on photosynthesis, a core was exposed to approximately 800 μmol m^{-2} s^{-1} and the change in F, F_m' and $\Delta F/F_m'$ was followed (Figure 6). The increase in F and F_m' was due to upward vertical migration. $\Delta F/F_m'$ did not decrease, but actually increased a little. If the cells had experienced CO_2 limitation, caused

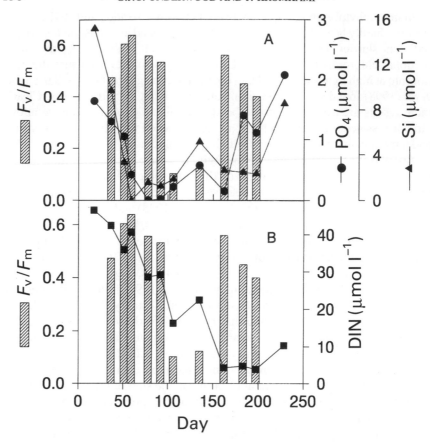

Fig. 5. Maximum quantum efficiency of PSII (F_v/F_m) of natural phytoplankton populations and (A) phosphate and silicate and (B) DIN concentrations of inner (eastern) station in the Eastern Scheldt station. From Kromkamp and Peene (1997).

by the high rates of photosynthesis, a decrease in $\Delta F/F_m'$ would have been likely. Also, photoprotection against high irradiance would induce non-photochemical quenching (thermal energy dissipation), lowering $\Delta F/F_m'$. This was also not observed. It was therefore concluded by Kromkamp *et al.* (1998) that during the overall upward vertical migration there was a vertical micro-migration within the first millimetre of the sediment, where algae at the surface migrated to deeper layers in order to prevent photoinhibition and CO_2 limitation, whilst being replaced by others. In a separate study on the effect of UVB radiation on microphytobenthos, vertical migration in *G. balticum* mats away from high midday irradiances resulted in significant decreases in F_s and F_m'. With increased UVB treatment, diatom mats

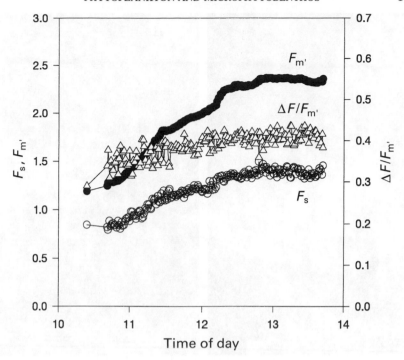

Fig. 6. Steady-state fluorescence (F_s), maximum fluorescence (F'_m) and effective quantum efficiency $\Delta F/F'_m$ of microphytobenthos from an intertidal mudflat in the central Western Scheldt. From Kromkamp *et al.* (1998), reproduced by permission of Inter-Research, Oldendorf, Germany.

exhibited significantly higher values of Φ_{PSII}, indicating that cells were migrating deeper into the sediments, where they were exposed to lower irradiances (Underwood *et al.*, 1999).

An exciting recent development is that variable fluorescence measurements can also be performed on single cells, using the PAM technique on a modified fluorescence microscope and image analysis techniques (Oxborough and Baker, 1997; Oxborough *et al.*, unpublished results). The advantage of this is that the photosynthetic response of single cells within a mixed population can be measured (Figure 7). Oxborough *et al.* (unpublished results) have shown significant differences in the efficiency of light usage between diatoms and euglinid algae within intertidal sediments, with euglinids maintaining a higher efficiency at high irradiances. This technique has important implication for biodiversity studies and will be able to image short-term changes in position and efficiency of individual cells within a biofilm.

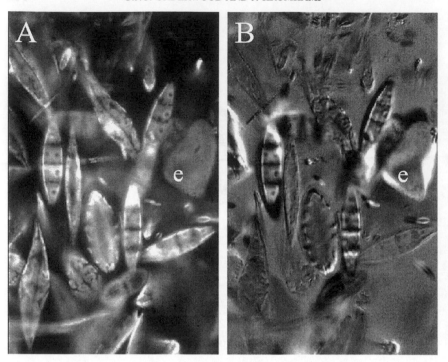

Fig. 7. Example of fluorescence yields (F'_m) and effective quantum efficiency ($\Delta F/F'_m$ or ϕ_{PSII}) in single cells using the PAM approach in combination with a modified fluorescence microscope in a microphytobenthic biofilm acclimated to 200 µmol m^{-2} s^{-1} irradiance (Oxborough, Hanlon, Underwood and Baker, unpublished results). (A) An F'_m image, showing a number of different diatom taxa, (B) ϕ_{PSII} image, showing differences between cells: the brighter the image, the higher the ϕ_{PSII}. The low F'_m yield of the euglenid (e) in (A), and its higher ϕ_{PSII} value in (B) relative to the diatoms is clearly visible.

In conclusion, it can be said that variable fluorescence methods are a valuable new tool in microalgal ecology because they can measure rapidly and non-intrusively, and provide data on the physiological condition of the algae. More research is needed between the relationship of C fixation or oxygen exchange methods and PSII electron flow.

ACKNOWLEDGEMENTS

The authors thank Nancy de Bakker, Alison Miles and David Paterson for providing some of the data and discussion, J.E. Cloern and B.E. Cole for providing San Francisco Bay data, and Astrid Hanlon for the images in Figure 7. Part of this work was funded by the UK Natural Environment Research

Council, grant numbers GR3/8907, GST/02/1562 and GT22/96/ENVD/5.This is publication number 2500 of The Netherlands Institute of Ecology, Centre for Estuarine and Coastal Ecology.

REFERENCES

Abercrombie, M., Hickman, C.J. and Johnson, M.L. (1973). *A Dictionary of Biology.* Penguin Books, Harmondsworth, UK.

Admiraal, W. (1984). The ecology of estuarine sediment-inhabiting diatoms. In: *Progress in Phycological Research* (Ed. by F.E. Round and D.J. Chapman), vol. 3, pp. 269–322. Biopress, Bristol.

Admiraal, W. and Peletier, H. (1979). Sulphide tolerance of benthic diatoms in relation to their distribution in an estuary. *Br. Phycol. J.* **14**, 185–196.

Admiraal, W. and Peletier, H. (1980). Distribution of diatom species on an estuarine mudflat and experimental analysis of the selective effect of stress. *J. Exp. Mar. Biol. Ecol.* **46**, 157–175.

Admiraal, W., Peletier, H. and Zomer, H. (1982). Observations and experiments on the population dynamics of epipelic diatoms from an estuarine mudflat. *Estuar. Coast. Shelf Sci.* **14**, 471–487.

Alpine, A.E. and Cloern, J.E. (1988). Phytoplankton growth rates in a light-limited environment, San Francisco Bay. *Mar. Ecol. Progr. Ser.* **44**, 167–173.

Baillie, P.W. and Welsh, B.L. (1980). The effect of tidal resuspension on the distribution of intertidal epipelic algae in an estuary. *Estuar. Coast. Mar. Sci.* **10**, 165–180.

Bakker, C., Herman, P.M.J. and Vink, M. (1994). A new trend in the development of the phytoplankton in the Oosterschelde (SW Netherlands) during and after the construction of a storm-surge barrier. *Hydrobiologia* **282/283**, 79–100.

Banse, K. (1993). On the dark bottle in the ^{14}C method for measuring marine phytoplankton production. In: *Measurement of Primary Production from the Molecular to the Global Scale* (Ed. by K.W. Li and S.Y. Maestrini), ICES Marine Science Symposia (197), pp. 132–140. International Council for the Exploration of the Sea, Copenhagen.

Barranguet, C., Kromkamp, J. and Peene, J. (1998). Factors controlling primary production and photosynthetic characteristics of intertidal microphytobenthos. *Mar. Ecol. Prog. Ser.* **173**, 117–126.

de Beer, D, Glud, A., Epping, E. and Kühl, M. (1997). A fast-responding CO_2 microelectrode for profiling sediments, microbial mats, and biofilms. *Limnol. Oceanogr.* **42**, 1590–1600.

Blanchard, G.F. and Cariou-Le Gall, V. (1994). Photosynthetic characteristics of microphytobenthos in Marennes-Oléron Bay, France: preliminary results. *J. Exp. Mar. Biol. Ecol.* **182**, 1–14.

Blanchard G.F. and Guarini, J.M. (1999). Temperature effects on microphytobenthic productivity in temperate, intertidal mudflats. *Vie Milieu* **48**, 271–284.

Blanchard, G.F., Guarini, J.M., Richard, P., Gros, P. and Mornet, F. (1996). Quantifying the short-term temperature effect on light-saturated photosynthesis of intertidal microphytobenthos. *Mar. Ecol. Progr. Ser.* **134**, 309–313.

Boyer, J.N., Stanley, D.W. and Christian, R.R. (1994). Dynamics of NH_4^+ and NO_3^- uptake in the water column of the Neuse River estuary, North Carolina. *Estuaries* **17**, 361–371.

Boynton, W.R., Kemp, W.M. and Keefe, C.W. (1982). A comparative analysis of nutrients and other factors influencing estuarine phytoplankton production. In: *Estuarine Comparisons* (Ed. by V.S. Kennedy), pp. 69–90. Academic Press, New York.

Bratbak, G., Levasseur, M., Michaud, S., Cantin, G., Fernández, E., Heimdal, B. and Heldal, M. (1995). Viral activity in relation to *Emiliania huxleyi* blooms: a mechanism of DMSP release? *Mar. Ecol. Progr. Ser.* **128**, 133–142.

Brotas, V. and Catarino, F. (1995). Microphytobenthos primary production of Tagus Estuary intertidal flats (Portugal). *Neth. J. Aquat. Ecol.* **29**, 333–339.

Brotas, V., Cabrita, T., Portugal, A., Serôdio, J. and Catarino, F. (1995). Spatio-temporal distribution of the microphytobenthic biomass in the intertidal flats of Tagus Estuary (Portugal). *Hydrobiologia* **300/301**, 90–104.

Bruno, S.F., Staker, R.D. and Sharma, G.M. (1980). Dynamics of phytoplankton productivity in the Peconic Bay estuary, Long Island. *Est. Coast. Mar. Sci.* **10**, 247–263.

Bruno, S.F., Staker, R.D., Sharma, G.M. and Turner, J.T. (1983). Primary productivity and phytoplankton size fraction in a temperate North Atlantic estuary. *Estuaries* **6**, 200–211.

Brussaard, C.P.D., Riegman, R., Noordeloss, A.A.M., Cadée, G.C., Witte, H. Kop, A.J., Nieuwland, G., van Duyl, F.C. and Bak, R.P.M. (1995). Effects of grazing, sedimentation and phytoplankton cell lysis on the structure of a coastal pelagic food web. *Mar. Ecol. Progr. Ser.* **123**, 259–271.

Brussaard, C.P.D., Kempers, R.S. Kop., A.J. Riegman, R. and Heldal, M. (1996). Virus-like particles in a summer bloom of *Emiliania huxleyi* in the North Sea. *Aq. Microbiol. Ecol.* **10**, 105–113.

Burkholder, J.M. and Glasgow, H.B.J. (1997). *Pfiesteria piscicida* and other *Pfiesteria*-like dinoflagellates: behaviour, impacts, and environmental controls. *Limnol. Oceanogr.* **42**, 1052–1075.

Cadée, G.C. (1986). Increased phytoplankton primary production in the Marsdiep area (Western Dutch Wadden Sea). *Neth. J. Sea Res.* **20**, 285–290.

Cadée, G.C. and Hegeman, J. (1974). Primary production of the benthic microflora living on tidal flats in the Dutch Wadden sea. *Neth. J. Sea Res.* **8**, 260–291.

Cadée G.C. and Hegeman, J. (1977). Distribution of primary production of the benthic microflora and accumulation of organic matter on a tidal flat area, Balgzand, Dutch Wadden Sea. *Neth. J. Sea. Res.* **11**, 24–41.

Cadée, G.C. and Hegeman, J. (1993). Persisting high levels of primary production at declining phosphate concentrations in the Dutch coastal area (Marsdiep). *Neth. J. Sea Res.* **31**, 147–152.

Cammen, L.M. (1982). Effect of particle size on organic content and microbial abundance within four marine sediments. *Mar. Ecol. Prog. Ser.* **9**, 273–280.

Cammen, L.M. (1991). Annual bacterial production in relation to benthic microalgal production and sediment oxygen uptake in an intertidal sandflat and an intertidal mudflat. *Mar. Ecol. Prog. Ser.* **71**, 13–25.

Cloern, J.E. (1982). Does the benthos control phytoplankton biomass in south San Francisco Bay? *Mar. Ecol. Prog. Ser.* **9**, 191–202.

Cloern, J.E. (1991). Annual variations in riverflow and primary production in the South San Francisco Bay Estuary (USA). In: *Estuaries and Coasts: Spatial and Temporal Intercomparisons* (Ed. by M. Elliot and J.-P. Ducrotoy), pp. 91–96. Olsen and Olsen, Fredensborg.

Cloern, J.E. (1996). Phytoplankton bloom dynamics in coastal ecosystems: a review with some general lessons from sustained investigation of San Francisco Bay, California. *Rev. Geophys.* **34**, 127–168.

Cloern, J.E (1999). The relative importance of light and nutrient limitation of phyto-plankton growth: a simple index of coastal ecosystem sensitivity to nutrient enrichment. *Aq. Ecol.* (in press).

Cloern, J.E., Alpine, A.E., Cole, B.E., Wong, R.L.J., Arthur, J.F. and Ball, M.D. (1983). River discharge control phytoplankton dynamics in the Northern San Francisco Bay estuary. *Estuar. Coast. Shelf Sci.* **16**, 415–429.

Cloern, J.E., Powell, T.M. and Huzzey, L.M. (1989). Spatial and temporal variability in South San Francisco Bay (USA). II. Temporal changes in salinity, suspended sediments, and phytoplankton biomass and productivity over tidal time scales. *Estuar. Coast. Shelf Sci.*, **28**, 599–613.

Cloern, J.E., Grenz, C. and Videgar-Lucas, L. (1995). An empirical model of the phytoplankton chlorophyll: carbon ratio—the conversion factor between productivity and growth rate. *Limnol. Oceanogr.* **40**, 1313–1321.

Cole, B.E. and Cloern, J.E. (1987). An empirical model for estimating phytoplankton productivity in estuaries. *Mar. Ecol. Prog. Ser.* **36**, 299–305.

Cole, B.E., Cloern J.E. and Alpine A.E. (1986). Biomass and productivity of three phytoplankton size classes in San Francisco Bay. *Estuaries* **9**, 117–126

Cole, J.J. (1989). Temporal and spatial patterns of phytoplankton production in Tomales Bay, California, USA. *Estuar. Coast. Shelf Sci.* **28**, 103–115.

Cole, J.J. and Cloern, J.E. (1984). Significance of biomass and light availability to phytoplankton productivity in San Francisco Bay. *Mar. Ecol. Prog. Ser.* **17**, 15–24.

Cole, J.J., Caraco, N.F. and Peierls, B.L. (1992). Can phytoplankton maintain a positive carbon balance in a turbid, freshwater, tidal estuary? *Limnol. Oceanogr.* **37**, 1608–1617.

Colijn, F. (1983). *Primary Production in the Ems-Dollard Estuary*. PhD thesis, University of Groningen, The Netherlands.

Colijn, F. and de Jonge, V. (1984). Primary production of microphytobenthos in the Ems-Dollard Estuary. *Mar. Ecol. Prog. Ser.* **14**, 185–196.

Colijn, F. and Dijkema, K. S. (1981). Species composition of benthic diatoms and the distribution of chlorophyll *a* on an intertidal flat in the Dutch Wadden sea. *Mar. Ecol. Prog. Ser* **4**, 9–21.

Collos, Y. and Slawyk, G. (1996). ^{13}C and ^{15}N uptake by marine phytoplankton. IV. Uptake ratios and the contribution of nitrate to the productivity of Antarctic waters (Indian Ocean sector). *Deep Sea Res.* **33**, 1039–1051.

Dau, H. (1994). Molecular mechanisms and quantative models of variable photosystem II fluorescence. *Photochem. Photobiol.* **69**, 1–23.

Dauchez, S., Legendre, L. and Fortier, L. (1995). Assessment of simultaneous uptake of nitrogenous nutrient (^{15}N) and inorganic carbon (^{13}C) by natural phytoplankton populations. *Mar. Biol.* **123**, 651–666.

Davies, A.G., Madariaga, I., de Bautista, B. and Fernández, E. (1992). The ecology of a coastal *Phaeocystis* bloom in the north-western English Channel in 1990. *J. Mar. Biol. Assoc. UK* **72**, 691–708.

Delgado, M., de Jonge, V.N. and Peletier, H. (1991). Experiments on resuspension of natural microphytobenthos populations. *Mar. Biol.* **108**, 321–328.

Dore, J.E. and Karl, D.M. (1996). Nitrification in the euphotic zone as a source for nitrite, nitrate, and nitrous oxide at station ALOHA. *Limnol. Oceanogr.* **41**, 1619–1628.

Dortch, Q. (1990). The interaction between ammonium and nitrate uptake in phytoplankton. *Mar. Ecol. Prog. Ser.* **61**, 183–201.

Dugdale, R.C. and Goering, R.F. (1967). Uptake of new and regenerated forms of nitrogen in primary productivity. *Limnol. Oceanogr.* **12**, 196–206.

Dugdale, R.C. and Wilkerson, F.P (1986). The use of ^{15}N to measure nitrogen uptake in eutrophic oceans; experimental considerations. *Limnol. Oceanogr.* **31**, 673–689.

Dugdale, R.C., Wilkerson F.P. and Morel, A. (1990). Realisation of new production in coastal upwelling areas: a means to compare relative performance. *Limnol. Oceanogr.* **35**, 822–829.

Edgar, L. A. and Pickett-Heaps, J.D. (1984) Diatom locomotion. In: *Progress in Phycological Research* (Ed. by F.E. Round and D.J. Chapman), vol 3, pp. 47–88. Biopress, Bristol.

Epping, E.H.G. and Jørgensen, B.B (1996). Light-enhanced oxygen respiration in benthic phototrophic communities. *Mar. Ecol. Prog. Ser.* **139**, 193–203.

Eppley, R.W. and Peterson, B.J. (1979). Particulate organic matter flux and planktonic new production in the deep ocean. *Nature* **282**, 677–679.

Estep, K.W. and MacIntyre, F. (1989). Taxonomy, life cycles, distribution and dasmotrophy of *Chrysochromulina*: a theory accounting for scales, haptonema, muciferous bodies and toxicity. *Mar. Ecol. Prog. Ser.* **57**, 11–21.

Falkowski, P.G. and Raven, J.A., (1997). *Aquatic Photosynthesis*. Blackwell Science, Oxford.

Falkowski, P.G., Dubinsky, Z. and Wyman, K. (1985). Growth-irradiance relationships in phytoplankton. *J. Plankton Res.* **30**, 311–321.

Falkowski, P.G., Wyman, K., Ley, A.C. and Mauzerall, D.C. (1986). Relationship of steady-state photosynthesis to fluorescence in eucaryotic algae. *Biochim. Biophys. Acta* **849**, 183–192.

Falkowski, P.G., Zieman, D., Kolber, Z. and Bienfang, P.K. (1991). Role of eddy pumping in enhancing primary production in the ocean. *Nature* **352**, 55–58.

Falkowski, P.G., Greene, R.M. and Geider, R.J. (1992). Physiological limitation on phytoplankton productivity in the ocean. *Oceanography* **5**, 84–91.

Fielding, P., Damstra, K. and Branch, G. (1988). Benthic diatom biomass, production, and sediment chlorophyll in Langebaan Lagoon, South Africa. *Estuar. Coast. Shelf Sci.* **27**, 413–426.

Fisher, T.R., Harding, L.W. Jr., Stanley, D.W. and Ward, L.G. (1988). Phytoplankton, nutrients, and turbidity in the Chesapeake, Delaware, and Hudson Estuaries. *Estuar. Coast. Shelf Sci.* **27**, 61–93.

Flameling, I. A. (1998). *Growth and Photosynthesis of Eukaryotic Microalgae in Fluctuating Light Conditions, Induced by Vertical Mixing*. PhD thesis, Catholic University Nijmegen, The Netherlands.

Flameling, I.A. and Kromkamp, J. (1998). Light dependence of quantum yields for PSII charge separation and oxygen evolution in eucaryotic algae. *Limnol. Oceanogr.* **43**, 284–297.

Fraser, A.J., Sargent, J.R., Gamble, J.C. and Seaton, D.D. (1989). Formation and transfer of fatty acids in an enclosed marine food chain comprising phytoplankton, zooplankton and herring (*Clupea harengus* L.) larvae. *Mar. Chem.* **27**, 1–18.

Freeberg, L.R., Marshall, A. and Heye, M. (1979). Interrelationships of *Gymnodinium breve* (Florida red tide) within the phytoplankton community. In: *Anonymous Toxic Dinoflagellate Blooms* (Ed. by D.L. Taylor and H.H. Seliger), pp. 139–144. Elsevier, New York.

French, F.W. and Hargraves, P.E. (1980). Physiological characteristics of plankton diatom resting spores. *Mar. Biol. Lett.* **1**, 185–195.

Gallegos, C.L. and Jordan, T.E. (1997). Seasonal progression of factors limiting phytoplankton pigment biomass in the Rhode River estuary, Maryland (US). I. Controls on phytoplankton growth. *Mar. Ecol. Prog. Ser.* **161**, 185–198.

Geel, C., Versluis, W. and Snel, J.H. (1997). Estimation of oxygen evolution by marine phytoplankton from measurements of the efficiency of photosystem II electron flow. *Photosynth. Res.* **51**, 61–70.

Geider, R.J. and Osborne, B.A. (1986). Light absorption, photosynthesis and growth of *Nannochloris atomus* in nutrient-saturated cultures. *Mar. Biol.* **93**, 351–360.

Genty, B., Briantais, J.-M. and Baker, N.R. (1989). The relationship between the quantum yield of photosynthetic electron transport and quenching of chlorophyll fluorescence. *Biochim. Biophys. Acta* **990**, 87–92.

Glud, R.N., Ramsing, N.B. and Revsbech, N.P. (1992). Photosynthesis and photosynthesis-coupled respiration in natural biofilms quantified with oxygen microsensors. *J. Phycol.* **28**, 51–60.

Goosen, N.K., Van Rijswijk, P., Kromkamp, J. and Peene, J. (1997). Regulation of annual variation in heterotrophic bacterial production in the Schelde estuary (SW Netherlands). *Aquat. Microb. Ecol.* **12**, 223–232.

Grande, K., Williams, P.J.LeB., Marra, J., Purdie, D.A., Heinemann, K., Eppley, R.W. and Bender, M.L. (1989). Primary production in the North Pacific gyre: a comparison of rates determined by the ^{14}C, O_2 concentration and ^{18}O methods. *Deep Sea Res.* **36**, 1621–1634.

Grobbelaar, J.U. (1990). Modelling phytoplankton productivity in turbid waters with small euphotic to mixing depth ratios. *J. Plankton Res.* **12**, 926–931.

Guarini J.M., Blanchard, G.F., Bacher, C., Riera, P. and Richard, P. (1998). Dynamics of spatial patterns of microphytobenthic biomass: inferences from a geostatistical analysis of two comprehensive surveys in Marennes-Oleron bay (France). *Mar. Ecol. Prog. Ser.* **166**, 131–141.

Guarini, J.M., Blanchard, G.F., Gros, P. and Harrison S.J. (1997). Modelling the mud surface temperature on intertidal flats to investigate the spatio-temporal dynamics of the benthic microalgal photosynthetic capacity. *Mar. Ecol. Prog. Ser.* **153**, 25–36.

Hallegraeff, G.M. (1993). A review of harmful algal blooms and their apparent global increase. *Phycologia* **32**, 79–99.

Harding, L.W. Jr. (1994). Long-term trends in the distribution of phytoplankton in Chesapeake bay: roles of light, nutrients and streamflow. *Mar. Ecol. Prog. Ser.* **104**, 267–291.

Hargrave, B.T., Prouse, N.J., Phillips, G.A. and Neame, P.A. (1983). Primary production and respiration in pelagic and benthic communities at two intertidal sites in the upper Bay of Funday. *Can. J. Fish. Aquat. Sci.* **40**(suppl.1), 229–243.

Harrison, W.G. and Platt, T. (1986). Photosynthesis–irradiance relationships in polar and temperature phytoplankton populations. *Polar Biol.* **5**, 153–164.

Hartig, P., Wolfstein, K., Lippemeier, S. and Colijn, F. (1998). Photosynthetic activity of natural microphytobenthos populations measured by fluorescence (PAM) and ^{14}C-tracer methods: a comparison. *Mar. Ecol. Prog. Ser.* **166**, 53–62.

Hay S.I., Maitland T.C. and Paterson, D.M. (1993). The speed of diatom migration through natural and artificial substrata. *Diatom Res.* **8**, 371–384.

Hecky, R.E. and Kilham, P. (1988). Nutrient limitation of phytoplankton in freshwater and marine environments: a review of recent evidence on the effects of enrichments. *Limnol. Oceanogr.* **33**, 796–822.

Heip, C.H.R., Goosen, N.K., Herman, P.M.J., Kromkamp, J., Middleburg, J.J. and Soetaert, K. (1995). Production and consumption of biological particles in temperate tidal estuaries. *Oceanogr. Mar. Biol. Annu. Rev.* **33**, 1–149.

Herman, P.M.J. and Scholten, H. (1990). Can suspension-feeders stabilise estuarine ecosystems? In: *Trophic Relationships in the Marine Environment* (Ed. by M. Barnes and R.N. Gibson), pp. 104–116. Aberdeen University Press, Aberdeen.

Hoagland, K.D., Rosowski, J.R., Gretz, M.R. and Roemer, S.C. (1993). Diatom extracellular polymeric substances: function, fine structure, chemistry and physiology. *J. Phycol.* **29**, 537–566.

Hofstraat, J.W., Peeters, J.H.C., Snel, J.H. and Geel, C. (1994). Simple determination of photosynthetic efficiency and photoinhibition of *Dunaliella tertiolecta* by saturating pulse fluorescence measurements. *Mar. Ecol. Prog. Ser.* **103**, 187–196.

Irigoien, X. and Castel, J. (1997). Light limitation and distribution of chlorophyll pigments in a highly turbid estuary: the Gironde (SW France). *Estuar. Coast. Shelf Sci.* **44**, 507–517.

Joint, I.R. (1978). Microbial production of an estuarine mudflat. *Estuar. Coast. and Mar. Sci.* **7**, 185–195.

Joint, I.R. and Pomroy, A.J. (1981). Primary production in a turbid estuary. *Estuar. Coast. Shelf Sci.* **13**, 303–316.

de Jong, D.J and de Jonge, V.N. (1995). Dynamics and distribution of microphyto-benthic chlorophyll-*a* in the Western Scheldt estuary (SW Netherlands). *Hydrobiologia.* **311**, 21–30.

de Jong, D.J., Nienhuis, P.H. and Kater B.J. (1994). Microphytobenthos in the Oosterschelde estuary (The Netherlands), 1981–1990—consequences of a changed tidal regime. *Hydrobiologia.* **283**,183–195.

de Jonge, V. (1992). *Physical Processes and Dynamics of Microphytobenthos in the Ems Estuary (The Netherlands).* PhD thesis, University of Groningen, The Netherlands.

de Jonge, V.N. (1997). High remaining productivity in the Dutch western Waddensea despite decreasing nutrient inputs from riverine sources. *Mar. Pollut. Bull.* **34**, 427–436.

de Jonge, V.N. and van Beusekom, J.E.E. (1992). Contributions of resuspended microphytobenthos to total phytoplankton in the Ems estuary and its possible role for grazers. *Neth. J. Sea Res.* **30**, 91–105.

de Jonge, V.N. and van Beusekom, J.E.E. (1995). Wind- and tide-influenced resuspension of sediment and microphytobenthos from tidal flats in the Ems estuary. *Limnol. Oceanogr.* **40**, 766–778.

de Jonge, V.N. and van Raaphorst, W. (1995). Eutrophication of the Dutch Wadden Sea (western Eutrope) and estuarine area controlled by the river Rhine. In: *Eutrophic Shallow Estuaries and Lagoons* (Ed. by A.J. McComb), pp. 129–142. CRC Press, London.

de Jonge, V.N., Bakker, J.F. and van Stralen, M. (1996). Recent changes in the contributions of River Rhine and North Sea to the eutrophication of the western Dutch Wadden Sea. *Neth. J. Aquat. Ecol.* **30**, 27–39.

Jönsson, B. (1991). A ^{14}C incubation technique for measuring microphytobenthic primary productivty in intact sediment cores. *Limnol. Oceanogr.* **36**, 1485–1492.

Jönsson, B., Sundbäck, K. and Nilsson, C. (1994). An upright life-form of an epipelic motile diatoms: on the behaviour of *Gyrosigma balticum. Eur. J. Phycol.* **29**, 11–15.

Kana, T.M. (1992). Relationship between photosynthetic oxygen cycling and carbon assimilation in *Synechococcus* WH7803 (Cyanophyta). *J. Phycol.* **28**, 304–308.

Kana, T.M., Darkangelo, C., Hunt, M.D., Oldham, J.B., Bennet, G.E. and Cornwell, J.C. (1994). Membrane inlet mass spectrometer for rapid high-precision determination of N_2, O_2, and Ar in environmental water samples. *Anal. Chem.* **66**, 4166–4170.

Kinney, E.H. and Roman, C.T. (1998). Response of primary producers to nutrient enrichment in a shallow estuary. *Mar. Ecol. Prog. Ser.* **163**, 89–98.

Kirchman, D.L., Moss, L. and Keil, R.G. (1992). Nitrate uptake by heterotrophic bacteria: does it change the f-ratio? *Arch. Hydrobiol. (Beih. Erg. Limnol. Adv)* **37**, 129–138.

Kolber, Z. and Falkowski, P.G. (1993). Use of active fluorescence to estimate phytoplankton photosynthesis in situ. *Limnol. Oceanogr.* **38**, 1646–1665.

Kolber, Z., Zehr, J. and Falkowski, P.G. (1988). Effects of growth irradiance and nitrogen limitation on photosynthetic energy conversion in photosystem II. *Plant Physiol.* **88**, 923–929.

Kolber, Z., Barber, R.T., Coale, K.H. *et al.* (1994). Iron limitation of phytoplankton photosynthesis in the equatorial Pacific Ocean. *Nature* **371**, 145–149.

Krall, J.P. and Edwards, G.E. (1991). Environmental effects on the relationship between the quantum yield of carbon assimilation and *in vivo* PSII electron transport in maize. *Aust. J. Plant Physiol.* **18**, 579–588.

Krause, G.H. and Weis, E. (1991). Chlorophyll fluorescence and photosynthesis: the basics. *Annu. Rev. Plant Physiol. Plant Mol. Biol.* **42**, 313–349.

Kromkamp, J. and Peene, J. (1995a). Possibility of net phytoplankton primary production in the turbid Schelde estuary (The Netherlands). *Mar. Ecol. Prog. Ser.* **121**, 249–259.

Kromkamp, J. and Peene, J. (1995b). On the net growth of phytoplankton in two Dutch estuaries. *Wat. Sci. Tech.* **32**, 55–58.

Kromkamp, J. and Peene, J. (1997). Photosynthesis and nutrient limitation of phytoplankton in the Oosterschelde estuary estimated using PSII quantum efficiency and electron flow. *Proceedings of the second network meeting: Integrated Marine System Analysis* (Ed. by F. Dehairs, M. Elskens and L. Goeyens), VUB Brussels, pp 3–27. Free University of Brussels, Brussels.

Kromkamp, J., Barranguet, C. and Peene, J. (1998). Determination of microphytobenthos PSII quantum efficiency and photosynthetic activity by means of variable chlorophyll fluorescence. *Mar. Ecol. Prog. Ser.* **162**, 45–55.

Kromkamp, J., Peene, J., Van Rijswijk, P., Sandee, A. and Goosen, N. (1995). Nutrients, light and primary production by phytoplankton and microphytobenthos in the eutrophic, turbid Westerschelde estuary (The Netherlands). *Hydrobiologia* **311**, 9–19.

Kroon, B.M.A. (1994). Variability of photosystem II quantum yield and related processes in *Chlorella pyrenoidosa* (Chlorophyta) acclimated to an oscillating light regime simulating a mixed photic zone. *J. Phycol.* **30**, 841–852.

Kühl, M., Lassen, C. and Jörgensen, B.B. (1994). Optical measurements of microbial mats: light measurements with fibre-optic microprobes In: *Microbial Mats* (Ed. by L.J. Stal and P. Caumette), NATO ASI Series **35**, pp. 149–166. Springer-Verlag, Berlin.

Kühl, M., Glud, N.G., Ploug, H. and Ramsing, N.R. (1996). Microenvironmental control of photosynthesis and photosynthesis-coupled respiration in an epilithic cyanobacterial biofilm. *J. Phycol.* **32**, 799–812.

Laanbroek, H.J and Verplanke, J.C. (1986). Seasonal changes in percentages of attached bacteria enumerated in a tidal and a stagnant coastal basin: relation to bacterioplankton productivity. *FEMS Microbiol. Ecol.* **38**, 87–98.

Landry, M.R. and Hasset, R.P. (1982). Estimating the grazing impact of marine micro-zooplankton. *Mar. Biol.* **67**, 282–288.

Langdon, C. (1993). The significance of respiration production measurements based on both carbon and oxygen. In: *Measurements of Primary Production from the Molecular to the Global Scale* (Ed. by W.K.W. Li and S.Y. Maestrini), pp. 69–78. ICES MSS **197**. International Council for the Exploration of the Sea, Copenhagen.

Lassen, C., Ploug, H. and Jorgensen, B.B. (1992a). A fibre-optic scalar irradiance microsensor: application for spectral light measurements in sediments. *FEMS Microb. Ecol.* **86**, 247–254.

Lassen, C., Ploug, H. and Jorgensen, B.B. (1992b). Microalgal photosynthesis and spectral scalar irradiance in coastal marine sediments of Limfjorden, Denmark. *Limnol. Oceanogr.* **37**, 760–772.

Laws, E.A. and Bannister, T.T. (1980). Nutrient- and light-limited growth of *Thalassiosira fluviatilis* in continuous culture, with implications for phytoplankton growth in the ocean. *Limnol. Oceanogr.* **25**, 457–473.

Leach, J.H. (1970). Epibenthic algal production in an intertidal mudflat. *Limnol. Oceanogr.* **15**, 514–521.

Lee, S.Y. (1990). Primary productivity and particulate organic matter flow in an estuarine mangroe-wetland in Hong Kong. *Mar. Biol.* **106**, 453–463.

Lewitus, A.J., Koepfler, E.T. and Morris, J.T. (1998). Seasonal variation in the regulation of phytoplankton by nitrogen and grazing in a salt-marsh estuary. *Limnol. Oceanogr.* **43**, 636–646.

Lindeboom, H.J. and de Bree, B.H.H. (1982). Daily production and consumption in an eelgrass (*Zostera marina*) community in saline Lake Grevelingen: discrepancies between the O_2 and ^{14}C method. *Neth. J. Sea Res.* **16**, 362–379.

Ludden, E., Admiraal, W. and Colijn, F. (1985). Cycling of carbon and oxygen in layers of marine microphytes: a simulation model and its eco-physiological implications. *Oecologia* **66,** 50–59.

McCarthy, J.J., Taylor, W.R. and Taft, J.L. (1977). Nitrogen nutrition of the plankton in the Chesapeake Bay. 1. Nutrient availability and phytoplankton preferences. *Limnol. Oceanogr.* **22**, 996–1011.

MacIntyre, H.L. and Cullen, J.J. (1995). Fine-scale vertical resolution of chlorophyll and photosynthetic parameters in shallow-water benthos. *Mar. Ecol. Prog. Ser.* **122**, 227–237.

MacIntyre, H.L., Geider R.J. and Miller, D.C. (1996). Microphytobenthos: the ecological role of the "secret garden" of unvegetated, shallow-water marine habitats. I. Distribution, abundance and primary production. *Estuaries* **19**, 186–201.

de Madariaga, I., Orive, E. and Boalch, G.T. (1989). Primary production in the Gernika estuary during a summer bloom of the dinoflagellate *Peridinium quiquecorne* Abe. *Bot. Mar.* **32**, 159–165.

de Madariaga, I., Gonzalez-Azpiri, L., Villate, F. and Orive, E. (1992). Plankton responses to hydrological changes induced by freshets in a shallow mesotidal estuary. *Estuar. Coast. Shelf Sci.* **35**, 425–434.

McMahon, T.G., Raine, R.C.T., Fast, T., Kies, L. and Patching, J.W. (1992). Phytoplankton biomass, light attenuation and mixing in the Shannon estuary, Ireland. *J. Mar. Biol. Assoc. UK* **72**, 709–720.

McQuid, M.R. and Hobson, L.A. (1995). Importance of resting stages in diatom seasonal succession. *J. Phycol.* **31**, 44–50.

Mallin, M.A., Paerl, H.W., Rudek, J. and Bates, P.W. (1993). Regulation of estuarine primary production by watershed rainfall and river flow. *Mar. Ecol. Prog. Ser.* **93**, 199–203.

Malone, T.C. (1977). Environmental regulation of phytoplankton productivity in the lower Hudson Estuary. *Estuar. Coast. Mar. Sci.* **5**, 157–171.

Malone, T.C., Crocker, L.H., Pike, S.E. and Wendler, B.W. (1988). Influences of river flow on the dynamics of phytoplankton production in a partially stratified estuary. *Mar. Ecol. Prog. Ser.* **48**, 235–249.

Mann, K.H. and Lazier, J.R.N. (1991). *Dynamics of Marine Ecosystems— Biological–Physical Interactions in the Ocean.* Blackwell, Boston.

Marshall, H.G. and Alden, R.W. (1990). A comparison of phytoplankton assemblages and environmental relationships in three estuarine rivers of the lower Chesapeake Bay. *Estuaries* **13**, 287–300.

Marshall, H.G. and Alden, R.W. (1993). A comparison of phytoplankton assemblages in the Chesapeake and Delware estuaries (USA), with emphasis on diatoms. *Hydrobiologia* **269/270**, 251–261.

Monbet, Y. (1992). Control of phytoplankton biomass in estuaries: a comparative analysis of microtidal and macrotidal estuaries. *Estuaries* **15**, 563–571.

Montagna, P.A. (1984). *In situ* measurement of meiobenthic grazing rates on sediment bacteria and edaphic diatoms. *Mar. Ecol. Prog. Ser.* **18**, 119–130.

Montagna, P.A., Coull, B.C., Herring, T.L. and Dudley, B.W. (1983). The relationship between abundance of meiofauna and their suspected microbial food (diatoms and bacteria). *Estuar. Coast. Shelf Sci.* **17**, 381–394.

Morrisey, D.J. (1988). Differences in effects of grazing by deposit-feeders *Hydrobia ulvae* (Pennant) (Gastropoda: Prosobranchia) and *Corophium arenarium* Crawford (Amphipoda) on sediment microalgal populations. II. Quantitative effects. *J. Exp. Mar. Biol. Ecol.* **118**, 43–53.

Nagasaki, K. and Yamaguchi, M. (1997). Isolation of a virus infectious to the harmful bloom causing microalga *Heterosigma akashiwo* (Raphidophyceae). *Aquat. Microbiol. Ecol.* **13**, 135–140.

Nienhuis, P.H. and Smaal, A.C. (1994). The Oosterschelde estuary, a case study of a changing ecosystem: an introduction. *Hydrobiologia* **282/283**, 1–14.

Nienhuis, P.H., Daemen, E.A.M.J., De Jong, S.A. and Hofman, P.A.G. (1985) *Biomass and Production of Microphytobenthos, Progress Report 1985.* Delta Institute for Hydronbiological Research, The Netherlands.

Nilsson, C. and Sundbäck, K. (1991). Growth and nutrient uptake studied in sand–agar microphytobenthic communities. *J. Exp. Mar. Biol. Ecol.* **153**, 207–226.

Nilsson, P, Jönsson, B., Swanberg, I.L. and Sundbäck, K. (1991). Response of a marine shallow-water sediment system to an increased load of inorganic nutrients. *Mar. Ecol. Prog. Ser.* **71**, 275–290.

Nixon, S.W. (1995). Coastal marine eutrophication: a definition, social causes, and future concerns. *Ophelia* **41**, 199–219.

Ogilvie, B., Nedwell, D.B., Harrison, R.M., Robinson, A. and Sage, A. (1997). High nitrate, muddy estuaries as nitrogen sinks: the nitrogen budget of the River Colne estuary (United Kingdom). *Mar. Ecol. Prog. Ser.* **150**, 217–228.

Oxborough, K. and Baker, N.R. (1997). An instrument capable of imaging chlorophyll *a* fluorescence from intact leaves at very low irradiance and at cellular and subcellular levels of organisation. *Plant. Cell. Environ.* **20**, 1473–1483.

Paerl, H.W., Joye, S.B. and Fitzpatrick, M. (1993). Evaluation of nutrient limitation of CO_2 and N_2 fixation in marine microbial mats. *Mar. Ecol. Prog. Ser.* **101**, 297–306.

Paerl, H.W., Pinckney, J.L., Fear, J.M. and Peierls, B.L. (1998). Ecosystem responses to internal and watershed organic matter loading: consequences for hypoxia in the eutrophying Neuse River Estuary, North Carolina, USA. *Mar. Ecol. Prog. Ser.* **166**, 17–25.

Palmer, J.D. and Round, F.E. (1967). Persistent vertical-migration rhythms in benthic microflora. VI. The tidal and diurnal nature of the rhythm in the diatom *Hantzschia virgata*. *Biol. Bull.* **132**, 44–55.

Pamatmat, M.M. (1968). Ecology and metabolism of a benthic community on an intertidal sandflat. *Int. Rev. Ges.Hydrobiol.* **53**, 211–298.

Passow, U. and Wassmann, P. (1994). On the trophic fate of *Phaeocystis pouchetii* (hariot). 4. The formation of marine snow by *P. pouchetii*. *Mar. Ecol. Prog. Ser.* **104**, 153–161.

Paterson, D.M. (1986). The migratory behaviour of diatom assemblages in a laboratory tidal micro-ecosystem examined by low-temperature scanning electron microscopy. *Diatom Res.* **1**, 227–239.

Paterson, D.M. (1989). Short-term changes in the erodibility of intertidal cohesive sediments related to the migratory behaviour or epipelic diatoms. *Limnol. Oceanogr.* **34**, 223–234.

Paterson, D.M., Crawford, R.M. and Little, C. (1990). Sub-aerial exposure and changes in stability of intertidal estuarine sediments. *Estuar. Coast. Shelf Sci.* **30**, 541–556.

Paterson, D.M., Underwood, G.J.C., Miles, A.C., Szyszko, P. and Davidson, I. (1997). Investigation of primary productivity by microphytobenthos on intertidal mudflats by *in situ* and laboratory studies. Final report to N.E.R.C., GR3/8907. Natural Environment Research Council.

Paterson, D.M., Wiltshire, K.H., Miles, A., Blackburn, J., Davidson, I., Yates, M.G., McGrorty, S. and Eastwood, J.A. (1998). Microbiological mediation of spectral reflectance from intertidal cohesive sediments. *Limnol. Oceanogr.* **43**, 1207–1221.

Peduzzi, P. and Weinbauer, M.G. (1993). The submicron size fraction of seawater containing high numbers of virus particles as bioactive agent in unicellular plankton community successions. *J. Plankton Res.* **15**, 1375–1386.

Peletier, H. (1996). Long-term changes in intertidal estuarine diatom assemblages related to reduced input of organic waste. *Mar. Ecol. Prog. Ser.* **137**, 265–271.

Pennock, J.R. (1985). Chlorophyll distributions in the Delaware estuary: regulation by light-limitation. *Estuar. Coast. Shelf Sci.* **21**, 711–725.

Pennock, J.R. (1987). Temporal and spatial variability in phytoplankton ammonium and nitrate uptake in the Delaware estuary. *Estuar. Coast. Shelf Sci.* **24**, 841–857.

Pennock, J.R. and Sharp, J.H. (1986). Phytoplankton production in the Delaware estuary: temporal and spatial variability. *Mar. Ecol. Prog. Ser.* **34**, 143–155.

Pennock, J.R. and Sharp, J.H. (1994). Temporal alternation between light- and nutrient-limitation of phytoplankton production in a coastal plain estuary. *Mar. Ecol. Prog. Ser.* **111**, 275–288.

Pinckney, J.L. (1994). Development of an irradiance-based ecophysiological model for intertidal benthic microalgal production. In: *Biostabilization of Sediments* (Ed. by W.E. Krumbein, D.M. Paterson and L.J. Stal), pp 55–84. Bibliotheks and Informations system der Carl von Ossietzky Universität, Oldenburg.

Pinckney, J. and Sandulli, R. (1990). Spatial autocorrelationanalysis of meiofaunal and microalgal populations on an intertidal sandflat: scale linkage between consumers and resources. *Estuar. Coast. Shelf Sci.* **30**, 341–353.

Pinckney, J. and Zingmark, R.G. (1991). Effects of tidal stage and sun angles on intertidal benthic microalgal productivity. *Mar. Ecol. Prog. Ser.* **76**, 81–89

Pinckney, J. and Zingmark, R.G. (1993). Biomass and production of benthic microalgal communities in estuarine habitats. *Estuaries* **16**, 887–897.

Pinckney, J., Paerl, H.W. and Fitzpatrick, M. (1995). Impacts of seasonality and nutrients on microbial mat community structure and function. *Mar. Ecol. Prog. Ser.* **123**, 207–216.

Pinckney, J.L., Paerl, H.W., Harrington, M.B. and Howe, K.E. (1998). Annual cycles of phytoplankton community-structure and bloom dynamics in the Neuse River estuary, North Carolina. *Mar. Biol.* **131**, 371–381.

Platt, T. and Sathyendranath, S. (1993). Fundamental issues in measurement of primary production. In: *Measurements of Primary Production from the Molecular to the Global Scale* (Ed. by W.K.W. Li and S.Y. Maestrini), pp. 3–8. ICES MSS **197**. International Council for the Exploration of the Sea, Copenhagen.

Platt, T., Lewis, M. and Geider, R. (1984). Thermodynamics of the pelagic ecosystem: elementary closure conditions for biological production in the open ocean. In: *Flow of Energy and Materials in Marine Ecosystems* (Ed. by M.J.R. Fasham), pp. 48–84. Plenum Press, New York.

Platt, T., Jauhari, P. and Sathyendranath, S. (1992). The importance and measurement of new production. In: *Primary Productivity and Biogeochemical Cycles in the Sea* (Ed. by P.G. Falkowski and A.D. Woodhead), pp. 273–284. Plenum Press, New York.

Ploug, H., Lassen C. and Jørgensen, B.B. (1993). Action spectra of microalgal photosynthesis and depth distribution of spectral scalar irradiance in a coastal marine sediment of Limfjorden, Denmark. *FEMS Microbiol. Ecol.* **12**, 69–78.

Pomeroy, L.R. (1959). Algal productivity in salt marshes of Georgia. *Limnol. Oceanogr.* **4**, 386–397.

Post, A.F., de Wit, R. and Mur, L.R. (1985). Interactions between temperature and light intensity on growth and photosynthesis of the cyanobacterium *Oscillatoria agardhii*. *J. Plankton Res.* **7**, 487–495.

Prasil, O., Kolber, Z., Berry, J. and Falkowski, P.G. (1996). Cyclic electron flow around photosystem II *in vivo*. *Photosynth. Res.* **48**, 395–410.

Ragueneau, O.B., Quiguineret, B. and Truguer, P. (1996). Contrast in biological responses to tidally-induced vertical mixing for two macrotidal ecosystems of Western Europe. *Estuar. Coast. Shelf Sci.* **42**, 645–665.

Randall, J.M. and Day, J. (1987). Effects of river discharge and vertical circulation on aquatic primary production in a turbid Louisiana (USA) estuary. *Neth. J. Sea Res.* **21**, 231–242.

Rasmussen, M.B., Henriksen, K. and Jensen, A. (1983). Possible causes of temporal fluctuation in primary production of the microphytobenthos in the Danish Wadden Sea. *Mar. Biol.* **73**, 109–114.

Raven, J.A. (1984). A cost–benefit analysis of photon absorption by photosynthetic unicells. *New Phytol.* **98**, 593–625.

Raven, J.A. (1997). Inorganic carbon acquisition by marine autotrophs. *Adv. Bot. Res.* **27**, 86–209.

Reise, K. (1985). *Tidal Flat Ecology: An Experimental Approach to Species Interactions.* Springer, Berlin.

Reisser, W. (1993). Viruses and virus-like particles of freshwater and marine eukaryotic algae—a review. *Arch. Protistenkd.* **143**, 257–265.

Rengefors, K. (in press). Seasonal succession of dinoflagellates coupled to benthic cyst dynamics in Lake Erken, Sweden. *Arch. Hydrobiol. Special Issues Adv. Limnol.*

Revsbech, N.P. and Jørgensen, B.B. (1983). Photosynthesis of benthic microflora measured with high spatial resolution by the oxygen microprofile method: capabilities and limitations of the method. *Limnol. Oceanogr.* **28**, 749–756.

Reynolds, C.S., Thompson, J.M., Ferguson, A.J.D. and Wiseman, S.W. (1982). Loss processes in the population dynamics of phytoplankton maintained in closed systems. *J. Plankton Res.* **4**, 561–600.

Riegman, R. and Mur, L.R. (1984). Regulation of phosphate uptake kinetics in *Oscillatoria agardhii. Arch. Microbiol.* **139**, 28–32.

Riegman, R., Noordeloos, A.M.N. and Cadée, G.C. (1992). *Phaeocystis* blooms and eutrophication of the continental coastal zones of the North Sea. *Mar. Biol.* **112**, 479–484.

Riegman, R., Kuipers, B.R., Noordeloos, A.A.M. and Witte, H.J. (1993). Size-differential control of phytoplankton and the structure of plankton communities. *Neth. J. Sea Res.* **31**, 255–265.

Rijstenbil, J.W. (1991). Indications of salinity stress in phytoplankton communities from different brackish-water habitats. In: *Estuaries and Coasts: Spatial and Temporal Intercomparisons* (Ed. by M. Elliot and J.-P. Ducrotoy), pp. 107–110. Olsen and Olsen, Fredensborg.

150 G.J.C. UNDERWOOD AND J. KROMKAMP

Round, F.E. (1981). *The Ecology of the Algae*. Cambridge University Press, Cambridge.

Saburova, M.A., Polikarpov, I.G. and Burkovsky, I.V. (1995). Spatial structure of an intertidal sandflat microphytobenthic community as related to different spatial scales. *Mar. Ecol. Prog. Ser.* **129**, 229–239.

Santos, P.J.P., Castel, J. and SouzaSantos, L.P. (1997). Spatial distribution and dynamics of microphytobenthic biomass in the Gironde estuary (France). *Oceanol. Acta* **20**: 549–556.

Schreiber, R.A. and Pennock, J.R. (1995). The relative contribution of benthic microalgae to total microalgal production in a shallow subtidal estuarine environment. *Ophelia* **42**, 335–352.

Schreiber, U., Schliwa, U. and Bilger, W. (1986). Continuous recording of photochemical and non-photochemical chlorophyll fluorescence quenching with a new type of modulation fluorometer. *Photosynth. Res.* **10**, 51–62.

Schreiber, U., Endo, T., Mi, H. and Asada, A. (1995). Quenching analysis of chlorophyll fluorescence by the saturation pulse method: particular aspects relating to the study of eukaryotic algae and cyanobacteria. *Plant Cell Physiol.* **36**, 873–882.

Seaton, G.C.R. and Walker, D.A. (1990). Chlorophyll fluorescence as a measure of photosynthetic carbon assimilation. *Proc. R. Soc. Lond.* **B252**, 29–35.

Serôdio, J., da Silva, J.M. and Catarina, F. (1997). Nondestructive tracing of migratory rhythms of intertidal benthic microalgae using *in vivo* chlorophyll a fluorescence. *J. Phycol.* **33**, 542–553.

Shaffer, G.P. and Onuf, C.P. (1985). Reducing the error in estimating annual production of benthic microflora: hourly to monthly rates, patchiness in space and time. *Mar. Ecol. Prog. Ser.* **26**, 221–231.

Smaal, A.C. and Prins, T.C. (1993). The uptake of organic matter and the release of inorganic nutrients by bivalve feeder beds. In: *Bivalve Filter Feeders in Estuarine and Coastal Ecosystem Processes* (Ed. by R.F. Dame), NATO series G VOL. **33**, pp. 271–298. Springer, Berlin.

Small, L.F., McIntire, C.D., MacDonald, K.B., Lara-Lara, J.R., Frey, B.E., Amspoker, M.C. and Winfield, T. (1990). Primary production, plant and detrital biomass and particle transport in the Columbia River Estuary. *Progr. Oceanogr.* **25**, 175–210.

Smayda, T.J. (1990). Novel and nuisance phytoplankton blooms in the sea: evidence for a global epidemic. In: *Anonymous Toxic Marine Phytoplankton*, pp. 29–40. Elsevier, New York.

Smayda, T.J. (1997). Harmful algal blooms: their ecophysiology and general relevance to phytoplankton blooms in the sea. *Limnol. Oceanogr.* **42**, 1137–1153.

Smetacek, V. and Passow, U. (1990). Spring bloom initiation and Sverdrup's critical-depth model. *Limnol. Oceanogr.* **35**, 228–234.

Smith, D.J. and Underwood, G.J.C. (1998). Exopolymer production by intertidal epipelic diatoms. *Limnol. Oceanogr.* **43**, 1578–1591.

Smith, D., Hughes, R.G. and Cox, E.J. (1996). Predation of epipelic diatoms by the amphipod *Corophium volutator* and the polychaete *Nereis diversicolor*. *Mar. Ecol. Prog. Ser.* **145**, 53–61.

Smith, R.E.H, Stapleford, L.C. and Ridings, R.S. (1994). The acclimated response of growth, photosynthesis, composition and carbon balance to temperature in the psychrophilic ice diatom *Nitzschia seriata*. *J. Phycol.* **30**, 8–16.

Smith, S.V. and Hollibaugh, J.T. (1993). Coastal metabolism and the oceanic organic carbon balance. *Rev. Geophys.* **31**, 75–89.

Smith, S.V., Hollibaugh, J.T., Dollar, S.J. and Vink, S. (1989). Tomales Bay, California: a case for carbon-controlled nitrogen cycling. *Limnol. Oceanogr.* **34**, 37–52.

Soetaert, K. and Herman, P.M.J. (1995). Nitrogen dynamics in the Westerschelde estuary (SW Netherlands) estimated by means of the ecosystem model MOSES. *Hydrobiologia* **311**, 225–246

Soetaert, K., Herman P.M.J. and Kromkamp, J. (1994). Living in the twilight: estimating net phytoplankton growth in the Westerschelde estuary (The Netherlands) by means of a global ecosystem model (MOSES). *J. Plankton Res.* **16**, 1277–1301.

Stal, L.J. (1994). Tansley Review No. 84. Physiological ecology of cyanobacteria in microbial mats and other communities. *New Phytol.* **131**, 1–32.

Stolte, W., McCollin, T., Noordeloos, A.A.M. and Riegman, R. (1994). Effect of nitrogen source on the size distribution within marine phytoplankton populations. *J. Exp. Mar. Biol. Ecol.* **184**, 83–97.

Sullivan, M.J. (1978). Diatom community structure: taxonomic and statistical analysis of a Mississippi salt marsh. *J. Phycol.* **14**, 468–475.

Sullivan, M.J. (1999). Applied diatom studies in estuarine and shallow coastal environments. In: *The Diatoms: Applications for the Environmental and Earth Sciences* (Ed. by E.F. Stoermer and J.P. Smol), pp. 334–351. Cambridge University Press, Cambridge.

Sullivan, M.J. and Moncreiff, C.A. (1988). Primary production of edaphic algal communities in a Mississippi salt marsh. *J. Phycol.* **24**, 49–58.

Sundbäck, K. and Jönsonn, B. (1988). Microphytobenthic productivity and biomass in sublittoral sediments of a stratified bay, southeastern Kattegat. *J. Exp. Mar. Biol. Ecol.* **122**, 63–81.

Sundbäck, K. and Snoeijs, P. (1991). Effects of nutrient enrichment on microalgal community composition in a coastal shallow-water sediment system: an experimental study. *Bot. Mar.* **34**, 341–358.

Sundbäck, K., Carlson, L., Nilsson, C., Jönsonn, B., Wulff, A. and Odmark, S. (1996). Response of benthic microbial mats to drifting green algal mats. *Aquat. Microb. Ecol.* **10**, 195–208.

Sundbäck, K., Odmark, S., Wulff, A., Nilsson, C. and Wängberg S.-Å. (1997). Effects of enhanced UVB radiation on a marine benthic diatom mat. *Mar. Biol.* **128**, 171–179.

Suttle, C.A. and Chan, A.M. (1993). Marine cyanophages infecting oceanic and coastal strains of *Synechococcus*: abundance, morphology, cross-infectivity and growth characteristics. *Mar. Ecol. Prog. Ser.* **92**, 99–109.

Suttle, C.A., Chan, A.M., Feng, C. and Garza, D.R. (1993). Cyanophages and sunlight: a paradox. In: *Trends in Microbial Ecology* (Ed. by R. Guerrero and C. Pedrós-Alió), pp. 303–307. Spanish Society for Microbiology, Barcelona.

Sverdrup, H.U. (1953). On conditions for the vernal blooming of phytoplankton. *J. Conseil* **18**, 287–295.

Tett, P. (1990). The photic zone. In: *Light and Life in the Sea* (Ed. by P.J. Herring, A.K. Campbell, M. Whitfield and L. Maddock), pp. 59–87. Cambridge University Press, Cambridge.

Thingstad, T.F. and Sakshaug, E. (1990). Control of phytoplankton growth in nutrient recycling ecosystems. Theory and terminology. *Mar. Ecol. Prog. Ser.* **63**, 261–272.

Thornton, D.C.O., Underwood, G.J.C. and Nedwell, D.B. (1999). Effect of illumination and emersion period on the exchange of ammonium across the estuarine sediment-water interface. *Mar. Ecol. Prog. Ser.* (in press).

Trimmer, M., Nedwell, D.B., Sivyer, D.B. and Malcolm, S.J. (1998). Nitrogen fluxes through the lower estuary of the river Great Ouse, England: the role of the bottom sediments. *Mar. Ecol. Prog. Ser.* **163**, 109–124.

Tripos (1997). Biomonitoring van fytoplankton in de Nederlandse zoute en brakke wateren 1996. In: Opdracht van: Rijkswaterstaat, Rijksinstituut voor Kust en Zee (RIKZ). Rapportnummer 97.0017–1a.

Turner, J.T. and Tester, P.A. (1997). Toxic marine phytoplankton, zooplankton grazers, and pelagic foodwebs. *Limnol. Oceanogr.* **42**, 1203–1214.

Underwood, G.J.C. (1994). Seasonal and spatial variation in epipelic diatom assemblages in the Severn estuary. *Diatom Res.* **9**, 451–472.

Underwood, G.J.C. (1997). Microalgal colonization in a saltmarsh restoration scheme. *Est. Coast. Shelf Sci.* **44**: 471–481.

Underwood G.J.C. and Paterson, D.M. (1993a). Recovery of intertidal benthic diatoms from biocide treatment and associated sediment dynamics. *J. Mar. Biol. Assoc. UK* **73**, 25–45.

Underwood, G.J.C. and Paterson, D.M. (1993b) Seasonal changes in diatom biomass, sediment stability and biogenic stabilization in the Severn Estuary. *J. Mar. Biol. Assoc. UK* **73**, 871–887.

Underwood, G.J.C. and Smith, D.J. (1998a). Predicting epipelic diatom exopolymer concentrations in intertidal sediments from sediment chl. *a. Microb. Ecol.* **35**, 116–125.

Underwood, G.J.C. and Smith, D.J. (1998b). *In situ* measurements of exopolymer production by intertidal epipelic diatom-dominated biofilms in the Humber estuary. In: *Sedimentary Processes in the Intertidal Zone* (Ed. by K.S. Black, D.M. Paterson and A. Cramp), pp. 125–134. Special Publication 139. Geological Society, London.

Underwood, G.J.C., Phillips, M. and Saunders K. (1998). Distribution of estuarine benthic diatom species along salinity and nutrient gradients. *Eur. J. Phycol.* **33**, 173–183.

Underwood, G.J.C., Nilsson, C., Sundbäck, K. and Wulff, A. (1999). Short-term effects of UVB radiation on chlorophyll fluorescence, biomass, pigments and carbohydrate fractions in a benthic diatom mat. *J. Phycol.* (in press).

Vadeboncoeur, Y. and Lodge, D.M. (1998). Dissolved inorganic carbon sources for epipelic algal production: sensitivity of primary production estimates to spatial and temporal distribution of C-14. *Limnol. Oceanogr.* **43**, 1222–1226.

van Boekel, W.H.M., Hansen, F.C., Riegman, R. and Bak, R.P.M. (1992). Lysis-induced decline of a *Phaeocystis* spring bloom and coupling with the microbial foodweb. *Mar. Ecol. Prog. Ser.* **81**, 269–276.

van Es, F.B. (1982). Community metabolism of intertidal flats in the Ems-Dollard estuary. *Mar. Biol.* **66**, 95–108.

van Etten, J.L., Lane, L.C. and Meints, R.H. (1991). Viruses and virus-like particles of eukaryotic algae. *Microbiol. Rev.* **55**, 586–620.

van Raalte, C.D., Valiela, I. and Teal, J.M. (1976). Production of benthic salt marsh algae: light and nutrient limitation. *Limnol. Oceanogr.* **21**, 862–872.

van Spaendonk, J.C.M., Kromkamp, J.C. and de Visscher, P.R.M. (1993). Primary production of phytoplankton in a turbid coastal plain estuary, the Westerschelde (The Netherlands). *Neth. J. Sea Res.* **31**, 267–292.

Varela, M. and Penas, E. (1985). Primary production of benthic microalgae in an intertidal sand flat of the Ria de Arosa, NW Spain. *Mar. Ecol. Prog. Ser.* **25**, 111–119.

Viles, H. and Spencer, T. (1995) *Coastal Problems, Geomorphology, Ecology and Society at the Coast*. Edward Arnold, London.

Weger, H.G., Herzig, R., Falkowksi, P.G. and Turpin, D.H. (1989). Respiratory losses in the light in a marine diatom: measurements by short-term mass spectrometry. *Limnol. Oceanogr.* **34**, 1153–1161.

Weisse, T., Tande, K., Verity, P., Hansen, F. and Gieskes, W. (1994). The trophic significance of *Phaeocystis* blooms. *J. Mar. Sys.* **5**, 67–79.

Wetsteyn, L.P.M.J. and Kromkamp, J.C. (1994). Turbidity, nutrients and phytoplankton primary production in the Oosterschelde (The Netherlands) before, during and after a large-scale coastal engineering project (1980–1990). *Hydrobiologia* **282/292**, 61–78.

Wiggert, J., Dickey, T. and Granata, T. (1994). The effect of temporal undersampling on primary production estimates. *J. Geophys. Res.* **99(C2)**, 3361–3371

Wilhelm, C. (1990). The biochemistry and physiology of light-harvesting processes in chlorophyll *b*- and *c*-containing algae. *Plant Physiol. Biochem.* **28**, 293–306.

Williams P.J.leB. (1993a). On the definition of plankton production terms. In: *Measurements of Primary Production from the Molecular to the Global Scale* (Ed. by W.K.W. Li and S.Y. Maestrini), pp. 9–19. ICES MSS 197.

Williams, P.J.leB. (1993b). Chemical and tracer methods of measuring plankton production. In: *Measurements of Primary Production from the Molecular to the Global Scale* (Ed. by W.K.W. Li and S.Y. Maestrini). ICES MSS **197**, 20–36. International Council for the Exploration of the Sea, Copenhagen.

Williams, P.J.leB. and Lefèvre, D. (1996). Algal ^{14}C and total carbon metabolisms. *J. Plankton Res.* **18**, 1941–1959.

Wiltshire, K.H., Schroeder, F., Knauth, H.-D. and Kausch, H. (1996). Oxygen consumption and production rates and associated fluxes in sediment-water systems: a combination of microelectrode, incubation and modelling techniques. *Arch. Hydrobiol.* **137**, 457–486.

Wiltshire, K.H., Blackburn, J. and Paterson, D.M. (1997). The cryo-lander: a new method for *in situ* sampling of unconsolidated sediments minimising the distortion of sediment fabric. *J. Sediment Res.* **67**, 980–984.

Windust, A.J., Wright, J.L.C. and Mclachlan, J.L. (1996). The effects of the diarrhetic shellfish poisoning toxins, okadaic acid and dinophysistoxin-1, on the growth of microalgae. *Mar. Biol.* 19–25.

Wofsy, S.C. (1983). A simple model to predict extinction coefficients and phytoplankton biomass in eutrophic waters. *Limnol. Oceanogr.* **28**, 1144–1155.

Yallop, M.L., de Winder, B., Paterson, D.M. and Stal, L.J. (1994). Comparative structure, primary production and biogenic stabilisation of cohesive and non-cohesive marine sediments inhabited by microphytobenthos. *Estuar. Coast. Shelf Sci.* **39**, 565–582.

Yin, K., Goldblatt, R.H., Harrison, P.J., John, A.S., Clifford, P.J. and Beamish, R.J. (1997). Importance of wind and river discharge in influencing nutrient dynamics and phytoplankton production in summer in the central Strait of Georgia. *Mar. Ecol. Prog. Ser.* **161**, 173–183.

Zevenboom, W., bij de Vaate, A. and Mur, L.R. (1982). Assessment of factors limiting growth rate of *Oscillatoria agardhii* in hypertrophic Lake Wolderwijd, 1978, by use of physiological indicators. *Limnol. Oceanogr.* **27**, 39–52.

Water Flow, Sediment Dynamics and Benthic Biology

D.M. PATERSON AND K.S. BLACK

I. SUMMARY

The influence of water flow is an important, and sometimes dominant, factor shaping the ecology of aquatic systems. However, the integration of scientific studies required for the understanding of the influence of flow on benthic ecology is still relatively underdeveloped. This chapter introduces some of the hydrodynamic and sedimentological concepts that are relevant to the study of benthic systems. The nature of the flow, whether laminar or

ADVANCES IN ECOLOGICAL RESEARCH VOL. 29
ISBN 0–12–013929–4

turbulent, and how this influences the stress experienced by the bed is considered. The importance of the balance between viscous and inertial forces and the significance of the Reynolds number are described. Particular reference is made to skin friction, which initiates sediment erosion from relatively smooth beds. The complex erosional behaviour of cohesive sediments is described including the recent classification of sediment erosion into type I, type II and possible intermediate behaviours.

The influence of biofilm or microbial mat development on the erosional properties of sediment had become a topic of considerable interest. The mechanism of this effect is thought to be mainly through the secretion of polymeric substances by cells in the biofilm. Biofilm assemblages can form highly structured communities at the sediment surface, and evidence from the literature suggests that components of the biofilm can be detected in suspension before erosion of the sediment bed, emphasizing the biological nature of the bed surface. Other benthic organisms and assemblages also utilize or mediate the flow environment to their possible advantage. Selected examples are given including sea grass assemblages and infaunal deposit-feeding bivalves. Future lines of research including the state-of-the-art in terms of modelling sediment dynamics, particularly with respect to biological properties, and the recent use of remote sensing techniques to classify habitat type and make inferences about sediment behaviour are considered.

II. INTRODUCTION

The nature and experience of water flow is a central structuring factor in the ecology of aquatic systems. The unusual physical properties of water, which allow it to become less dense as it freezes, to have the greatest surface tension of any known liquid and to act as a widely effective solvent, are central to the metabolic processes that support life on earth. All life requires water and, where it is scarce, effective mechanisms have evolved to conserve and acquire the water needed for survival and reproduction. The properties of water within organisms have been studied widely both as a structural and chemical requirement for the functioning of cells and as a medium for nutrient and waste transport. Viewed from this perspective, it is easy to forget the tremendous power and destructive force inherent in the mass transport of water and the financial expenditure required to protect the coastline from erosion and the catastrophic consequences tidal insurgencies. However, although water flows are not so extreme under normal circumstances, there is increasing recognition of the importance of the fine structure of flow in the nature of the environment and in the adaptation of living organisms to their surroundings. Examples are the use of rising vortices created and used by the blackfly larvae (Simulidae) in lotic environments to enhance capture of detrital material, and the fine structure of shark or dolphin skin to reduce drag and enhance locomotion (Mullins, 1997). Thus, reasons for the physical orientation, habitat selection and efficiency of an organism become clear, in ecological terms, only when the fine structure of the flow environments is understood (Vogel, 1994). The improved interpretation of the ecological influence of flow is due partly to technological advances in the fine-scale measurement of flow, but also as a result of a conceptual change exemplified in the tendency toward interdisciplinary research. Where biologists and physicists have worked together in the study of the influences of water flow on the ecology and physical dynamics of ecosystems, a great deal of new information has been gained and important new research questions raised. This chapter introduces some of the concepts of flow dynamics relevant to estuarine benthic systems and reviews some of the more recent work on the interactions and consequences of flow on the life strategy of estuarine organisms, with examples ranging from microphytobenthos to aquatic macrophytes. This is not intended to be an exhaustive review of the literature but to introduce concepts, terms and consequences of flow particularly with respect to cohesive (muddy) substrata using a limited number of examples.

As a final introductory remark, it is important to recognize that we do not yet understand many of the ways in which organisms experience and adapt to flow or even the fine-scale structuring of the flow itself. Equally, organisms are not simply the passive "recipients" of flow or stress but living components of the environment which shape and mediate flow in a manner that often enhances their prospects of survival, nourishment and reproductive success (Darwinian fitness). Where we are dealing with flow on a micro-scale, our intuition may let us down. One of the greatest proponents of the physical determination of biological form was D'Arcy Thompson (1860–1948); he propounded many ideas on this approach in his seminal work *On Growth and Form*, which was first published in 1917 and has been reprinted regularly since (Thompson, 1961). Today, some of these ideas seem extreme, but they are not necessarily at odds with Darwinian thinking, although the action of the "structuring forces" must now be seen through our knowledge of a hereditary influence that Thompson did not recognize. However, as we understand more about the interaction between physics and biology, Thompson's insights seem all the more remarkable and he clearly recognized the fact that intuition may fail when understanding is limited. He explained that, as we examine properties at scales beyond our limited senses: "We have come to the edge of a world of which we have no experience, and where all our preconception must be recast" (D'Arcy Thompson, 1917). These thoughts are still relevant today despite the use of laser technology, biomolecular techniques, electron microscopy and our computer modelling capacity.

III. THE INFLUENCE OF FLOW

A. Hydrodynamic Forces

In freshwater ecology, aquatic environments are broadly described as "lentic" where water is confined to a basin, lake or pool, or as "lotic" where the water is horizontally displaced as within a river or stream (Moss, 1998). No such simple terms exist for estuarine or marine environments. Even in freshwater systems, the use of the terms lentic and lotic can be misleading as these simple terms disguise a great variety of conditions (and consequent hydrodynamic forces) ranging between static (stagnant), laminar, transitional and fully turbulent flow (Figure 1). In nature, most relevant flow is turbulent (wherein the flow is irregular and chaotic), and laminar flow can almost be considered a laboratory artefact. Flow in estuaries is more complicated than within fresh-water systems due to tidal effects, the influence of surface wave action and the consequence of mixing flows of varying salt concentration and density (Dyer, 1997). In addition, the usually turbid nature of estuarine water, particularly very close to the sediment–water interface, can have a significant effect on the turbulence and viscosity of the flow (Gust, 1976). Thus, it is important to consider the effects of sediment transport, both in terms of the suspended pool of sediments as well as the traction load that rolls and jumps along the sea floor, in the nature of the flow conditions.

There are four major hydrodynamic forces that objects experience in flow: *pressure drag,* such as occurs over a rippled seabed; *acceleration effect,* such as occurs under oscillatory waves; *lift,* which is of increasing importance where the bottom particle size is relatively large (e.g. over cobbles and boulders); and *skin friction* (friction drag), which is the force acting directly on the surface of grains (Soulsby, 1999). Clearly, the balance and importance of these forces depend on the nature of the organism (shape, size and orientation) and its environment. Skin friction is significant and can predominate over other hydrodynamic forces where the surface is a smooth bed (e.g. mudflat) or a submerged flat plate, oriented parallel with the flow. Skin friction is the dominant force responsible for the erosion of muddy sediments. A full description of estuarine flow is not attempted here (Dyer, 1998, presents a thorough overview) but of chief concern is the stress (τ_0) induced on the bed by the overlying flow and its measurement.

B. Reynolds Number

The importance of the difference between laminar and turbulent flow has been outlined. The transition from laminar to turbulent flow can be sudden and is influenced by a number of factors. As the velocity or density of a fluid increases, or as the viscosity declines, the transition to turbulence is promoted. The advent of turbulent flow can be seen as the dominance of inertial forces

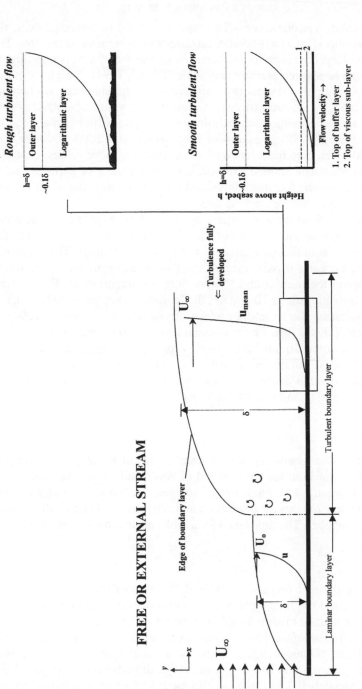

Fig. 1. Transition of flow condition over a flat boundary. Initial laminar condition becomes transitional and then fully turbulent with distance from the leading edge of the plate. Velocity profiles (u) are given for laminar and turbulent boundary layers. Details of the near-bed velocity profile under smooth and rough turbulent conditions are also given. Note absence of the viscous sublayer under rough turbulent flow. Adapted from Allen (1977).

over more sedate viscous forces. The ratio of inertial to viscous forces, the Reynolds number (Re), is a fundamental descriptive character of the flow. The utility of the Reynolds number is increased by the fact that it is non-dimensional and can be applied across many scales (Vogel, 1994). A high Reynolds number implies rapid, turbulent flow, whereas a low number suggests laminar flow, governed by viscous forces. A transitional value of 2000 is often suggested as the Reynolds threshold value above which flow become turbulent. Massey (1989) gives a more cautious range of 2000–4000 for flow in pipes. However, care must be taken when applying these values to natural systems. Where unconstrained flow occurs over a variety of surfaces and at varying velocities, the transition threshold can vary by an order of magnitude and these figures cannot be applied directly.

Organisms of differing sizes experience local flow in quite different ways. For small organisms, the effects of fluid viscosity are dominant, and inertial forces, derived from the momentum of the mass, are small. The Reynolds number for an organism swimming through water is, again, an expression of a ratio between the inertial and viscous forces it experiences. For bacteria swimming through fluid at 10 μm s^{-1}, the Reynolds number is in the order of 1×10^{-5}, indicating the large dominance of viscous forces (Vogel, 1994). A consequence of this is that, for small organisms, viscous forces dominate and as soon as the force driving movement stops, so does the organism. Small cells cannot "coast" along. For a fish swimming at 10 m s^{-1}, the value of the Reynolds number can exceed 1×10^7, indicating the greater importance of inertial forces as size and velocity increase.

1. Transport Processes

For organisms that inhabit aquatic systems, a critical feature is the transport capacity of the medium for the removal of waste and the uptake of nutrients (including oxygen). The transport of oxygen is a good example: aerobic organisms require oxygen for respiration and can be limited in their distribution by restricted supplies. The time taken for a molecule to diffuse over distance x is

$$x^2/2D_m$$

where x is the distance (m) and D_m is the diffusion coefficient (m^2 s^{-1}).

In a theoretical stagnant system, transport processes depend on molecular diffusion. For aquatic organisms this can be a very serious problem since the diffusion coefficient of oxygen in water, for example, is 10 000 times smaller than that in air (Denny, 1993). Thus, on a small spatial scale (microns) diffusion may seem rapid (for oxygen at 20°C, diffusive transport over 1 μm takes approximately 0.3 ms) but as scales increase the rate of diffusion rapidly becomes limiting (10 cm would take 25 days).

In nature, almost no aquatic systems are stagnant or limited to strictly laminar flow patterns where molecular diffusion controls transport. Most natural flow is turbulent and transport processes become dominated by the structure of the flow. The term "turbulence" expresses the variation in flow velocity around a mean value, and the "intensity" of the turbulence is expressed as the root mean square value of the fluctuations (Massey, 1989). Material, and energy, is distributed by turbulent flow in the form of eddies, an expression for "packets" of water, moving with their own momentum and velocity within the flow. This process depends on the energy of the flow, and transport rates are expressed in terms of the "eddy diffusivity", in some ways analogous to, but much more rapid than, the molecular diffusion coefficient. However, note that molecular diffusion acts in all directions from a source and, since flow *per se* is not involved, it is non-vectorial (independent of direction). Eddy diffusion may have different values in each of the x, y and z directions but is highly dependent on direction of flow. The nature of turbulent flow in systems is sometimes expressed in terms of the turbulent kinetic energy (TKE). In general, large eddies are associated with higher levels of mass transport and energy, whereas smaller eddies have less energy. The balance of turbulent versus laminar flow and the nature of the turbulent flow is an area where organisms have the potential to change and mediate the flow within their environment and increase their ecological "fitness". Aquatic organisms are often dependent on critical levels of flow and may select their local environment to fulfil flow criteria (Denny, 1993; Giller and Malmqvist 1998). Too little flow may limit transport processes below those required to maintain metabolism or nutrient supply, while too rapid flow may lead to drag and shear beyond the abilities of the organism, or the environment, to resist. However, living forms can also moderate flow to their advantage, as will become apparent.

IV. THE ACTION OF FLOW ON SEDIMENTS

The way in which a bed of sediment experiences stress from water flow depends on the nature of the sediment. The bed may be hydraulically rough or smooth, an expression of the "roughness" or relief of the surface and how the bed elements interact with the flow. Where the surface does not interfere with boundary flow, for example under "smooth turbulent flow", there is a viscous (sometimes also called "laminar") sublayer adjacent to the bed in which viscous forces dominate (Denny, 1993). Here, the profile of mean longitudinal velocity is linear and the momentum transfer, responsible for grain movement, is by molecular viscosity. The extent of the viscous sublayer depends on the depth, velocity and viscosity of the total flow, and ranges in thickness from a few millimetres over continental shelf sediments to a few centimetres in the deep sea. Where the elements of the bed disrupt flow, for example under "rough turbulent conditions", the constituent grains or elements protrude into the flow inducing turbulent conditions up to the boundary and destroy any viscous sublayer. The critical relief required to induce the development of rough turbulent conditions is principally dependent on the nature (speed) of the prevailing flow. At the boundary between these conditions, the flow is termed "transitional". Under rough turbulent conditions, momentum exchange is dominated by energetic turbulent eddy motions, and mixing and diffusion rates are consequently much greater under turbulent than laminar flow. The transfer of momentum from faster to slower moving levels in a moving flow (and thus towards the bed), through either turbulent eddies or viscous action, creates a tangential drag force, which is usually referred to as the bed shear stress (denoted τ_0). Synonymous terms found in the fluid dynamics literature are "surface", "wall" or "boundary" stress, or "friction" or "surface drag". Frictional drag, as one might expect, saps momentum, and if no external force were applied the flow would eventually lose all of its momentum and velocity would be reduced to zero (Komar, 1978). However, momentum at a large scale of TKE can be considerable. If the motivating force for the Gulf Stream were removed today, it would take years to become static.

Most of the existing literature on the biological implications of flow concentrates on the forces of drag and lift. It is more difficult to find reference to bed shear stress and the erosion process. Yet the process of sediment erosion is central to the ecology and behaviour of intertidal and subtidal organisms associated with cohesive sediments, and to the supply of organic material to (and from) the water column. The actual process of erosion is better understood for sandy non-cohesive sediments than for cohesive muddy sediments (Soulsby, 1999). This is because, without biological influence, a particle of sand from a non-cohesive bed can be considered as the unit of erosion. The size, density and shape of the grain are all important, but essentially only a small number of parameters is necessary to predict particle behaviour (Miller *et al.*, 1977). The erosion of cohesive sediment is a very different proposition. Grains behave in a cohesive or "collective" manner and cannot be viewed as individual units of erosion.

The principal unit of erosion is a floc, comprising a heterogeneous, poorly mixed array of silt and clay particles associated with and embedded within organic material (Figure 2). Flocs typically contain 10^5 to 10^6 individual particles (Pethick, 1986). The fluid force necessary to detach the floc from the bed cannot as yet be predicted from *a priori* knowledge of the bed properties. Therefore, all the features of the bed that affect the attraction between grains are influential in determining the nature of the erosion process. This includes changes in the electrostatic field surrounding clay minerals, which may be contingent upon porewater pH and ionic composition, the nature of the organic material present in the sediment and the temperature (and hence viscosity) of the eroding fluid. The GM6 programme (cited by Parker, 1997) indicated 32 *variables* of importance in determining the stability of a cohesive bed, while Black and Paterson (1996) indicated seven major *processes* that should be considered (Figure 3; updated by Tolhurst, 1999). These interacting factors together control the stability of the bed and the resistance of the surface to erosion through applied bed shear stress. It is notable that the conditions within the bed are greatly influenced by biological processes (Faas *et al.*, 1992). This biogenic mediation of cohesive sediment dynamics has only recently been acknowledged; indeed, the earliest systematic work on cohesive sediment erosion of the 1950s and 1960s completely ignored biological influence, and only in the last decade or so have we seen a dramatic shift in philosophy. A significant literature is now developing dealing with biogenic stabilization (Montague, 1986; Dade *et al.*, 1990; Paterson and Daborn, 1991; Krumbein *et al.*, 1994; Paterson, 1997, 1999; Black *et al.*, 1998; Eisma, 1998; Sutherland *et al.*, 1998a).

A. Bed Stress and Velocity Gradients

The tangential force experienced by a surface adjacent to a steady unidirectional flow is related to the nature of the flow environment. In theory, a static submerged surface hinders the flow of water, creating a velocity gradient away from the surface. The water molecules directly in contact with the surface are "held" static with respect to the surface (no-slip condition). The exact mechanism of this effect is still a matter of debate but the influence is the same whether the surface is hydrophilic, hydrophobic, rough or smooth. Because flow adjacent to the surface is essentially zero and molecules of water attract one another (viscosity), the layer adjacent to the layer of zero flow is slowed. The region over which flow "feels" the effect of the bed is known as the boundary layer and is arbitrarily defined as the region in which the velocity gradient from the bed extends to 99% of the natural or "free stream" velocity. Thus, whilst an identifiable region of retarded flow occurs adjacent to the seabed, the extent of this is a theoretical construction. Given the variability of natural flow, the boundary layer as defined above will vary considerably in time and space. Boundary layers, for instance, will form where flow encounters a bluff obstruction, such as over ripples on a sandy beach or even over the apex of a worm tube, and "grow" in thickness with distance downstream. However, flow within this ill-defined region of changing velocity is extremely important in benthic ecology and sediment dynamics. The force or stress that a bed "feels" as a result of flow is related to the rate of change in velocity in this region. In other words, bed shear stress is related to the rate of change of velocity in the boundary layer, and greater stresses are associated with steeper velocity gradients. As free stream velocity increases, the boundary layer becomes thinner and the rate of

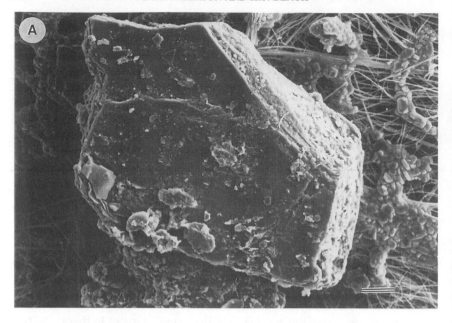

Fig. 2. (Facing page and above.) Low-temperature scanning electron micrographs of floc material from erosion experiments. (A) Large mineral particle; (B) composite floc including diatoms cells and organic material; (C) detail of composite flocs showing diatom debris as well as complete cells and spicules. Bars, 10 μm. Images courtesy of T. Tolhurst.

change of velocity increases since velocity at the bed is zero (the no-slip condition). The greater shear within thin boundary layers is also of ecological interest, since the higher level of turbulence will promote a greater mixing of nutrients, particles and resuspended biota. The physical processes in the region just above the bed are thus as important to benthic biota as the conditions within the bed.

The shape of the velocity profile within the boundary layer is also of importance, and is central to computation of the stress at the bed. Above the bed, a relatively well-defined logarithmic region can be present over both smooth and rough boundaries. Adjacent to the bed, the velocity profile varies over hydraulically smooth and hydraulically rough substrata. The presence of the viscous sublayer (and a transitional or "buffer" layer directly above) over smooth boundaries modifies the profile just above the bed from curvilinear to linear (see Figure 1). Under rough turbulent conditions, the velocity profile declines logarithmically down to the bed (strictly the topmost surfaces of the grains), and turbulence extends to the wall. This increases the stress experienced by benthic organisms and reduces the occurrence of hydraulic refugia (e.g. the laminar sublayer).

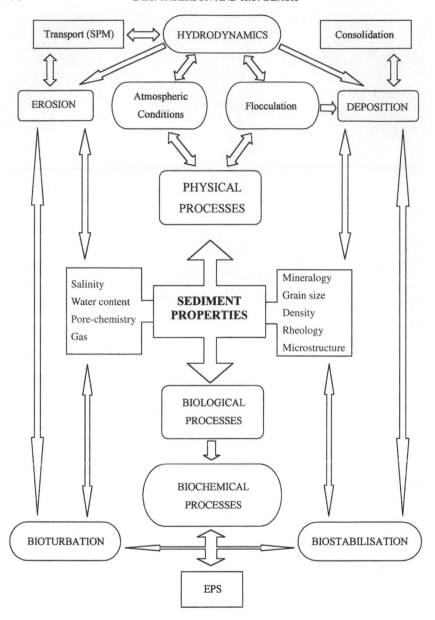

Fig. 3. Diagram of the sedimentological, physical and biological features that interact to control the properties of cohesive sediments. EPS, extracellular polymeric substances; SPM, suspended particulate matter. Figure updated by Tolhurst (1999).

B. Measurement of Bed Stress

It is perhaps rather ironic that, in spite of its key importance to both sediment transport and benthic ecology, we are unable *directly* to measure bed shear stress. There is no such thing as a bed stress gauge or instrument. It is a quantity that can be derived only from measurement of other variables, such as flow speed or water slope (in channel flow) and more recently heat dissipation (Gust, 1988). In its simplest form, and provided the flow is steady on the time-scale of 10–20 min, the mean stress experienced by the seabed can be computed from measurements of current speed at a single height (usually at 1 m) above the bed using the formula

$$\tau_0 = \rho_s C_{100}(U_{100})^2 \tag{1}$$

where ρ_s is the density of seawater, C_{100} is a drag coefficient and U_{100} is the current velocity 1.0 m above the seabed. This expression is known to marine geologists as the "quadratic stress law", and is economical in terms of both the necessary measurement hardware and of computation time. Unfortunately, it is also often misused in situations where wave oscillation causes variable flow or where flow is not even approximately steady. Clearly, however, evaluation of the bed stress is contingent upon using the appropriate drag coefficient (C_{100}), which can vary considerably according to the nature of the sea bottom (e.g. rough, smooth, muddy, sandy, rippled and so on) and the nature and distribution of the benthic biota. Choosing an appropriate value for C_{100} is difficult, not least because of the lack of direct experimental evidence. Soulsby (1983) reported, from a total of four observations, values of 0.0022 and 0.0030 for (presumably bioturbated) mud and muddy sand respectively, whereas Ariathurai and Krone (1976) cited a value of 0.0015 for a hydraulically smooth laboratory mud.

If the velocity profile is known, for example from an array of vertically spaced current meters, then the shear stress at the bed can be calculated based on the so-called universal von-Karmann–Prandtl "law of the wall". This simple expression relates flow velocity to distance above the bed and accounts for the nature of the bed (rough or smooth).

$$u = u_*/\kappa \, \ln(z/z_0) \tag{2}$$

where u is the current velocity (at a prescribed height), κ is von Karman's constant (usually taken as 0.4), z is height above the bed and z_0 is the roughness length. This equation can be treated graphically in order to calculate the friction velocity u_*. A plot of the mean velocity at height z ($u(z)$) against $\ln(z)$ gives a straight line with a slope of κ/u_* and an intercept of z_0. Once the friction velocity is known, bed shear stress can be calculated from

$$\tau_0 = \rho_s \, (u_*)^2 \tag{3}$$

This method has been used extensively by engineers and oceanographers in many laboratory and natural tidal situations (see Soulsby, 1999).

The velocity profile is a useful method when the near bed vertical current displays a logarithmic profile. However, in many natural geophysical flows, such as fast tidal flows across broad intertidal flats or rapidly accelerating flows found under waves, the boundary layer flow regime is more complex and commonly non-logarithmic. In these instances it is difficult, if not impossible, to derive accurate estimates of bed stress from velocity measurements. Many natural environments have complex flow patterns. Intertidal mudflats are a perfect example; aside from the fact that they are often remote and inhospitable (and dangerous to work on), the tidal flow across mudflats is often not rectilinear but rotary in nature (Dyer, 1998). Surface waves are rarely completely absent across tidal flats, and hence the seabed is influenced by a combined wave–current shear stress. Proximity to a river in the main body of the estuary and freshwater runoff through marsh channels creates a flow environment of considerable complexity.

A wide variety of instrumentation currently exists to measure fluid flow. In the sea, researchers have used either simple impeller and rotor-type current meters, or more advanced electromagnetic flow sensors and acoustic meters to determine flow speed and direction. In the laboratory, miniature impeller devices such as the Nixon Instruments Streamflo sensor, have been used for many years, although nowadays laser Doppler technology is routinely used to measure flow. Lasers permit flow and turbulence very close to the bed to be measured with considerable accuracy (Best *et al.*, 1997). The velocity of fluid flow has also been measured using the dissipation of heat from a thin wire (hot wire/film anemometry), which enables velocity profiles to be measured on a very fine spatial scale. More recently, flush-mounted configurations have been developed which have been calibrated directly against wall stress (Gust, 1988; Graham *et al.*, 1992). These represent a significant advance in wall stress measurement technology as they can, and have been, used in situations where logarithmic mean velocity profiles do not exist (e.g. Gust and Morris, 1989).

C. Erosion Process for Cohesive Sediments

Sediment erosion is often described as taking place when the hydrodynamic stress exceeds the inherent strength or resistance to erosion of the bed. If this is really the case, then the sediments exhibit a threshold stress that must be exceeded for erosion to begin. The stress at the point just when erosion begins, incipient erosion, is called the critical bed shear stress ($\tau_{0\,crit}$). Once this threshold is exceeded, erosion may occur by a number of mechanisms. Several classifications of the erosion process have been suggested (Mehta, 1991; Amos, 1995) and there is a danger of some confusion. Two main types of erosion have been noted since early in the description of cohesive sediment erosion: surface and mass (or later bulk) erosion (Mehta, 1991). These have been recently described as type I and type II erosion and also as "benign" and "chronic" erosion (Amos, 1995). Type I erosion occurs where particles are eroded from the surface as flocs once the threshold is exceeded (Figure 4). Erosion peaks rapidly and then decreases with time. This appears to be the dominant mode of erosion under natural field settings (Amos *et al.*, 1998). Under type II, erosion may be rapid at first, as with type I, but the bed continues to erode (Figure 4). Type II erosion tends to occur when the stress on the bed greatly exceeds the erosion threshold ($\tau_0 \gg \tau_{0\,crit}$).

It seems that observation and description of the erosion process is by no means settled. In recent observations of the erosion properties of the Fraser River delta (Canada), Amos *et al.* (1997) noted that the first stage of erosion could be through the resuspension of a surface fluff layer. This material was hardly bound to the bed and resuspended by very low applied shear. Amos *et al.* (1997) described this phenomenon as type Ia erosion. These workers also

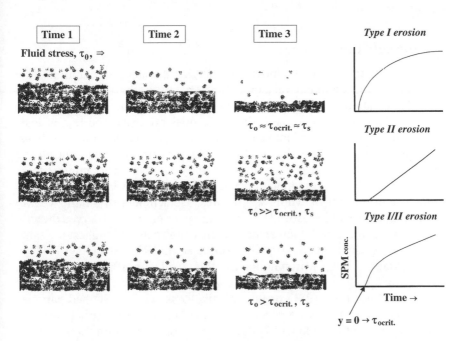

Fig. 4. Classification of cohesive sediment erosion. Type I erosion decreases over time toward a maximum suspended particulate matter concentration. Erosion ceases at the point where the fluid shear stress (τ_0) is equal to the bed shear strength (τ_s). Type I erosion can be subdivided into types Ia and Ib (see text). Type II erosion does not reach a steady state and erosion continues from the bed. Type I/II erosion is considered to share elements of both type I and type II behaviour, whereby erosion is initially rapid and declines with time but never quite reaches zero.

noted in this and previous studies (Amos *et al.*, 1992) that often there was no clear distinction between type I and type II erosion, confirming earlier comments by Mehta (1991) that erosion "can occur in several ways which are not wholly distinct and independent of each other". In some cases, erosion rate was initially high and declined, as with conventional type I erosion, but then the erosion continued at a reduced but steady rate. They considered this behaviour as an intermediate stage of erosion, between types I and II, and described it as transitional erosion (type I/II). We are left, therefore, with four types of erosion, each with differing characteristics of floc detachment and bed behaviour (Figure 4).

The nature of the erosion is critical for infauna but less so in terms of surface microbial biofilms. Under both type I and type II erosion, biofilms will be removed rapidly. Infauna will be affected by type II erosion only if a sustained period of stress, unlikely in the intertidal environment, can be maintained. Therefore, redistribution of infauna by sediment erosion takes place largely on an episodic basis in response to unusual high-energy conditions.

1. Threshold Conditions

In the modelling of sediment movement, a critical boundary condition is the "threshold" for sediment resuspension. However, the existence of the threshold condition for muddy sediments is still a contentious issue. Lavelle and Mofjeld (1987) argue that a threshold does not exist, and that at least some particles will be lifted into suspension by even very slow moving flows. On a micro-hydrodynamic basis, this is likely to be true since not all the subunits within a floc will be bound as tightly as others. One might expect floc disintegration or disaggregation and the breaking of weak primary bonds prior to entrainment of whole aggregates. Detachment of small, weakly bonded subunits within suspended floc aggregates has been observed directly (e.g. Al Ani *et al.*, 1991). Increasing evidence now exists to suggest the same is true for deposited muddy deposits (Peirce *et al.*, 1970; Black, 1991; Duck and McManus, 1991; Mirtskhoulava, 1991; Tolhurst, 1999). In addition, observations such as increased settling velocity of eroded flocs at increased bed stress (Lee *et al.*, 1981) or the negative-ϕ skewing of the bed particles under constant bed stress (Kusuda *et al.*, 1982) have been attributed to this initial detachment. Unsold and Walger (1987) coined the term "sub-critical" to describe the small patches and ephemeral low rates of erosion during their experiments. These terms may be synonymous with the Amos type Ia erosion (Table 1 in Amos *et al.*, 1997).

In purely statistical terms, therefore, there is no such thing as a critical erosion bed stress. There will always be a finite, albeit very low, rate of sediment transport for a finite fluid flow for an abiotic bed. However, biological colonization may influence this situation. Where the sediment surface is covered by a microbial mat, none of the mineral components of the bed is actually exposed to the flow. The argument may then be based at a cellular or even molecular level whereby colloidal polymers may leach from the bed (Tolhurst, 1999). For practical purposes, the subcritical levels of erosion are often below the resolution of the measurement technique or are lost in the noise of ambient turbidity. The measurable change in erosion profile around some critical value is still of practical interest. The definition of critical erosion stress is often based on a *time-averaged reference sediment transport rate*, which is then compared with a *reference bed stress* (Johnston, 1943; Mehta and Partheniades, 1982; Parchure and Mehta, 1985; Gross and Dade, 1991). For tidally deposited estuarine and coastal mud, reference bed stresses range between 0.1 and 1.0 Nm^{-2}, although all too often the minimum entrainment rate is not quoted in published research and is assumed to be zero (although correction may be made for ambient turbidity). Cohesive sediment mechanics awaits the arrival of a universal, well-defined and repeatable operational definition of the reference bed stress.

2. Shear Strength and Shear Stress

A number of publications have outlined the importance of microbiota in the physical structuring and erosional behaviour of sediments. Paterson (1997) has reviewed the mechanisms of this influence. The two major effects are the direct tensile influences of the organisms (usually filamentous forms) on sediment shear strength and the resistance to shear stress caused by the secretion of organic materials by organisms within the sediment. It is worth distinguishing clearly between shear strength and shear stress, terms often confused in the biological literature. The bed shear strength of sediment (denoted τ_s in soil mechanics) is a measurement of the internal strength of the bed. Bed shear strength is often measured as a resistance to torque, or turning movement, as in the Torr vane (vane shear strength), although this is known to be unreliable for unconsolidated sediments (Kravitz, 1970). Shear stress (τ_0) is quite different and refers not to an inherent property of the bed but to the force experienced at the interface due to water flow.

The important concept for erosional studies is the putative "operational" relationship between critical shear stress ($\tau_{0\,crit}$) (noting the problems with this definition outlined above) in relation to the magnitude of the measurable shear strength (τ_s; see Mehta, 1991). The Hydraulics Research *Estuarine Muds Manual* (Delo, 1988) gives an equation for the conversion of shear strength to the critical shear stress required for erosion. There are a number of difficulties with this relationship. The benthic boundary layer has often been described as an area supporting intense gradients around the surface region (e.g. Davison *et al.*, 1997; Taylor and Paterson, 1998). These gradients occur over relatively small spatial scales and have not yet been fully investigated in terms of the process of sediment erosion and transport. The occurrence of a biofilm or mat has an impact of the erodibility of the sediment but does not alter resistance to torque at a depth of several centimetres into the bed, the scale usually used to define τ_s. This is a problem pointed out by Amos (1995). A number of anomalous results in terms of sediment stability can be explained by the development of surface assemblages, for example the buoyant layer found by Sutherland *et al.* (1998a). Thus, a relationship between the critical shear stress and shear strength may be reasonable where a sediment has been prepared and maintained in the laboratory but is less likely to be a useful predictive tool under natural conditions (Amos, 1995).

V. THE LIVING SEDIMENT

A. The Influence of Organic Material

Almost all sediments in natural waters show evidence of biological activity. Even sediments deep within the ocean bed have shown evidence of bacterial activity (Cragg *et al.*, 1990). The importance of cell exudates was first recorded experimentally by Holland *et al.* (1974), who noted that stabilization of sediment was more pronounced by cultures of diatoms that produced copious quantities of polymer (mucilage). This link has also been demonstrated for other diatom species (Sutherland *et al.*, 1998b) and for marine bacteria (Dade *et al.*, 1990, 1996; Decho, 1994). However, scholars had previously considered the potential importance of organic material in marine sediments. Paracelsus (1493–1541) commented on mucilage and its role in the colonization of surfaces, while in the last century Huxley actually classified deep-sea mucilaginous material as a separate life form (*Bathybius haeckelii*, cited in Krumbein *et al.*, 1994).

It is no longer contentious that cellular exudates produced by microbial metabolism, referred to as extracellular polymeric substances (EPS; Decho, 1990), can influence the stability of sediment. It is still unclear as to the generic significance of these stabilizing effects. The ecological importance of EPS is varied. Polymers can act as collating agents sequestering xenobiotic compounds and may thus protect cells from toxins. Diatoms have been shown to produce more EPS under conditions of raised metal concentration (Miles, 1994). Field studies, conducted after the 1989 mining waste spill into Restronguet Creek in Cornwall, showed that intertidal diatom assemblages survived the contamination despite highly increased levels of cadmium and zinc (Miles, 1994). EPS is largely composed of water and may protect cells against desiccation. In diatoms (Edgar and Pickett-Heaps, 1984), polymeric material is also exuded as a mechanism of locomotion. Diatoms, bacteria and cyanobacteria also produce polymers which can attach cells to the substratum

(Wetherbee *et al.*, 1998). Therefore, the evolutionary advantages of EPS production may be manifest, but we are still relatively naive about the structure and properties of microbial exudates in aquatic environments (Decho, 1990; Taylor and Paterson, 1998; Taylor *et al.*, 1999). In addition to local influences, organic material on the bed may influence the succession of the system since the organic material and associated biofilms act as cues for the development of further natural assemblages of organisms through the inhibition and facilitation of settlement (Wieczorek and Todd, 1998). Polymers can be extremely influential in altering the flow characteristics of a fluid or suspension. Polymers on the bed may enhance particle cohesion but also reduce the roughness of the bed (Paterson, 1997), which reduces the surface shear for a given flow. The literature on the influence of organic content is, however, extremely varied and the magnitude of the reported effects varies from considerable to marginal or even destabilizing (Paterson, 1997; Tolhurst, 1999).

Polymers are abundant in seawater (Aluwihare *et al.*, 1997) and can alter the viscosity of the fluid; where viscosity is reduced, the shear stress at the surface also declines. Where polymers are concentrated near a surface, the effect may be more pronounced and polymer production has been suggested as a mechanism to reduce drag in fish swimming (Daniel, 1981). The influence of this effect is not always clear-cut given the energy required for mucilage production and the extent of the drag reduction created (see Vogel, 1994).

B. Erosion of Biological Components

The sedimentological approach to the erosion of cohesive beds deals with the movement of particles or flocs. However, the bed is a complex matrix of material with varying properties. There is no *a priori* reason to assume that all components of the bed behave in the same way. Recent evidence suggests that the erosion process may occur in discrete stages. Several authors have begun to examine each component of the bed in terms of its behaviour under pre- and post-threshold conditions (Wiltshire *et al.*, 1998). These studies have shown that biological components and markers (e.g. EPS and pigments) may have measurable erosion thresholds that differ slightly from the threshold for the bed *per se*. Wiltshire *et al.* (1998) showed that the erosion of chlorophyll *a* preceded the accumulation of sediment in the water column. Sutherland (1996) also defined a separate $\tau_{0\,crit}$ for chlorophyll *a*. The interface between the sediment and the water or air is the region of great microbial activity (Characklis and Wilderer, 1989; Krumbein *et al.*, 1994; Stal, 1995; Paterson, 1999) where microbial mats and biofilms tend to form. This creates a structure on a microscale within the sediment (Yallop *et al.*, 1994; Paterson, 1995; Défarge *et al.*, 1996). Examination of the erosion process reveals the sequential removal of components of this structure, often before the sediment bed erodes. This is not surprising, given observations on the resuspension of diatom assemblages (de

Jonge and Van den Bergs, 1987). Admiraal (1984) put forward the hypothesis that some benthic algae follow a facultative phytoplanktonic existence whereby cells are deposited on the ebb tide, resuspended on the flood tide, and deposited once more as the tide recedes. The ecological advantage seems obvious. Cells on the surface of the sediment have adequate light for photosynthesis but once covered by water, especially turbid water, light will become limiting. Cells adopting a facultative phytoplanktonic lifestyle can continue to photosynthesize in the water column. This proposed cycle requires further investigation and it is unclear how the flow dynamics might operate without active behaviour on the part of the cells. Once resuspended, the potential for cells to be swept into the main channel along with the sediment is high. Alternatively, cells might be deposited at the high-water mark and exposed for an extended period over the spring/neap cycle. Such a facultative lifestyle requires the cells to avoid these adverse pathways. Cells must be at the sediment surface and be resuspended easily as the flood tide arrives, but return to a suitable part of the bed for the exposure period. The suggestion that cells might alter adhesion by varying the structure and properties of their polymers under varying environmental conditions has some support (Peterson, 1987; Ruddy et al., 1998a). Therefore, a mechanism to explain the dynamics of facultative phytoplankton can be suggested, whereby cells increase their adhesive qualities during the late ebb to be retained on the bed, while reducing adhesion before the flood to ensure resuspension to the water column. The turbulence induced in the ebb and flood tides then becomes important in promoting the chance encounter of the cells with the bed, a concept similar to the chance distribution of aquatic propagules and gametes (Denny, 1993). Cells suggested to follow this cycle tend to be lightly silicified (e.g. *Cylindrotheca closterium*) and hence have greater buoyancy than most epipelic forms.

These unproven observations would have interesting ecological implications. The consequence in terms of aquaculture is that, on approaching threshold conditions, a layer of highly organic and nutrient-rich material is swept into the water column. This burst of nutrition is relatively concentrated and a valuable resource for benthic filter feeders. The variation in the nature of material in suspension over the tidal cycle is now receiving renewed attention (Edelvang and Austen, 1997; Urrutia et al., 1997; Roegner, 1998; Tolhurst, 1999) and perhaps the facultative theory will be put to the test.

C. The Infauna

It has been common to regard the infauna as passive colonizers of sediments, responding to episodic erosion/deposition events, with little emphasis on the way in which these organisms structure and alter their own environments. This perspective is changing and many studies now address the biological mediation of benthic habitats (de Boer, 1981; Montague, 1986; Meadows and Tait,

1989; Meadows *et al.*, 1990, 1998a,b; Meadows and Meadows, 1991; Hall, 1994; Paterson, 1997). The impact of biota on the habitat can be relatively obvious. Tube-forming organisms may stabilize the bed and alter the flow pattern above the interface. The understanding of these effects also requires a basic appreciation of benthic hydrodynamics. For example, protruding tubes change the near bed flow but such changes are density dependent. Single tubes create turbulence and local erosion (scour) but if the density of the tube field increases, the effect is to reduce the erosive shear stress at the bed and displace the velocity gradient to the region above the tube field (Luckenbach, 1986). The construction of tubes that do not protrude may also have an effect in that they increase the shear strength of the bed and this may impact bed stability and erosion (Meadows and Tait, 1989).

Infauna can also be destabilizing. In a recent study, Widdows *et al.* (1998) concluded that the erodibility of natural intertidal sediments from the Humber Estuary was related to the density (and bioturbatory activity) of the benthic bivalves *Macoma balthica* and *Cerastoderma edule*. The authors concluded that sedimentological measures were of little use in predicting sediment stability. In addition, they noted that biodeposition (material deposited as faeces or pseudo-faeces) greatly increased the clearance rates of the water above the bed. Therefore, the infauna decreased stability but increased deposition. At first, this may seem counter-intuitive since the bivalves require the substratum and, by destabilizing the sediment, they increase the likelihood of sediment erosion and loss of habitat. Taking the example of *M. balthica*, the smaller size classes (< 5 mm) are found within 4 cm of the surface while some larger individuals can be found down to > 15 cm (Davey and Partridge, 1998). Assuming type I erosion (Amos *et al.*, 1998), a sustained period of increasing hydrodynamic stress would be required to threaten the habitat of the bivalves, even at a limited depth of a few centimetres. To expose any bivalves (making some assumptions about sediment properties), 76 000 tonnes of sediment would have to be eroded from the Humber Skeffling mudflat on a single tidal cycle. The changes in stability induced at the surface by the feeding activity are unlikely to induce more catastrophic type II erosion. In addition, activity at the surface may serve to increase the fitness of the bivalves by optimizing their foraging potential. It has been noted in laboratory experiments that deposit-feeding bivalves may cease to feed after a few days in captivity (Zwarts *et al.*, 1994). The reasons for this may be related to the quality of food available within the confined system or inappropriate culture conditions (e.g. wrong type of substratum). The addition of fresh sediment induces feeding to begin again. Therefore renewal of the surface sediment is important for the bivalves. Feeding destabilizes the bed and promotes erosion. Sediment is redistributed in the water column and material is deposited on the bed through the action of both biodeposition and natural sedimentation. Thus, the surface of the bed is continually being reworked, which appears to be to the advantage of the deposit feeders. The deposit feeders mediate the stability of the sediment and possibly gain in terms of food supply. The interaction between the behaviour of the organisms and the dynamics of the system could be interpreted as leading to an improved fitness for the assemblage (Figure 5). This kind of feedback mechanism requires further study.

D. Macrophytes

Depositional systems tend to be dynamic and unstable, and are generally unsuitable habitats for rooted aquatic macrophytes. However, under the correct condition of flow, colonization can take place leading to the development of significant swards of plants. One of the first *in situ* flume experiments to be

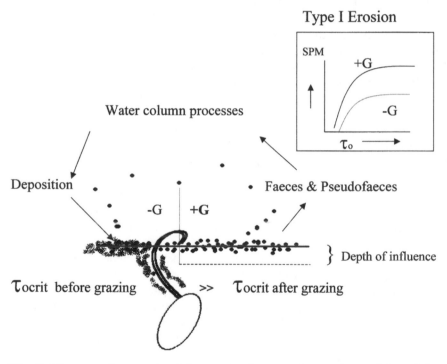

Fig. 5. Diagram of a postulated interaction between the grazing activity of the common bivalve, *Macoma balthica,* and sediment dynamics. The surface layer of sediment is disturbed by the activity of *M. balthica* and $\tau_{0\,crit}$ is reduced. The influence is mainly at the sediment surface, inducing type I (possibly type Ia) erosion, at a lower threshold stress. Material (sediment and faeces) is resuspended in the water column and redistributed. This leads to greater turnover rates for surface sediment material and potential benefits in terms of food supply to surface deposit feeders.

conducted confirmed the influence of vegetation on the reduction of flow and the trapping and binding of sediments (Scoffin, 1970). Since this early work several groups have investigated sediment stabilization by sea grasses (e.g. Fonseca *et al.*, 1982) and improved technology has allowed detailed study of the microstructure of flow above and within the beds. The reciprocal effects of the grasses on water flow (Fonseca *et al.*, 1982, 1983; Fonseca, 1989; Koch, 1996) and how the flow may benefit the plants have also been investigated.

Sea grasses require light for photosynthesis and therefore suspended material (sediments, phytoplankton), which reduces light penetration in the water column, can limit growth. However, growth also requires the transport of nutrients to and waste from the plants. The energy of the flow can be expressed in terms of the turbulent kinetic energy (TKE). Large eddies (high TKE) can maintain sediments is suspension leading to turbid flow. Less energetic, smaller eddies (low TKE) can still efficiently disperse nutrients in solution but have insufficient energy to maintain large particles in suspension. Within the seagrass beds, large eddies are damped leading to sediment deposition, while the blades, and the associated epiphytes, create or enhance smaller low-energy eddies (Koch, 1996). Thus, the beds moderate the flow dynamics across several scales to create conditions that may enhance macrophyte growth and development. Turbulence induced by the epiphytes also enhances photosynthesis by decreasing the thickness of the diffusive boundary layer around the fronds (Koch, 1994), allowing more efficient uptake of carbon. In addition, the velocity profile may be inverted by the influence of the macrophyte stands. The narrow stalks of the plants retard the flow adjacent to the bed but these are less effective than the expanded upper portion of the blades. Thus flow within the canopy may be impeded more than the flow adjacent to the bed. This has been noted for *Zostera marina* (Fonseca *et al.*, 1982; Ackerman and Okubo, 1993) and for saltmarsh plant such as *Spartina anglica* (Shi *et al.*, 1995).

Flow within the bed also has importance consequences for organisms inhabiting the sediments, for example enhanced deposition, protection from high-energy bursts and longer periods of sediment stability. The examination of the flow structure has enhanced the understanding of the interaction between the grasses and their local environment. This may also lead to improved efforts to protect these often-threatened sea grass meadows. Macrophytes are important stabilizing agents for estuarine sediments. Fluctuation in seagrass populations have been recorded since the "wasting disease" of the 1930s, but it is also likely that global change and sea levels will threaten or change their distribution (Glemarec *et al.*, 1997). Anthropogenic influences such as yachting, xenobiotics, exploitation for fishing, recreation and dredging also damage the natural habitat of the sea grasses, and efforts to preserve their habitats are now underway.

VI. REMOTE SENSING AND MULTIDISCIPLINARY STUDIES

Patterns of change on the surface of intertidal mud and sand flats reflect temporal and seasonal variations, but also change in response to anthropogenic influences, for example the growth and development of nuisance swards of green algae in response to increasing nutrient loading (Eisma 1998; Raffaelli et al., 1998). However, it is difficult to gather evidence on the changing state of intertidal systems. Traditional ground surveys are time consuming, expensive and usually provide only limited spatial coverage. Additionally, the collection of long-term data sets is difficult to maintain given the expense and labour required for each survey. Recent advances in the remote sensing (RS) of intertidal deposits may provide the solution.

Synoptic mapping of intertidal systems (Guillaumont et al., 1997) and estuaries can be used to characterize the bed in terms of biological components and grain size distribution. Synoptic maps of sediment type have been produce by a number of workers (Doerffer and Murphy, 1986; Yates et al., 1993). The information that can be gained from the surface depends on the ability to separate the influence of the substratum and the biota from the signal. In the example of spectral reflectance data, recent research has shown that the optical properties of the surface are influenced by the distribution of microbial biomass (photosynthetic pigments). The accumulation of pigments alters the absorption characteristics of the surface, and therefore the spectral reflectance properties. Using multi-wavelength radiometry (e.g. the compact airborne spectrographic imager) the spectral reflectance signal from the surface can be analysed to determine the spatial coverage of the pigment and hence the algae. Such studies are fraught with logistic difficulties, such as the timing of flight lines, atmospheric interference with the signal, correction for scatter and reflectance. In addition to physical difficulties, it has been shown that, for muddy intertidal flats, the reflectance signal can change within minutes owing to the migratory patterns of epipelic diatoms (e.g. Hay et al., 1993) and other microphytobenthos. The RS implications of this have been noted by Paterson et al. (1998), while the effects of cell migration have been confirmed using non-destructive measurement of the fluorescence signal from the surface of sediments (Serôdio et al., 1997).

Developments in RS may provide increasing discrimination of sediment type and the distribution of biota. However, this is potentially only the first step. Many of the predictions regarding the ecophysiology and behaviour of intertidal sediments are now linked to the biology of the system (Daborn et al., 1993; Eisma, 1998). Knowledge of the distribution and patchiness of the biomass over the intertidal may allow other inferences to be drawn. The cascade effect, noted by Daborn et al. (1993), provides an example of the complexity of the linkages thorough the system, as predicted by Montague

(1986). Daborn *et al.* noted that the stability of their study site on the Bay of Fundy was altered by the passage of large flocks of semi-palmated sandpipers. They postulated the production of EPS by diatoms increased the stability of the sediment and this was promoted by the removal of the invertebrate, *Corophium volutator*, by the grazing birds. This finding was possible only because of the wide and interdisciplinary nature of the research being carried out (Daborn, 1989).

Similar projects have evolved along this theme leading to other interdisciplinary programmes (e.g. the LISP-UK programme; Black and Paterson, 1996, Black *et al.*, 1998). The potential to follow cascade effects through remote sensing has been suggested by Riethmüller *et al.* (1998). These workers postulated a link between tidal flat colour (pigment distribution) and sediment erosion threshold. If such links are proven to be robust, given adequate ground truthing, then the prospect that many other links and cascade effects could be unravelled is a realistic one. The distribution of pigment (largely chlorophyll *a*) is used a biomass proxy for many ecological studies of depositional systems and can be linked directly to EPS distribution under the correct conditions (Underwood and Smith, 1998a). Biomass is a central parameter in many models of system dynamics (Pinckney, 1994) and the prospect of delimiting pigment distribution on a large scale makes the application of such models more feasible over a tidal basin scale. This is an area of research that is likely to be exploited in the future as RS imagery and analysis becomes more accessible and reliable.

VII. MODELLING

The ability to predict the movement of shoreline sediments under a given set of circumstances is undoubtedly a useful management and decision-making tool for coastal engineers, and may increase our understanding of the ecology of coastal ecosystems. Numerical mathematical models are used routinely by scientists to examine a wide variety of environmental fluid dynamics problems, and the present-day computational capabilities of computers allow fairly sophisticated models to be run on relatively short time-scales. However, the utility of a numerical model is dependent on our understanding of the processes under consideration, and for this reason models predicting the behaviour of cohesive sediments are not yet adequate or realistic.

Large-scale, estuary-wide models of water and (fine) sediment movement are currently available for a number of specific environments (e.g. Cheng *et al.*, 1993; Teisson, 1997; Wang *et al.*, 1998). Generic, yet sophisticated models, such as the UK-based EcoS model (Environmental Contaminant Simulation, from Plymouth Marine Laboratory), the MIKE3MT code of the Danish Hydraulics Institute, the DELFT3D-Sed model from WL Delft Hydraulics in the Netherlands, and the DIVAST model developed jointly at Bradford

University, UK, and the Katholieke Universiteit Leuven, are increasingly being used as tools to represent sediment transport processes in a wide variety of different estuarine environments. In the context of our discussion, the chief variable is the bed stress (τ_0) which parameterizes bottom friction. Coupled water–sediment models and morphodynamic models need to prescribe a threshold value for τ_0. To predict the entrainment of bottom sediments, and where vertical flux through time is also of concern, the rate of sediment erosion (ϵ) must also be known. A great deal of effort usually accompanies the development of numerical models (e.g. Cassuli and Cheng, 1992); however, poor parameterization of both τ_0 and ϵ will be disastrous in terms of the utility of the model. The current lack of a detailed understanding of the variability in both space and time of both these variables is at the centre of our inability to model the dynamics of muddy sediments correctly.

A. Temporal Variability

A bewildering variety of factors influences cohesive sediment behaviour (Black and Paterson, 1996), and natural mudflats are an extremely complex hydrodynamic–sedimentary environment (see Figure 1). Tidal energy and oscillatory wave motions, desiccation, rain impaction, biostabilization and bioturbation, fluid mud formation, consolidation, even ice in some cases, act together to mediate the threshold erosion stress and erosion rate. It is beyond the scope of this discussion to provide an overview of the myriad interactions here, but several reviews exist (e.g. Rhoads and Boyer, 1982; Paterson, 1997). Also, because most of these factors can vary through time on a variety of scales (e.g. minutes, days, months, years and longer), threshold erosion stress and erosion rate must be represented within models as time-variable parameters on the appropriate scale. Currently, very few models consider this, even on an exploratory basis, and these parameters are treated as constants.

Intertidal regions are unique in that the sediments are periodically exposed to the atmosphere for part of the tidal cycle. In many English and Welsh estuaries (except the Thames) the tidal phasing is such that exposure (low-water spring tides) consistently occurs between 0900 and 1500 hours (Dyer, 1998). Upper flat sediments can be exposed for the greater portion of the tidal cycle, whereas the lower flat exposure may vary from 0 to about 2.5 h. During exposure, the sediment surface may undergo heating and drying out by evaporation caused by sunlight and wind. This decreases the surface moisture content giving rise to an increase in the threshold erosion stress (Amos et al., 1988). Paterson et al. (1990), for example, quoted a significant increase in the stability of mudflat sediments from the Tamar and Severn estuaries, UK, during exposure, which was more extreme towards the upper intertidal region. In contrast, they also showed that, when sediments were exposed, rainfall may increase the surface moisture content and substantially reduce the threshold stress, or even induce erosion directly through pitting and runoff. Clearly, and in contrast to subtidal mud, intertidal mudflat sediments can experience a change in erosion resistance even whilst the tide is out. Intratidal variation in erosion resistance is central to the short–medium term development of mudflats, since these influence resuspension of bed sediments by subsequent tides (Black, 1998). In terms of modelling sediment transport in these environments, the temporal variation in τ_0 (and ϵ) throughout the *entire* tidal cycle needs to be considered. However, to the authors' knowledge such formulations have yet to be included within numerical models.

Changes through time influence the biological processes that mediate the behaviour and stability of cohesive mud (Paterson, 1997). A principal driving function of all biological processes is temperature. This has been recently re-examined in terms of microphytobenthos dynamics and productivity (Blanchard *et al.*, 1996; Blanchard and Guarini, 1999) and can now be linked to processes such as EPS production (Underwood and Smith, 1998b), reproduction and biomass production. The summer stabilization of intertidal mudflat surfaces is now firmly established, and it is known that sediments accumulate on intertidal banks and shoals during the summer and erode during the winter months (Frostick and McCave, 1979). Time—and biology—is thus a key factor relating to the behaviour of cohesive sediment systems, and appreciation of the time-scales of variability is central to any numerical representation of mudflat mechanics if there is to be a reasonable degree of realism. This contrasts with non-cohesive systems such as sandy beaches, in which single time-independent prescriptions of τ_0 and ϵ are sufficient.

B. Spatial Variability

Most numerical models represent a mudflat surface as a smooth, two-dimensional, uniform and homogeneous (in terms of cohesive strength) substratum (e.g. Hamm *et al.*, 1997; Wood *et al.*, 1998), but even casual visual observation reveals that this is a gross oversimplification (Meadows *et al.*, 1998b). The natural topographic variability associated with ridge-runnel features, dendritic drainage networks and "bio-roughness", as well as the natural patchiness of carbon, nutrients and biota in mudflat sediments (e.g. Schroder and van Es, 1982; Meadows and Tufail, 1986; Eisma 1998), creates substantial spatial variation in τ_0 and ϵ on a series of scales both across and down the shore. Modelling that does not parameterize this variability will not resolve the erosion flux across the mudflat accurately. Since mudflats fringing estuaries act as both a source and a sink for suspended particles, inaccurate computation of the intertidal–subtidal fine-sediment flux may have serious consequences for the estuary sediment budget as a whole.

Kilometre-scale variability in cohesive strength has been noted from several macro-tidal estuaries, arising primarily because of the variable exposure time of different parts of the mudflat, although this can also be attributed to variations in sediment texture (Anderson, 1983). Widdows *et al.* (1996), for example, recorded a general seaward decrease in threshold erosion stress ($\tau_{0\,crit}$) from 0.70 Nm^{-2} to about 0.32 Nm^{-2} on Skeffling clays in the Humber estuary, UK. Amos *et al.* (1997) reported a similar, although reverse, trend for submerged benthic sediments along a transect on the Fraser River delta ($\tau_{0\,crit}$ 0.11–0.50 Nm^{-2}), and noted an inverse relationship of $\tau_{0\,crit}$ with erosion rate (ϵ).

Metre-scale variability in erosion resistance, associated with shore-normal bedform features similar to ridge-runnel topography found on sandy beaches, is also apparent. On the crests, the sediment is much drier owing to better drainage. In the troughs seawater becomes ponded following tidal emergence, and the surficial sediments have very little or no measurable strength. Widdows *et al.* (1998) cited field measurements of 0.18 Nm^{-2} and 0.50 Nm^{-2} for trough and crest sediments respectively. Superimposed on these larger scales of interest are even smaller scales of heterogeneity in cohesive strength. S. Shayler (unpublished data), as part of the recent European Union-funded INTRMUD project, documented significant differences in the stability of small ripple peaks and troughs using the Cohesive Strength Meter (Tolhurst *et al.*, 1999). Heterogeneity, for whatever reason and on whichever scale, is a predominant characteristic of muddy tidal flat environments. However, dealing with such widespread heterogeneity from a modelling (and measurement) viewpoint is problematic. Currently the best practical approach is to use a method of spatially averaging the distribution of values, and then a sensitivity and validation analysis to establish the predictability of the model and to provide error bounds on model predictions (Teisson, 1997). The actual method used is heavily contingent on the aims and objectives of the model (e.g. whether a water-quality or morphodynamic model), as well as the resolution of the model grid and computing power. What is certain, however, is the necessity within models to incorporate data on $\tau_{0\,crit}$ and ϵ obtained from field measurements and from as many measurement locations as possible.

C. Modelling Approaches

Modelling approaches to these problems have differed historically, and two rather separate camps have arisen. The first of these comprises the system-wide engineering modellers who *still* treat cohesive sediment dynamics in estuaries rather empirically, obtaining values for the various parameters from laboratory experiments or the literature (e.g. this criticism can be levelled at the DELFT3D-Sed model), or worse, setting a value to some parameters (critical shear stress for deposition, for instance) without any information on its value for the site under study. Sometimes, the intertidal regions within an estuary are ignored altogether in estuary models. Alternatively, there are approaches that focus only upon the bottom boundary and associated processes. One might suppose that only when these two approaches are unified will there be a more realistic, less empirical and (hopefully) less site-specific model output.

Few researchers have undertaken the second path, in spite of evidence that this is precisely where further research is most urgently required. Part of the problem is that, in addition to the aforementioned spatial hetero-geneity, a great many dynamic linkages exist amongst the biotic and abiotic components of muddy sediments, and the majority of these remain, even today, poorly understood (Parker, 1997). In addition, factors may vary in importance between sites and in relation to one another. The field situation is rather too complex for an experimental or modelling approach. However, two of the most recent studies have arisen from the UK government-funded LISP (Littoral Investigation of Sediment Properties) project (see Black and Paterson, 1996), concerning the muddy intertidal of the Humber estuary, UK. Both studies examined the role of selective biological processes as mediators of sediment stability.

Willows et al. (1998) conducted a flume study to assess the influence of the small clam, M. balthica, on sediment erosion. Experiments indicated that the density of individuals at the mud surface directly influenced the amount of sediment eroded (erosion rate), up to a maximum clam density. This phenomenon was modelled using a model based on the Weibull function, and included critical erosion current velocity, clam density, erosion rate and maximum suspended sediment concentration as a function of both current velocity and clam density, in addition to several experimental constants. Field data supported the laboratory-determined functional relationship between Macoma density and the quantity of sediment resuspended at environmentally realistic maximum current velocities, and provided field-based parameter esti-mates for modelling the erosion of in situ sediment. Using predicted flood–ebb tidal current velocities along a shore-normal transect, they esti-mated that natural densities of Macoma increase the quantity of sediment resuspended over that compared to non-bioturbated mud by approximately 0.42 kg m^{-2} per tide.

This study was pioneering in terms of providing modellers with a straight-forward means of quantifying the complex effects of biotic components on rates of sediment erosion. Inclusion of the type of functional relation advocated in this study into algorithms describing the bottom boundary condition of larger-scale sediment transport models is the next step. In addition, research of a similar nature ought to be encouraged on the other principal biotic influences on sediment stability: microbial mats and tube construction by worms.

As part of the LISP study, Ruddy *et al.* (1998a,b) outlined a cohesive sediment transport model which incorporated a highly interdependent community of benthic algae, bacteria and macro-heterotrophs with small-scale nitrogen cycling and the algal secretion of "excess" carbon as sediment-binding polymers. Muddy estuarine silts comprise a dynamic and otherwise favourable environment for various biota. Given the known tendency for benthic microbiota to stabilize mud surfaces through exudation of EPS, one might expect the overall sediment dynamics to be a complex function of ecological and biogeochemical processes as well as physical processes. Using a simplified virtual sediment–water interface, they showed that under "normal" conditions, a biolayer composed of unicellular algae can form at the sediment–water interface. Rapid algal growth generates regularly occurring nitrogen-limited conditions in the biolayer. In the model, algal carbohydrate polymers produced and exuded during nitrogen limitation profoundly influence sediment dynamics by adjusting bed cohesion. Stable, but dynamic, sediment surfaces result in the model from a combination of EPS secretion, bacterial remineralization of exuded carbohydrate and grazing of the algae by bioturbatory macroheterotrophs.

The model predictions compared favourably with the large quantity of field data originating from the LISP project (Black *et al.*, 1998), indicating that the conceptual ideas within the model may realistically represent physical and biological processes occurring *in situ*. For example, model runs were iterated through a sequence of tides and light under conditions of zero biological influence and under conditions of bioturbation by *M. balthica* (which produces significant destabilization). A combination of exudation and bioturbation lead to a dynamic, but stable, surface sediment layer. The model, whilst undoubtedly complex, provides compelling evidence that ecological and chemical dynamics within the thin surface layer of intertidal mud can mediate or even dominate sediment transport processes. The ability of the model to simulate the LISP field observations demonstrates that the nitrogen-limited exudation by algae in a dynamic context might be a sufficiently robust self-regulating mechanism to withstand factors such as diurnal variation, intense grazing, sediment–water exchange processes, sediment burial and sediment erosion.

Parker (1997) asked several perceptive questions in his preface to the Fourth Nearshore and Estuarine Cohesive Sediment Transport Conference (Burt *et al.*, 1997):

(1) Are the conclusions, rates or coefficients arising from supposedly abiotic experiments, or experiments that ignore microbiological impacts, vulnerable to alternative formulation?

(2) How long will it be before we see included in the constitutive equations describing fine sediment transport, terms or groups balancing bacterial metabolism and mucous production, microbiological grazing and digestion, bio-roughness production and pelletization?

Ruddy and his colleagues have provided an innovative approach and their modelling philosophy is most certainly pioneering. Not only does it offer a greater understanding of the factors of importance with respect to sediment erosion through identification of the key biogeochemical and biophysical linkages, it promises to provide hard numbers (on the relevant temporal scale) for the dynamic variables of use to system modellers, $\tau_{0\,crit}$ and ϵ. However, a cautionary note is needed: the model is still largely unproven and even the central tenent of nitrogen-limited EPS secretion has not been shown to occur in these systems. The approach, however, may provide a conceptual framework for future research both in terms of the modelling and in understanding the detail of the model parameters. In addition, different benthic assemblages, different tidal rhythms or different nutrient loading may produce somewhat different dynamics. Site specificity has always plagued cohesive sediment systems, and there is a pressing need to examine other depositional environments using Ruddy's philosophy.

These examples illustrate some recent advances in our understanding of, and ability to model, muddy sediments. It is not difficult to appreciate how the structural and compositional complexity of these sediments has been a major impediment to progress in the modelling (as well as measurement) field. Uncertainty in model output increases proportionately with the number of variables, which often leads researchers to simplify some aspects of the system or ignore others (Dyke 1996). In biologically active and chemically reactive mud (which includes almost all naturally found muds in marine and fluvial environments), as we have seen, oversimplification can be a dangerous indulgence. Over the past 10 years many cohesive sediment transport models have been developed for a variety of applications, but Teisson's recent remarks are a timely reminder: "Because of [these] limitations, cohesive sediment transport models will never be ready-to-use models, and call on the expertise of the modeller gained in this field" (Teisson, 1997).

There is still much to learn regarding the basic behaviour of natural mud. For example, most mud in estuaries and on the continental shelf contains some sand, and yet only now are we beginning to understand the erosion of multi-class sediments (Williamson and Ockenden, 1997). In practice, the future of cohesive sediment research will be best served through an alliance of observation and experimentation with numerical modelling.

VIII. FUTURE LINES OF ENQUIRY

The physical and sedimentological properties of sediment systems were formerly regarded to be the realm of geologists and civil engineers. Holistic studies have shown that the physical dynamics and biology of depositional systems are intimately coupled. The ecology of intertidal systems is receiving increasing attention because of IPCC (Intergovernmental Panel on Climate

Change) predictions of global change scenarios, which will have a major impact on coastal systems. The current state-of-the-art is such that many of the measurements required could now be made, but greater integration between disciplines is still required. Ecology is the study of organisms in their environment and includes both biotic and abiotic components. It is becoming more accepted that, rather than simply responding to physical forces, organisms can strongly mediate their physical environment. It is our belief that the conceptual linkages between the disciplines, perhaps to be expressed in the form of testable hypotheses linked to biosedimentary models and cascade predictions, will provide the most intriguing beneficial results in the future.

ACKNOWLEDGEMENTS

Work referred to in this paper has been supported by Natural Environment Research Council awards GR3/8907, GST/02/759, GST/02/787 and by the EU through award MAS3-CT97–0158 (BIOPTIS). This funding is gratefully acknowledged. Trevor Tolhurst kindly supplied Figures 2 and 3, while Dr S. Hagarthey, Dr G. Underwood and Professor D. Nedwell made valuable comments on the manuscript.

REFERENCES

Ackerman, J.D. and Okubo, A. (1993). Reduced mixing in a marine macrophyte canopy. *Funct. Ecol.* **7**, 305–309.

Admiraal, W. (1984). The ecology of estuarine sediment-inhabiting diatoms. In: *Progress in Phycological Research* (Ed. by F.E. Round and D.J. Chapman), vol. 3, pp. 269–322. Biopress, Bristol.

Al Ani, S., Dyer, K.R. and Huntley, D.A. (1991). Measurement of the influence of salinity on floc density and strength. *Geomar. Lett.* **11**, 154–158.

Allen, J.R.L. (1977). *Physical Processes of Sedimentation.* Allen and Unwin, London.

Aluwihare, L.I., Repeta, D.J. and Chen, R.F. (1997). A major biopolymeric component to dissolved organic carbon in surface seawater. *Nature*, **387**, 166–169.

Amos, C.L. (1995). Siliclastic tidal flats. In: *Geomorphology and Sedimentology of Estuaries. Development in Sedimentology* (Ed. by G.M. Perillo), vol. 53, pp. 273–306. Elsevier, Amsterdam.

Amos, C.L., Wagoner, N.A. and Daborn, G.R. (1988). The influence of subaerial exposure on the bulk properties of fine-grained intertidal sediment from the Minas Basin, Bay of Fundy. *Estuar. Coast. Mar. Sci.* **27**, 1–13.

Amos, C.L., Atkinson, A., Daborn, G.R. and Robertson, A. (1992). The nature of the erosion of fine-grained sediment deposits in the Bay of Fundy. *Mar. Geol.* **108**, 175–196.

Amos, C.L., Feeney, T., Sutherland, T.F. and Luternauer, J.L. (1997). The stability of fine-grained sediments from the Fraser River delta. *Estuar. Coast. Shelf Sci.* **45**, 507–524.

Amos, C.L., Brylinsky, M., Sutherland, T.F., O'Brien, D., Lee, S. and Cramp, A.C. (1998) The stability of a mudflat in the Humber estuary, South Yorkshire, UK. In *Sedimentary Processes in the Intertidal Zone* (Ed. by K.S. Black, D.M.

Paterson and A. Crump), pp. 24–43. Special Publication no. 139. Geological Society, London.

Anderson, F.E. (1983). The northern muddy intertidal, a seasonally changing source of suspended sediment to estuarine waters—a review. *Can. J. Fish. Aquat. Sci.* **40**(Suppl. 1), 143–159.

Ariathurai, R. and Krone, R.B. (1976). Finite element model for cohesive sediment transport. *J. Hydraul. Div. Proc. ASCE* **102**(Hy 3), pp. 323–338.

Best, J.L., Bennett, S., Bridge, J,S and Leeder, M.R. (1997). Turbulence modulation and particle velocities over flat sand beds at low transport rates. *J. Hydraul. Eng.* **123**, 1118–1129.

Black, K.S. (1991). *The Erosion Characteristics of Cohesive Estuarine Sediments, Some In Situ Experiments and Observations.* PhD thesis, University of Wales.

Black, K.S. (1998). Suspended sediment dynamics and bed erosion in the high shore mudflat region of the Humber estuary, UK. *Mar. Pollut. Bull.* **37**, 120–131.

Black, K.S. and Paterson, D.M. (1996). LISP-UK, an holistic approach to the interdisciplinary study of tidal flat sedimentation. *Terra Nova* **8**, 304–308.

Black, K.S., Paterson, D.M. and Cramp, A. (1998). *Sedimentary Processes in the Intertidal Zone.* Special Publication no. 139. Geological Society, London.

Blanchard, G. and Guarini, J.-M. (1999). Temperature effects on microphytobenthic productivity in temperate intertidal mudflats. *Vie Milieu* **48**, 271–284.

Blanchard, G.F., Guarini, J.-M., Richard, P. Gros, P. and Mornet, F. (1996). Quantifying the short-term temperature effect on light-saturated photosynthesis of intertidal microphytobenthos. *Mar. Ecol. Prog. Ser.* **134**, 309–313.

de Boer, P.L. (1981). Mechanical effects of micro-organisms on intertidal bedform migration. *Sedimentology* **28**, 129–132.

Burt, N., Parker, R. and Watts, J. (1997). *Cohesive Sediments.* Wiley Interscience, Chichester.

Casulli, V. and Cheng, R.T. (1992). Semi-implicit finite difference methods for three-dimensional shallow water flow. *Int. J. Num. Methods. Fluids* **15**, 629–648.

Characklis, W.G. and Wilderer, P.A. (1989). *Structure and Function of Biofilms.* John Wiley, Chichester, UK.

Cheng, R.T., Casulli, V. and Gartner, J.W. (1993). Tidal, residual, intertidal mudflat (TRIM) model and its applications to San Francisco Bay, California. *Estuar. Coast. Shelf Sci.* **36**, 235–280.

Cragg, B.A., Parkes, R.J., Fry, J.C., Herbert, R.A., Wimpenny, J.W.T. and Gettliff, J.M. (1990). Bacterial biomass and activity profiles within deep sediment-layers. *Proceedings of the Ocean Drilling Programme* **112**, 607–619.

Daborn, G.R. (1989). *Littoral Investigation of Sediment Properties.* Publication 17. Acadia Centre for Estuarine Research, Nova Scotia.

Daborn, G.R., Amos, C.L., Berlinsky, M., Christian, H., Drapeau, G., Faas, R.W., Grant, J., Long, B., Paterson, D.M., Perillo, G.M.E. and Piccolo, M.C. (1993). An ecological "cascade" effect. Migratory birds affect stability of intertidal sediments. *Limnol. Oceanogr.* **38**, 225–231.

Dade, W.B., Davis, J.D., Nichols, P.D., Nowell, A.R.M., Thistle, D. Trexler, M.B. and White, D.C. (1990). Effects of bacterial exopolymer adhesion on the entrainment of sand. *Geomicrobiol. J.* **8**, 1–16.

Dade, W.B., Self, R.L., Pellerin, N.B., Moffet, A., Jumars, P.A. and Nowell, A.R.M. (1996). The effects of bacteria on the flow behavior of clay seawater suspensions. *J. Sediment. Res.* **66**, 39–42.

Daniel, T.L. (1981). Fish mucus. *In situ* measurements of polymer drag reduction. *Biol. Bull.* **160**, 376–382.

Davey, J.T. and Partridge, V.A. (1998). The macrofaunal communities of the Skeffling muds (Humber estuary), with special reference to bioturbation. In: *Sedimentary Processes in the Intertidal Zone* (Ed. by K.S. Black, D.M. Paterson and A. Cramp), pp. 125–134. Special publication no. 139. Geological Society, London.

Davison, W., Fones, G.R. and Grime, G.W. (1997). Dissolved metals in surface sediment and a microbial mat at 100 mm resolution. *Nature* **387**, 885–888.

Decho, A.W. (1990). Microbial exopolymer secretions in ocean environments, their role(s) in food webs and marine processes. *Oceanogr. Mar. Biol. Annu. Rev.* **28**, 73–153.

Decho, A.W. (1994). Molecular-scale events influencing the macro-scale cohesiveness of exopolymers. In: *Biostabilization of sediments* (Ed. by W.E. Krumbein, D.M. Paterson and L. Stahl), pp. 135–148. BIS, Oldenburg, Germany

Défarge, C., Trichet, J., Jaunet, A., Robert, M., Tribble, J. and Sansone, F.J. (1996). Texture of microbial sediments revealed by cryo-scanning electron microscopy. *J. Sed. Res.* **66**, 935–947.

Delo, E.A. (1988). *Estuarine Muds Manual*. Report SR164. Hydraulics Research Ltd, Wallingford, UK.

Denny, M.W. (1993). *Air and Water: The Biology and Physics of Life's Media*. Princeton University Press, Princetown, New Jersey.

Doerffer, R. and Murphy, D. (1986). Factor analysis and classification of remotely sensed data for monitoring tidal flats. *Helgoland. Meeres.* **43**, 275–293.

Duck, R.W. and McManus, J. (1991). Cohesive sediments in Scottish freshwater lochs and reservoirs. *Geomar. Lett.* **11**, 1127–1131.

Dyer, K.R. (1997). *Estuaries, A Physical Introduction*. Wiley Interscience, Chichester.

Dyer, K.R. (1998). The typology of intertidal mudflats. In: *Sedimentary Processes in the Intertidal Zone* (Ed. by K.S. Black, D.M. Paterson and A. Cramp), pp. 11–24. Special Publication no. 139. Geological Society, London.

Dyke, P. (1996). *Modelling Marine Processes*. Prentice Hall, Englewood Cliffs.

Edelvang, K. and Austen, I. (1997). The temporal variation of flocs and fecal pellets in a tidal channel. *Estuar. Coast. Shelf Sci.* **44**, 361–367.

Edgar, L.A. and Pickett-Heaps, J.D. (1984). Diatom locomotion. In: *Progress in Phycological Research* (Ed. by F.E. Round and D.J. Chapman), vol. 3, pp. 47–88. Biopress, Bristol.

Eisma, D. (1998). *Intertidal Deposits*. CRC Press, Boca Raton.

Faas, R.W., Christian, H.A. and Daborn, G.R. (1992). Biological control of mass properties of surficial sediments, an example from Starr's Point tidal flat, Minas Basin, Bay of Fundy. *Nearshore and Estuarine Cohesive Sediment Dynamics, American Geophysical Union*, vol. 42, pp. 360–377. Springer-Verlag, Berlin.

Fonseca, M.S. (1989). Sediment stabilization by *Halophita decipiens* in comparison to other seagrasses. *Estuar. Coast. Shelf Sci.* **29**, 501–507.

Fonseca, M.S., Fisher, J.S., Zeiman, J.C. and Thayer, G.W. (1982). Influence of the seagrass *Zostera marina* L. on current flow. *Estuar. Coast. Shelf Sci.* **15**, 351–364.

Fonseca, M.S., Zeiman, J.C., Thayer, G.W. and Fisher, J.S. (1983). The role of current velocity in structuring seagrass meadows. *Estuar. Coast. Shelf Sci.* **17**, 367–380.

Frostick, L.E. and McCave, I.N. (1979). Seasonal shifts of sediment within an estuary mediated by algal growth. *Estuar. Coast. Mar. Sci.* **9**, 569–576.

Giller, P.S. and Malmqvist, B. (1998). *The Biology of Streams and Rivers*. Oxford University Press, Oxford.

Glemarec M., Lefaou, Y. and Cuq, F. (1997). Long-term changes of seagrass beds in the Glenan Archipelago (South Brittany). *Oceanol. Acta* **20**, 217–227.

Graham, D.I., James, P.W., Jones, T.E.R., Davies, J.M. and Delo, E.A. (1992). Measurement and prediction of surface shear stress in an annular flume. *J. Hydraul. Eng. ASCE* **118**, 1270–1286.

Gross, T.F. and Dade, W.B. (1991). Suspended sediment storm modelling. *Mar. Geol.* **99**, 343–360.

Guillaumont, B., Bajjouk, T. and Talec, P. (1997). Seaweed and remote sensing, a critical review of sensors and data processing. In: *Progress in Phycological Research* (Ed. by F.E. Round and D.J. Chapman), vol. 12, pp. 213–282. Biopress, Bristol.

Gust, G. (1976). Observations on turbulent drag reduction in a dilute suspension of clay in seawater. *J. Fluid Mech.* **75**, 29–47.

Gust, G. (1988). Skin friction probes for field applications. *J. Geophys. Res.* **93**, 14121–14132.

Gust, G. and Morris, M.J. (1989). Erosion threshold and entrainment rates of undisturbed *in situ* sediments. *J. Coastal Res.*, *Special Issue 5, High Concentration Cohesive Sediment Transport*, pp. 87–100.

Hall, S.J. (1994). Physical disturbance and marine benthic communities, Life in unconsolidated sediments. *Oceanogr. Mar. Biol. Annu. Rev.* **32**, 179–239.

Hamm, L., Chesher, T., Fettweis, M., Pathirana, K.P.P. and Peltier, E. (1997). An intercomparison exercise of cohesive sediment transport numerical models. In: *Cohesive Sediments* (Ed. by N. Burt, R. Parker and J. Watts), pp. 449–458. Wiley Interscience, Chichester.

Hay, S.I., Maitland, T.C. and Paterson, D.M. (1993). The speed of diatom migration through natural and artificial substrata. *Diatom Res.* **8**, 371–384.

Holland, A.F., Zingmark, R.G. and Dean, J.M. (1974). Quantitative evidence concerning the stabilization of sediments by marine benthic diatoms. *Mar. Biol.* **27**, 191–196.

Johnston, J.W. (1943). *Laboratory Investigations of Bedload Transportation and Bed Roughness.* US Dept. Agric. Soil Conserv. Service, Minnesota, SCS-TP-50.

de Jonge, D.J. and Van den Bergs, J. (1987). Experiments on the resuspension of estuarine sediments containing benthic diatoms. *Estuar. Coast. Shelf Sci.* **24**, 725–740.

Koch, E.W. (1994). Hydrodynamics, diffusion-boundary layers and photosythesis of the seagrasses *Thalassia testudinum* and *Cymodocea nodaosa*. *Mar. Biol.* **118**, 767–776.

Koch, E.W. (1996). Hydrodynamics of a shallow *Thalassia testudinum* bed in Florida, USA. In: *Sea Grass Biology. Proceedings of an International Workshop, Rottnesy Island, Australia* (Ed. by J. Kuo, R.C. Phillips, D.I. Walker and H. Kirkman), pp. 105–110. The University of Western Australia, Nedlands.

Komar, P.D. (1978). Boundary layer flow under steady unidrectional currents. In: *Marine Sediment Transport and Environmental Management* (Ed. by D.J. Swift and D.J. Stanley), pp. 107–126. Wiley, New York.

Kravitz, J.M. (1970). Repeatability of three instruments used to examine the undrained shear strength of extremely weak saturated cohesive sediments. *J. Sediment. Petrol.* **40**, 1026–1037.

Krumbein, W.E., Paterson, D.M. and Stal, L.J. (1994). *Biostabilization of Sediments.* Carl von Ossietzky Universität, Oldenburg, Germany.

Kusuda, T., Umita, T. and Awaya, Y. (1982). Erosional process of fine cohesive sediments. *Memoirs of the Faculty of Engineering, Kyushu University* **42**, 317–333.

Lavelle, J.W. and Mofjeld, H.O. (1987). Do critical erosion stresses for incipient motion and erosion really exist? *J. Hydraul. Eng.* **113**, 370–393.

Lee, D., Lick, W. and Kang, S.W. (1981). The entrainment and deposition of fine-grained sediments in Lake Erie. *J. Great Lakes Res.* **7**, 224–233.

Luckenbach, M.W. (1986). Sediment stability around animal tubes. The roles of hydrodynamic processes and biotic activity. *Limnol. Oceanogr.* **31**, 779–787.

Massey, B.S. (1989). *The Mechanics of Fluids*, 6th edn. Von Nostrand Reinhold, New York.

Meadows, P. and Meadows, A. (1991). The environmental impact of burrowing animals and animal burrows. *Zool. Soc. Lond. Symp.* **63**, 157–181.

Meadows, P. and Tait, J. (1989). Modification of sediment permeability and shear strength by two burrowing invertebrates. *Mar. Biol.* **101**, 75–82.

Meadows, P., Tait, J. and Hussain, S.A. (1990). Effects of estuarine infauna on sediment stability and particle sedimentation. *Hydrobiologia* **190**, 263–266.

Meadows, P.S. and Tufail, A. (1986). Bioturbation, microbial activity and sediment properties in an estuarine ecosystem. *Proc. R. Soc. Edinb.* **90**B, 129–142.

Meadows, P., Meadows, A., West, F.J.C., Shand, P.S. and Shaikh, M.A. (1998a). Mussels and mussel beds (*Mytilus edulis*) as stabilisers of sedimentary environments in the intertidal zone. In: *Sedimentary Processes in the Intertidal Zone* (Ed. by K.S. Black, D.M. Paterson and A. Cramp), pp. 331–349. Special publication no. 139. Geological Society, London.

Meadows, P., Murray, J.M.H., Meadows, A., Muir-Wood, D. and West, F.J.C. (1998b). Microscale biogeotechnical differences in intertidal sedimentary ecosystems. In: *Sedimentary Processes in the Intertidal Zone* (Ed. by K.S. Black, D.M. Paterson, and A. Cramp), pp. 349–366. Geological Society, London. Special publication no. 139.

Mehta, A.J. (1991). *Characterisation of Cohesive Soil Bed Surface Erosion, with Special Reference to the Relationship Between Erosion Shear Strength and Bed Density*. University of Florida, Gainesville, UFL/COEL/MP-91/4.

Mehta, A.J. and Partheniades, E. (1982). Resuspension of deposited cohesive sediment beds. *18th Conference on Coastal Engineering*, pp. 1569–1588.

Miles, A. (1994). *Heavy Metals in Estuarine Systems*. PhD Thesis, University of Bristol, Bristol.

Miller, M.C., McCave, I.N. and Komar, P.D. (1977). Threshold of sediment motion under unidirectional currents. *Sedimentology* **24**, 507–527.

Mirtskhoulava, T.E. (1991). Scouring by flowing water of cohesive and non-cohesive beds. *J. Hydraul. Res.* **29**(3), 341–354.

Montague, C.L. (1986). Influence of biota on the erodibility of sediments. In: *Lecture Notes on Coastal and Estuarine Studies* (Ed. by A.J. Mehta), vol. 14, pp. 251–269. Springer-Verlag, Berlin.

Moss, B. (1998). *Ecology of Fresh Waters*. Blackwell, Oxford.

Mullins, J. (1997). Secrets of a perfect skin. *New Sci.* **2065**, 28–31.

Parchure T.M. and Mehta, A.J. (1985). Erosion of soft cohesive sediment deposits. *J. Hydraul. Eng.* **111**, 1308–1326.

Parker, W.R. (1997). On the characterisation of cohesive sediments for transport modelling. In: *Cohesive Sediments* (Ed. by N. Burt, R. Parker and J. Watts). Wiley Interscience, Chichester.

Paterson, D.M. (1994). Microbiological mediation of sediment structure and behaviour. In: *Microbial Mats* (Ed. by P. Caumette and L.J. Stal), NATO ASI series, vol. 35, pp. 97–109. Springer-Verlag, Berlin.

Paterson, D.M. (1995). Biogenic structure of early sediment fabric visualized by low-temperature scanning electron microscopy. *J. Geol. Soc. Lond.* **152**, 131–140.

Paterson, D.M. (1997). Biological mediation of sediment erodibility, ecology and physical dynamics. In: *Cohesive Sediments* (Ed. by N. Burt, R. Parker, and J. Watts), pp. 215–229. Wiley Interscience, Chichester.

Paterson, D.M. (1999). The fine-structure and properties of the sediment interface. In: *The Benthic Boundary Layer, Transport, Processes and Biogeochemistry* (Ed. by B.B. Boudreau and B.B. Jorgensen). Oxford University Press, Oxford.

Paterson, D.M. and Daborn, G.R. (1991). Sediment stabilisation by biological action; significance for coastal engineering. In: *Developments in Coastal Engineering* (Ed. by D.H. Peregrine and J.H. Lovelace), pp. 111–119. University of Bristol Press, Bristol.

Paterson, D.M., Crawford, R.M. and Little, C. (1990). Subaerial exposure and changes in the stability of intertidal estuarine sediments. *Est. Coast. Shelf Sci.* **30**, 541–556.

Paterson, D.M., Yates, M.G., Wiltshire, K.H., McGrorty, S., Miles, A., Eastwood, J.E.A., Blackburn, J. and Davidson, I. (1998). Microbiological mediation of spectral reflectance from intertidal cohesive sediments. *Limnol. Oceanogr.* **43**, 1207–1221.

Peterson, C.G. (1987). Influence of flow regime on development and desiccation response of lotic diatom communities. *Ecology* **68**(4), 946–954.

Pethick, J. (1986). *An Introduction to Coastal Geomorphology*, 2nd edn. Edward Arnold, New York.

Pierce, T.J., Jarman, R.T. and de Turville, C.M. (1970). An experimental study of silt scouring. *Proc. Inst. Civil Eng.* **45**, 231–243.

Pinckney, J. (1994). Development of an irradiance-based ecophysiologial model for inter-tidal benthic microalgal production. In: *Biostabilization of Sediments* (Ed. by W.E. Krumbein, D.M. Paterson, and L.J. Stal), pp. 55–83. BIS, Universität Oldenburg.

Raffaelli, D.G., Raven, J. and Poole, L. (1998). Ecological impact of green macroalgal blooms. *Annu. Rev. Mar. Biol. Oceanogr.* **36**, 97–125.

Rhoads, D.C. and Boyer, L. (1982). The effects of marine benthos on the physical properties of sediments, a successional perspective. In: *Animal–Sediment Relations, The Biogenic Alteration of Sediments* (Ed. by M.J.S. Tevesz and P.L. McCall), pp. 3–52. Plenum Press, New York.

Riethmüller, R., Hakvoort, H.M., Heineke, J.M., Heymann, K., Kühl, H. and Witte, G. (1998). Relating erosion shear stress to tidal flat surface colour. In: *Sedimentary Processes in the Intertidal Zone* (Ed. by K.S. Black, D.M. Paterson and A. Cramp), Special publication no. 139, pp. 283–293. Geological Society, London.

Roegner, C. (1998). Hydrodynamic control of the supply of suspended chlorophyll *a* to infaunal estuarine bivalves. *Estuar. Coast. Shelf Sci.* **47**, 369–384.

Ruddy, G., Turley, C. and Jones, T.E.R. (1998a). Ecological interaction and sediment transport on an intertidal mudflat I. Evidence for a biologically mediated sediment–water interface. In: *Sedimentary Processes in the Intertidal Zone* (Ed. by K.S. Black, D.M. Paterson and A. Cramp), Special Publication no. 139, pp. 135–148. Geological Society, London.

Ruddy, G., Turley, C. and Jones, T.E.R. (1998b). Ecological interaction and sediment transport on an intertidal mudflat. II. An experimental dynamic model of the sediment–water interface. In: *Sedimentary Processes in the Intertidal Zone* (Ed. by K.S. Black, D.M. Paterson and A. Cramp), Special publication no. 139, pp. 149–166. Geological Society, London.

Schroder, H.G.J. and van Es, F.B. (1982). Distribution of bacteria in intertidal sediments of the Ems-Dollard Estuary. *Neth. J. Sea Res.* **14**, 268–287.

Scoffin, T.P. (1970). The trapping and binding of subtidal carbonate sediments by marine vegetation in Bimini Lagoon, Bahamas. *J. Sediment. Petrol.* **40**, 249–273.

Serôdio, J., da Silva, J.M. and Catarino, F. (1997). Non-destructive tracing of migratory rhythms of intertidal benthic microalage using *in vivo* chlorophyll *a* fluorescence. *J. Phycol.* **33**, 542–553.

Shi, Z., Pethick, J.S. and Pye, K. (1995). Flow structure in and above the various heights of a saltmarsh canopy. A laboratory flume study. *J. Coast. Res.* **11**, 1204–1209.

Soulsby, R.L. (1983). The bottom boundary layer of shelf seas. In: *Physical Oceanography of Coastal and Shelf Seas* (Ed. by B. Johns), pp. 189–266. Elsevier, Amsterdam.

Soulsby, R.L. (1999). *Dynamics of Marine Sands*. DETR Publication. Thomas Telford, London.

Stal, L.J. (1995). Transley Review No.84. Physiological ecology of cyanobacteria in microbial mats and other communities. *New Phytol.* **131**, 1–32.

Sutherland, T.F. (1996). *Biostabilisation of Estuarine Subtidal Sediments*. DPhil thesis, Dalhousie University.

Sutherland, T.F., Amos, C.L. and Grant, J. (1998a). The effect of buoyant biofilms on the erodibility of sublittoral sediments of a temperate microtidal estuary. *Limnol. Oceanogr.* **43**, 225–235.

Sutherland, T.F., Grant, J. and Amos, C.L. (1998b). The effect of carbohydrate production by the diatom *Nitzschia curvilineata* on the erodibility of sediment. *Limnol. Oceanogr.* **43**, 65–72.

Taylor, I. and Paterson, D.M. (1998). Microspatial variation in carbohydrate concentrations with depth in the upper millimetres of intertidal cohesive sediments. *Est. Coast. Shelf Sci.* **46**, 359–370.

Taylor, I., Paterson, D.M. and Mehlert, A. (1999). Analysis of carbohydrates from estuarine sediments. *Biogeochemistry* **45**, 303–327.

Teisson, C. (1997). A review of cohesive sediment transport models. In: *Cohesive Sediments* (Ed. by N. Burt, R. Parker and J. Watts), pp. 367–381. Wiley Interscience, Chichester.

Thompson, D. (1961). *On Growth and Form*. Canto Edition, Cambridge (abridged).

Tolhurst, T.J. (1999). *Microbial Mediation of Intertidal Sediment Erosion*. PhD thesis, University of St Andrews.

Underwood, G.J.C. and Smith, D.J. (1998a). Predicting epipelic diatom exopolymer concentrations in intertidal sediments from sediment chlorophyll *a*. *Microb. Ecol.* **35**, 116–125.

Underwood, G.J.C. and Smith, D.J. (1998b). *In situ* measurement of exopolymer production by intertidal epipelic diatom dominated biofilms in the Humber estuary. In: *Sedimentary Processes in the Intertidal Zone* (Ed. by K.S. Black, D.M. Paterson and A. Cramp), Special publication no. 139, pp. 125–134.Geological Society, London.

Unsold, G. and Walger, E. (1987). Critical entrainment conditions of sediment transport. In: *Seawater–sediment Interactions in Coastal Waters: An Interdisciplinary Approach* (Ed. by J. Rumohr, E. Walger and B. Zeitzschel), pp. 328–338. Springer-Verlag, Berlin.

Urrutia, M.B., Iglesias, J.I.P. and Navarro, E. (1997). Feeding behaviour of *Cerastoderma edule* in a turbid environment; physiological adaptations and derived benefit. *Hydrobiologia* **355**, 173–180.

Vogel, S. (1994). *Life in Moving Fluids*, 2nd edn. Academic Press, Princeton.

Wang, P.F., Cheng, R.T., Richter, K., Gross, E.S., Sutton, D. and Gartner, J.W. (1998). Modeling tidal hydrodynamics of San Diego Bay, California. *J. Am. Water Res. Assoc.* **34**, 1123–1140.

Wetherbee, R., Lind, J.L. Burke, J. and Quatrano R.S. (1998). The first kiss, establishment and control of initial adhesion by raphid diatoms. *J. Phycol.* **34**, 9–15.

Widdows, J., Brinsley, M. and Elliot, M. (1996). Use of an *in situ* flume to investigate particle flux at the sediment–water interface in relation to changes in current velocity and macrofauna community structure. In: *LISP Preliminary Results.*

Interim Report (Ed. by K.S. Black and D.M. Paterson), Sediment Ecology Research Group, University of St Andrews, St Andrews.

Widdows, J., Brinsley, M. and Elliot, M. (1998). Use of an *in situ* flume to quantify particle flux (biodeposition rates and sediment erosion) for an intertidal mudflat in relation to current velocity and benthic macrofauna. In: *Sedimentary Processes in the Intertidal Zone* (Ed. by K.S. Black, D.M. Paterson and A. Cramp), Special publication no. 139, pp. 85–97. Geological Society, London.

Wieczorek, S.K. and Todd, C.D. (1998). Inhibition and facilitation of settlement of epifaunal marine invertebrate larvae by microbial biofilm cues. *Biofouling* **12**, 81–118.

Williamson, H. and Ockenden, M.C. (1997). Laboratory and field investigations of mud and sand mixtures. *Adv. Hydrosci. Eng.* **1**, 622–629.

Willows, R.L., Widdows, J. and Wood, R.G. (1998). Influence of an infaunal bivalve on the erosion of an intertidal cohesive sediment; a flume and modelling study. *Estuar. Coast. Shelf Sci.* **43**, 1332–1343.

Wiltshire K.H., Tolhurst, T., Paterson, D.M., Davidson, I. and Gust, G. (1998). Pigment fingerprints as markers of erosion and changes in cohesive sediment surface properties in simulated and natural erosion events. In: *Sedimentary Processes in the Intertidal Zone* (Ed. by K.S. Black, D.M. Paterson, and A. Cramp), Special publication no. 139, pp. 99–114. Geological Society, London.

Wood, R.G., Black, K.S. and Jago, C.F. (1998). Measurement and preliminary modelling of current velocity over an intertidal mudflat, Humber estuary, UK. In: *Sedimentary Processes in the Intertidal Zone* (Ed. by K.S. Black, D.M. Paterson and A. Cramp), Special Publication no. 139, pp. 167–175. Geological Society, London.

Yallop, M.L., de Winder, B., Paterson, D.M. and Stal, L.J. (1994). Comparative structure, primary production and biogenic stabilisation of cohesive and non-cohesive marine sediments inhabited by microphytobenthos. *Est. Coast. Shelf Sci.* **39**, 565–582.

Yates, M.G., Jones, A.R., McGrorty, S. and Goss-Custard, J.D. (1993). The use of satellite imagery to determine the distribution of intertidal surface sediments of The Wash, England. *Estuar. Coast. Shelf Sci.* **36**, 333–344.

Zwarts, L., Blomert, A.M., Spaak, P. and de Vries, B. (1994). Feeding radius, burying depth and siphon size of *Macoma balthica* and *Scrobicularia plana*. *J. Exp. Mar. Biol. Ecol.* **183**, 193–212.

Ecology of Estuarine Macrobenthos

P.M.J. HERMAN, J.J. MIDDELBURG, J. VAN DE KOPPEL AND C.H.R. HEIP

I. SUMMARY

Macrobenthos is an important component of estuarine ecosystems. Based on a cross-system comparison, we show that estuarine macrobenthos may directly process a significant portion of the system-wide primary production, and that estuarine macrobenthic biomass may be predicted from primary production data. At large scales, food may be the prime limiting factor for benthic biomass. Depending on the characteristics of the system, grazing by benthic suspension feeders may be the most important factor determining system dynamics.

The detailed spatial patterns and dynamics resulting from feeding interactions are discussed separately for suspension feeders and deposit feeders. The theory on local seston depletion and its consequences for spatial distribution of suspension feeders is compared critically with observed patterns of spatial distribution. It is concluded that additional non-linear interactions between

ADVANCES IN ECOLOGICAL RESEARCH VOL. 29
ISBN 0–12–013929–4

biomass of the benthos and water currents must exist to explain the observed patterns.

The relation between organic matter deposition fluxes and benthic community structure is discussed in the framework of the classical Pearson–Rosenberg paradigm. The importance of organic matter quality, in addition to quantity, is stressed. A simple model framework to investigate the relation between community structure and quantity of organic flux is proposed.

Internal dynamics of benthic food webs are characterized by a high degree of omnivory (feeding on different trophic levels). This feature is contrasted with published data on food webs in other systems. It is hypothesized that the high quality of marine detritus (compared with terrestrial detritus) is the prime factor explaining the differences. Since theoretical studies suggest that omnivory destabilizes food webs, a number of stabilizing mechanisms in benthic food webs are discussed. Problems and mechanisms that could be explored fruitfully in theoretical studies and field comparisons are identified.

II. INTRODUCTION

Estuaries, in general, are shallow, open and dynamic systems. The small volume of water per square metre of sediment surface, the presence of intertidal flats and very shallow subtidal areas, and the generally well-mixed nature of the water column are physical conditions that intensify the exchange of matter and energy between the water column and the sediment system. In deep, stratified marine systems, the benthos essentially receives the deposits of the pelagic export production and returns nutrients to the deeper water layers. It may take years before these nutrients can be utilized by photoautotrophs again. In estuarine systems, however, pelagic–benthic links are not only quantitatively more important, but also qualitatively different, since the benthos can be responsible for direct grazing on live phytoplankton, recycling of nutrients occurs within a growing season, and there is important *in situ* benthic primary production. Moreover, it is not possible to understand the dynamics, spatial distribution or trophic composition of the benthos without taking into account the general patterns of energy and matter flow in estuarine systems.

According to their size, benthic organisms are classified as microbenthos, meiobenthos and macrobenthos. Microbenthos (< 32 μm) is composed of bacteria and Protista. Many bacterial species are simultaneously present in sediments, in guilds that are metabolically dependent on one another. Total bacterial density is typically in the order of 10^9 cells per ml of sediment. The meiofauna (from 32 μm to 1 mm) is usually dominated by Nematoda, with densities in the order 10^6 m^{-2} (Heip *et al.*, 1995). Further dominant groups are Copepoda, Ostracoda, Turbellaria and Foraminifera. The macrofauna (> 1 mm) is composed of Mollusca, Polychaeta, Echinodermata, Crustacea and other groups. In estuaries typical densities are in the order of 10^4 m^{-2}, and biomass is usually in the range 1–100 g AFDW m^{-2} (AFDW = ash-free dry weight) (Heip *et al.*, 1995). Variability in macrofauna biomass of two to three orders of magnitude is typically found within most estuarine systems.

Extensive research has been conducted on the feeding habits of many macrofaunal species. Classifications of macrofaunal species into feeding groups have been made (e.g. Fauchald and Jumars, 1979) but considerable debate remains about these classifications. For many species, sufficient auto-ecological information is lacking. However, the accumulation of auto-ecological information for some species has rendered proper feeding group classification an even more difficult task. In this review, we will argue that the qualitative composition of the food of benthic animals fosters omnivory and generalism as a general strategy, precluding the use of a too narrow feeding group classification. We will, however, make a general distinction between two major feeding types: (1) *suspension feeders*, which filter their food directly from the water column, and (2) *deposit feeders*, which depend on the physical deposition of food particles on to the sediment surface, and (for "deep-deposit feeders") the subsequent incorporation of the food particles into the sediment matrix. Water and sediment have dramatically different properties for animals that want to extract food particles from them, and because of this we expect different processes to regulate the occurrence and dynamics of both groups. Again, individual species may show varying degrees of generalism, as is the case for "interface feeding" spionid polychaetes (Dauer *et al.*, 1981; Taghon and Greene, 1992).

In this review, we argue that the benthos is responsible for a large and predictable part of total system metabolism in estuaries, and that food is an important limiting and structuring factor in benthic communities. The particular characteristics of the sediment surface as a boundary layer and of the sediment as a habitat have consequences for the dynamics and spatial distribution of benthic communities, which we try to generalize. Our hope is that increased insight into the physical, geochemical and biological processes controlling benthic communities will facilitate the interpretation of community characteristics, spatial distribution and temporal evolution of benthic communities.

Macrofauna is an important food resource for epibenthic crustaceans, fish and birds. Humans also harvest many species of shellfish and crustaceans. Evaluation of the consequences of anthropogenically induced changes in a system will likely include the possible responses of the benthos. Moreover, the benthos is often monitored as an indicator of possible changes in the system. Being fixed in place and relatively long lived, the benthos integrates environmental influences at a particular place over a relatively long timespan. For the purposes of predictive modelling, and also for the *post hoc* interpretation of changed patterns of occurrence and altered community composition in monitoring studies, insight into the processes governing the temporal and spatial dynamics of benthic populations is needed. For this insight, we believe a consideration of the dynamics of the interaction between the benthos and its food is essential.

A. The Importance of System Dynamics for Macrobenthos

Heip *et al.* (1995), extending previous data compilations by Nixon (1981), Dollar *et al.* (1991) and Kemp *et al.* (1992), presented a general relation between system-averaged water depth and the fraction of total estuarine primary production mineralized in the sediment. The relation is given by the equation

$$\log_{10}(F) = 1.6 - 0.0146H \tag{1}$$

where F is the percentage of the system mineralized in the sediment and H denotes water depth (m).

The factor -0.0146 has the dimension m^{-1} and was interpreted by Heip *et al.* (1995) as representing the multiplicative effect of two factors: a degradability constant (time^{-1}) and an effective sinking speed (m time^{-1}). An estimate of the latter, taking degradability constants around 20 year^{-1} for fresh phytoplanktonic material (Westrich and Berner, 1984; Middelburg, 1989), is in the order or 2–3 m day^{-1}. This is an estimate for net effective sinking speed, taking into account the effects of resuspension cycles.

Drawing on the relation of eqn.(1), an order-of-magnitude estimate for sediment mineralization in shallow estuaries can be made. The fraction mineralized in shallow systems (approximately 10 m) is around 60% of the total primary production. Taking an estimate for yearly primary production of about 200 g C m^{-2} year^{-1}, one would arrive at approximately 120 g C m^{-2} year^{-1} for sediment mineralization. Direct observations on benthic mineralization rates show considerable variability. Heip *et al.* (1995) summarized values for 14 estuarine systems, estimated either from CO_2 production rates or O_2 consumption rates. Excluding the deep St Lawrence estuary, these values vary between 31 and 392 g C m^{-2} year^{-1}, with a median of 120 g C m^{-2} year^{-1}. Adding recently published values (Kristensen, 1993; Roden *et al.*, 1995; Middelburg *et al.*, 1996; Caffrey *et al.*, 1998; Rocha, 1998) and our own unpublished data for CO_2 production on a tidal flat in the Westerschelde, we obtain a total of 33 sites, with the frequency distribution shown in Figure 1A. The median value for this data set, which is probably biased by intertidal Westerschelde sediments, is 181 g C m^{-2} year^{-1}. The higher values in the data set typically come from estuaries (e.g. Westerschelde) with high exogenous carbon input.

It is difficult to establish which fraction of the carbon mineralized in estuarine sediments is directly attributable to the metabolism of the macrobenthic animals.

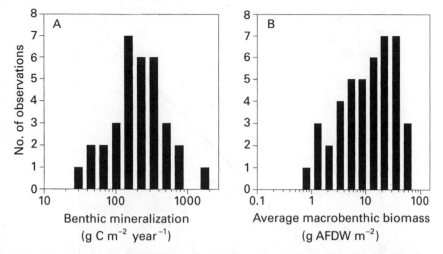

Fig. 1. Frequency distribution of observations of (A) total sediment mineralization (measured as either total inorganic carbon production or oxygen consumption) and (B) system-averaged benthic biomass. Most observations are listed in Tables 23 and 29 of Heip *et al.* (1995); for (A) additional data have been taken into account (see text).

At a system scale, one would need concurrent estimates of sediment respiration and of benthic production from a representative sampling of the different strata in the system. From the median figures on macrobenthic biomass and benthic mineralization it can, however, be qualitatively deduced that this fraction is not insignificant. Taking a median biomass of 15 g AFDW m^{-2} (Figure 1B), a carbon : AFDW ratio of 0.5, an annual production/biomass ratio (P/B) of 2 year^{-1} (Heip *et al.*, 1995) and a respiration : production ratio of 1.8 (Banse and Mosher, 1980), one would estimate a typical macrobenthic respiration rate in estuaries of the order of 25 g C m^{-2} year^{-1}, which constitutes around 15–20% of the median total respiration of the sediment. This estimate at the system scale is very similar to that based on concurrent biomass and sediment respiration data from single stations (Dauwe *et al.*, 1999b).

Parsons *et al.* (1977) showed a dependence of system-averaged benthic biomass on the magnitude of the spring phytoplankton bloom. This relation strongly suggests dependence between benthic biomass and pelagic primary productivity. In Figure 2 we assemble data on benthic biomass and system productivity. For the systems in the SW Netherlands (Westerschelde, Grevelingen, Oosterschelde and Veerse Meer; see Heip (1989) and Nienhuis (1992) for a description of the systems), benthic biomass was averaged from the large database available at the Netherlands Institute of Ecology. For other systems we relied on published estimates. Both our own data and the published results are not necessarily true system biomass averages, because

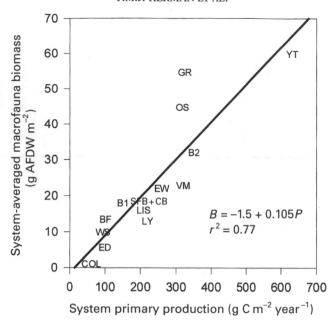

Fig. 2. Relation between system-averaged macrobenthic biomass and primary productivity of shallow well-mixed estuarine systems. Data points are indicated by the abbreviation of the name of the system. The regression line is a predictive linear least-squares line. System abbreviations and sources are: YT, Ythan estuary (Baird and Milne, 1981); GR, Grevelingen (benthic data from the authors' own database; primary production from de Vries, 1984); OS, Oosterschelde (benthic data from the authors' own database; primary production from Wetsteyn and Kromkamp, 1994); B1, Balgzand (Dutch Wadden Sea) in the 1970s (Beukema and Cadée, 1997); B2, Balgzand (Dutch Wadden Sea) in the 1980s (Beukema and Cadée, 1997); VM, Veerse Meer (benthic data from the authors' own database; primary production from Nienhuis, 1992); EW, Ems estuary near the Wadden Sea (benthic biomass from Meire *et al.*, 1991; primary production from Baretta and Ruardij, 1988); ED, Ems estuary, inner part ("Dollard") (benthic biomass from Meire *et al.*, 1991; primary production from Baretta and Ruardij, 1988); SFB, San Francisco Bay (benthic biomass from Nichols, 1977; primary production from Cole and Cloern, 1984); LY, Lynher estuary (Warwick and Price, 1975); WS, Westerschelde (benthic data from the authors' own database; primary production from Soetaert *et al.*, 1994); BF, Bay of Fundy (Wildish *et al.*, 1986); COL, Columbia River estuary (Small *et al.*, 1990); LIS, Long Island Sound (benthic biomass from Parsons *et al.*, 1977; primary production from Riley, 1956, cited in Sun *et al.*, 1994); CB, Chesapeake Bay (benthic biomass from Dauer, 1993; primary production from Harding *et al.*, 1986).

of bias in the sampling scheme. The same may be true for the primary production data, which in principle are net annual primary production figures including microphytobenthos and macrophytobenthos production, but in practice may more or less deviate from this ideal, depending on the data available. The resulting relation suggests that for these shallow estuarine systems between 5% and 25% of the annual primary production is consumed by macrobenthos respiration. On a system-averaged basis, suspension feeders are often the dominant component (with respect to biomass) of estuarine benthic assemblages. Their contribution to benthic biomass is 41%, 74%, 82% and 81% for the Westerschelde, Oosterschelde, Veerse Meer and Grevelingen data sets respectively. Suspension feeders are, both in absolute and relative terms, less dominant in the turbid Westerschelde, Ems-Dollard and Columbia River estuaries, all three of which are at the low end of the production spectrum. However, suspension feeders contribute 86% to the total benthic production in the Bay of Fundy, which is not very much more productive than the Westerschelde (Wildish *et al.*, 1986).

Suspension feeders typically occur in much higher local biomass than deposit feeders, which are much more equally spread over (generally lower) biomass classes (Figure 3). In the Oosterschelde, over 75% of the total biomass of suspension feeders was found at local biomass values of more than

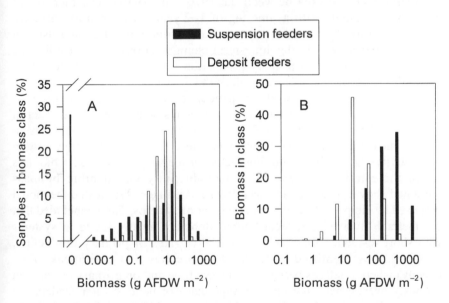

Fig. 3. (A) Frequency distribution of suspension and deposit feeder biomass in the Oosterschelde ($n = 2368$, data collected between 1990 and 1998) over logarithmic biomass classes. Suspension feeders are absent from a large number of samples and occur more frequently in very high local biomass classes. (B) Percentage of the total recorded biomass found at local biomass values in logarithmic classes. More than 75% of the recorded suspension feeder biomass is found at a local suspension feeder biomass of over 100 g. AFDW m^{-2}, whereas the largest fraction of deposit feeder biomass is within the 10–100-g AFDW m^{-2} class.

100 g AFDW m^{-2} (Figure 3B), whereas 53% of the samples contained less than 1 g AFDW m^{-2} of suspension feeders (Figure 3A) and 28% of the samples contained no suspension feeders at all (Figure 3A). For deposit feeders, only 15% of the system biomass was found at local biomass values of more than 100 g AFDW m^{-2} (Figure 3B) but only 20% of the samples contained less than 1 g AFDW m^{-2} of deposit feeders and there were virtually no samples without deposit feeders (Figure 3A).

Beukema and Cadée (1997) provided direct observational evidence for limitation of system-averaged benthic biomass by estuarine productivity. They showed that in the Balgzand area (Dutch section of the Wadden Sea) a substantial increase of pelagic primary production between the 1970s and the 1980s (nearly a doubling) was followed by a nearly proportional increase of system-averaged benthic biomass. The proportional increase was equally large for suspension and deposit feeders.

Field evidence for a direct coupling between benthic biomass and pelagic primary production was also produced by Josefson *et al.* (1993), who showed an increase in biomass and a change in community structure in the Skaggerak-Kattegat area between the 1970s and 1980s. The increase in biomass correlated (with a time lag of 1–2 years) with runoff variables known to be directly related to phytoplankton primary production. Austen *et al.* (1991) described similar long-term changes correlated to (indirectly determined) increases in primary production for two widely spaced sites, one in the Western North Sea and one in the Skagerrak.

Heip *et al.* (1995) showed a dependence of total system biomass of commercial benthic suspension feeders on the residence time of the water in the system. This pattern was modelled assuming that residence time was an inverse measure of food exchange with the coastal sea, and that system productivity was the basic limiting factor for the suspension feeder biomass (as in the relation between system productivity and benthic biomass presented above). Dame and Prins (1998) and Dame (1996) tested the model proposed by Heip *et al.* (1995) for a number of systems. They concluded that commercial bivalve filter-feeder populations are mostly found in systems with short residence times and high relative rates of primary production. However, they identified systems (Delaware Bay and Chesapeake Bay, notably) where relations based on residence time and on primary production seriously overestimate the relative importance of benthic filter feeders. As argued by Dame and Prins (1998), this may have historical reasons in over-fishing or eutrophication. In detailed ecosystem models of Marennes-Oléron and Carlingford Lough, Bacher *et al.* (1998) demonstrated the importance of physical transport and advection of food into coastal bays for the carrying capacity of bivalve filter feeders. In Marennes-Oléron, which is open to the import of phytoplankton from coastal waters, the carrying capacity is far higher than in Carlingford Lough.

The general conclusions are that a substantial fraction of the carbon flow in estuarine systems passes through macrobenthic populations, and that therefore macrobenthic populations at a system level may be limited by food fluxes to the sediments.

B. The Importance of Macrobenthos for System Dynamics

Deposit feeders transport particles and fluid during feeding, burrowing, tube construction and irrigation activity (Rhoads, 1974; Aller, 1988; Aller and Aller, 1998). By enhancing transport of (labile) particulate organic carbon to deeper sediment layers, these organisms stimulate anaerobic degradation processes and so can affect the form and rate at which metabolites are returned to the water column. Bioturbation causes an upward movement of reduced components such as sulfides, as a consequence of which rates of sedimentary oxygen uptake are higher and not necessarily coupled directly to organic carbon oxidation (Aller, 1994). As a result, there may be temporal uncoupling between benthic oxygen uptake and nutrient regeneration. Enhanced exchange between overlying water and pore fluids has conse-quences for sedimentary oxygen uptake, the extent that aerobic processes contribute to organic matter oxidation, the efficiency of reoxidation of reduced substances and the fraction of nutrients produced that escapes the sediments (Aller, 1988; Aller and Aller, 1998). The effect of bioturbating infauna on the benthic cycling of nitrogen, in particular the coupling between nitrification and denitrification, has been studied intensively (Aller, 1988; Pelegrí et al., 1994; Rysgaard et al., 1995). This loss of (fixed) nitrogen from estuaries due to benthic denitrification directly influences the availability of this limiting nutrient for phytoplankton production.

The importance of macrobenthic suspension feeders for total system dynamics has been the subject of a number of model studies. Officer et al. (1982) used an idealized Lotka–Volterra modelling approach to show that equilibrium between benthic suspension feeders and phytoplankton is reached when the grazing time-scale (τ_g) equals the production time-scale (τ_p):

$$\tau_g = \frac{H}{RB} = \tau_p = \frac{1}{\mu_p} \qquad (2)$$

in which H is system water depth (m), R is specific clearance rate of the benthos (m^3 g^{-1} day^{-1}), B is benthic biomass (g m^{-2}), and μ_p is specific phytoplankton net growth rate (day^{-1}). For this equilibrium to be reached at low phytoplankton biomass (a few milligrams of chlorophyll per cubic metre), the system must be shallow (a few metres) and benthic biomass relatively high (in the order of 100 g total fresh weight m^{-2}). This is a realistic situation for many well-mixed tidal estuaries. Using an ecosystem simulation model, Herman and Scholten (1990) showed how grazing by benthic suspension feeders effectively controls phytoplankton in the Oosterschelde. Herman (1993) extended the argument to show that the open nature of estuarine systems is essential: when nutrients accumulate in a system without effective removal mechanisms, one would expect ever-increasing levels of phytoplankton under eutrophication, despite high grazing pressure. Koseff *et al.* (1993) modelled the conditions under which phytoplankton can develop a bloom in the presence of benthic suspension feeders. In contrast to Officer *et al.* (1982), they explicitly included vertical mixing in the water column as a critical process. In non-stratified waters the possibility for a bloom depends in a non-linear way on two non-dimensional ratios of time scales: α' and K'. α' is the ratio of the production time-scale (τ_p) to the grazing time-scale (τ_g), essentially defined as in eqn. (2), but with vertical averaging of the phytoplankton growth rate, as this model explicitly resolves the vertical dimension. K' is defined as the ratio of the production time-scale to the vertical (turbulent) mixing time-scale H^2/K, where H is water depth and K is the turbulent diffusion coefficient. The possibility for the development of a bloom depends on the details of the model versions, but is qualitatively different in only two cases: when vertical sinking velocity exceeds a certain threshold, or when stratification develops. The most general case for a non-stratified system, with a vertically variable turbulent mixing coefficient depending on tidally varying current velocity and a moderate phytoplankton sinking velocity, is illustrated in Figure 4. Benthic grazing can limit bloom development provided that both K' values are high (vigorous vertical mixing and shallow water) and grazing is sufficiently intense compared with primary production (high α').

Lucas *et al.* (1998) investigated the role of stratification for this type of model in much more detail. With respect to the influence of benthic grazing on the occurrence of phytoplankton blooms, their results are qualitatively similar to those of Koseff *et al.* (1993): as soon as even a mild form of permanent vertical stratification affects the system, benthic grazing rates are no longer a major factor to be considered for the prediction of blooms. Benthic grazing is especially important in shallow well-mixed systems, although under favourable conditions of light and stratification, it may also be significant in deeper water columns of the order of 15 m.

The mechanisms relating macrobenthic energy flow to system productivity operate on a much smaller temporal and spatial scale than the scale of the system at which the correlation is observed. How do macrobenthic animals sequester their food, how do they influence the total flux rate of organic matter to the sediment, and what determines their share of the resources? These questions require the study of spatial and temporal distributions of physical, chemical and biological variables within the system. Numerous factors potentially influence the spatial distribution of benthic animals in estuarine systems (see Dame, 1996, for an extensive review on suspension feeders). We will concentrate our discussion here on food. If

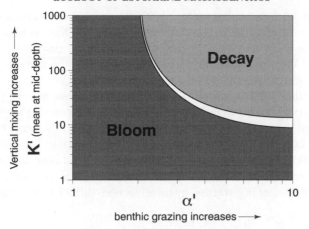

Fig. 4. Influence of (non-dimensionalized) benthic grazing and vertical mixing on the possibilities for the development of a bloom in a shallow tidal estuary. α' is the ratio of the grazing to the phytoplankton growth time scales. K' is the ratio of the production time-scale to the vertical mixing time-scale (see text for details). This figure summarizes model runs with tidally varying current velocity and vertically varying turbulent mixing. Parameter regions where blooms consistently develop are separated from regions where the phytoplankton decays by a transition zone characterized by oscillations in time. Adapted from Koseff *et al.* (1993).

food limitation sets the constraints for the system-integrated biomass of benthic animals, then a number of predictions can be made about the spatial distribution pattern of macrobenthos, based on the characteristics of the animals and of estuarine pelagic–benthic exchange.

In principle, the situation is different for benthic suspension feeders and deposit feeders. Suspension feeders have direct active access to pelagic production. In well-mixed estuaries, all suspension feeders are consequently to a certain extent in competition with one another (and with the pelagic herbivores), since their food resource is continuously redistributed over space, and food not taken by one animal remains available to the others. In contrast, deposit feeders are restricted to the food items deposited in a limited volume of sediment, even though they can also influence the sedimentation flux through their bioturbation (see section IV.B).

III.   SPATIAL DISTRIBUTION AND DYNAMICS OF SUSPENSION FEEDERS

A. The Sediment Surface as a Boundary Layer

Due to the mixing and the general availability of the food resources, one can assume that the biomass of suspension feeders will be distributed over space according to the suitability of the sites for capturing food. The best studied aspect (see Wildish and Kristmanson, 1997, for a recent review) of suitability is the relation with water flow velocity and, linked with that, vertical turbulent mixing of the water column. The filtration capacity of suspension feeders is high (several litres per hour for a 1g AFDW animal), the animals can occur locally in high biomass (several hundreds of g AFDW m^{-2}), and they are restricted to filtering the lowermost water layers. Consequently, the probability of refiltering previously filtered water increases with increasing local biomass. This may lead to local food depletion in the lower part of the water column, an effect that is counteracted only by vertical turbulent mixing. The latter effect is current dependent, since turbulence is generated by shear near the bottom.

In one (vertical) dimension, the concentration of a substance, say food particles, in the water is governed by an advection–diffusion–reaction equation

$$\frac{\delta C}{\delta t} = \frac{\delta}{\delta z}\left(K_z\frac{\delta C}{\delta z}\right) - \frac{\delta}{\delta z}(wC) + \text{sources} - \text{sinks} \tag{3}$$

where C is the concentration of the substance, t is time, z is the vertical coordinate, K_z is the depth-dependent turbulent mixing coefficient, w is the sinking rate, and sources and sinks are biological production or consumption reactions.

This equation can readily be extended to two (e.g. Verhagen, 1986; Fréchette *et al.*, 1989) or three dimensions, but we will concentrate here on the vertical dimension only. Eqn. (3) can be solved analytically or numerically to yield time- and depth-dependent values of the concentration, given appropriate boundary conditions and, for time-dependent solutions, initial conditions (e.g. Crank, 1986; Boudreau, 1997).

The depth-dependent values of K_z are of prime importance for the flux of food particles to benthic consumers. Mixing is generated by the generation of turbulent eddies from the shear (vertical gradient) near the bottom. The eddy diffusivity is characterized by a turbulent velocity scale and a length scale. In estuaries, the friction velocity u^* and the water depth H are used for these parameters. Using the logarithmic law velocity profile for flows with only one source of turbulence (the bottom shear), the vertical mixing coefficient for momentum and mass transport is then given by

$$K_z = \kappa u^* z\left(1 - \frac{z}{H}\right) \tag{4}$$

where κ is the von Karman constant (approximately 0.4), H is water depth and u^* is the friction velocity. This friction velocity is given by the relation

$$u(z) = \frac{u^*}{\kappa}\ln\left(\frac{z}{z_0}\right) \tag{5}$$

which describes the logarithmic profile of current velocity $u(z)$ as a function of distance to the bed. The parameter z_0, the roughness height, corresponds to the height above the bed where the extrapolated current velocity drops to zero. For rough turbulent flow, z_0 is determined by the height k_s of the roughness elements on the bed, with the approximate relationship $z_0 \approx k_s/30$. The friction velocity u^* is related to the bottom shear stress, a tangential force per unit area responsible (when a critical threshold is exceeded) for the movement of particles on the bed, by

$$\tau_0 = \rho(u^*)^2 \tag{6}$$

where ρ is the density of water. Since for every height in the water column $u(z)$ is proportional to u^*, with a proportionality coefficient dependent on height z (eqn (5)), the bottom shear stress can also be expressed as a quadratic function of $u(z)$ at a particular height. Conventionally, a height of 100 cm is used, with the relation

$$\tau_0 = \rho C_{100} (u_{100})^2 \tag{7}$$

where u_{100} is the velocity at 100 cm from the bed, and C_{100} is the drag coefficient at 100 cm. From the equations above, it can easily be derived that

$$C_{100} = \frac{\kappa}{\ln(100/z_0)} \tag{4}$$

B. Seston Depletion Above Suspension Feeder Beds

Wildish and Kristmanson (1979) have solved eqn. (3), while assuming steady state, neglecting particle sinking and production/consumption terms (except for the consumption of food by the benthos at the lower boundary) and adopting an average mixing coefficient (K_z) linearly related to the free-streaming current velocity. Under these assumptions, and taking typical values for the coefficients, the vertical flux of food (NA) at the lower boundary is given by

$$NA = \gamma u_{\text{free}} (C_0 - C') \tag{9}$$

where γ is a dimensionless hydrodynamic parameter related to bottom roughness, for which Wildish and Kristmanson give a value of 0.003, u_{free} is the current velocity at the top of the benthic boundary layer in m h^{-1}, C_0 is the upstream food concentration in g m^{-3} and C' is the effective food concentration at the intake in g m^{-3}.

This vertical flux of food is actually the food flux to the benthic grazers at the lower boundary of the model. It is the food that is actively filtered from the water. It can, therefore, also be expressed in terms of the suspension feeder biomass, its clearance rate and the effective food concentration

$$NA = BR\alpha C \tag{10}$$

where NA is the food flux in g m^{-2} h^{-1}, B is the biomass of suspension feeders in g m^{-2}, R is the clearance rate per gram of biomass in m^3 g^{-1} h^{-1}, and α is the filtration efficiency (dimensionless). From eqns (9) and (10), the following equation can be derived.

$$\frac{(C_0 - C')}{C'} = \frac{BR\alpha}{\gamma u_{\text{free}}} \tag{11}$$

in which the right-hand side is a dimensionless measure called the seston depletion index (SDI; Wildish and Kristmanson, 1997). It is the ratio of the filtration capacity to the turbulent mixing intensity. When it exceeds a certain threshold value, depletion effects can be assumed to occur.

Eqn (11) can be considered as a first-order approximation to the problem of seston depletion by suspension feeders. Verhagen (1986) and Fréchette *et al.* (1989) have presented more elaborate analytical and numerical solutions, including solutions of the two-dimensional problem. They show how the depletion effect gradually builds up as the current flows over a dense bed of suspension feeders, leading to relatively more favourable conditions at the edge of the bed. After several metres in the bed, the vertical distribution of food is reasonably described by eqn (11) (Wildish and Kristmanson, 1997).

Assuming that a minimum (fixed) value of C' is needed for growth and survival of the organisms, several conclusions can be drawn from this equation. For a fixed C' and a fixed C_0, the maximum allowable filtration capacity of the animals will increase linearly with current velocity if the roughness of the bottom is uniform. It will increase more steeply over rougher surfaces, which is important because the animals themselves may contribute considerably to the roughness (Green *et al.*, 1998; O'Riordan *et al.*, 1995). In systems with different "background" food concentrations' C_0, the maximum filtration capacity will be higher in the more productive systems. With species differing in their specific filtration capacity, those that filter more actively will be limited at a lower biomass than the less active ones.

In estuarine systems, omission of the advective term in eqn (3) (i.e. the term representing the sinking of the food particles) is not always warranted. Figure 5 illustrates the differences in sedimentation and resuspension patterns of chlorophyll between a tidal flat in the Westerschelde (total suspended matter (SPM) 80–250 mg l^{-1}) and one in the

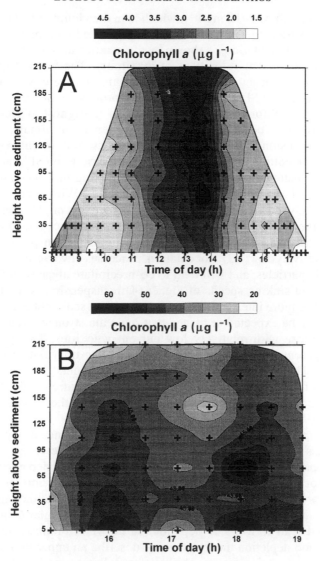

Fig. 5. Concentration contours of chlorophyll *a* in suspended matter during a tidal cycle over tidal flats in two estuaries differing in the total suspended matter concentration. Concentrations are expressed as a function of time of day (abscissa) and height above the sediment (ordinate). Crosses denote positions of samples in space and time. (A) Tidal cycle on a flat in the Oosterschelde, in the immediate vicinity of a mussel bed. (B) Tidal cycle on the Molenplaat (Westerschelde) near the centre of the flat. See text for discussion.

Oosterschelde (SPM 2–12 mg l^{-1}), which is in the vicinity of a large mussel bed. In the Westerschelde, the pattern is dominated by a pattern of resuspension and sedimentation, as shown by the maxima at flooding and ebbing tide, when the current velocity is at maximum. The vertical gradient is towards higher concentrations near the bottom, and there is a good correlation between total SPM and suspended chlorophyll *a* in the water. In the Oosterschelde, chlorophyll *a* concentration is highest at flood (when grazing pressure per unit volume of water is minimal), sedimentation–resuspension intensity is limited, concentration of chlorophyll *a* is lower near the bottom and there is no correlation between SPM and chlorophyll *a*. The pattern can be interpreted when taking into account the relatively high SPM content of the Westerschelde. With increasing SPM content of the water, a number of correlated processes might be expected to occur: (1) the productivity of the pelagic algae decreases due to light limitation; (2) as a consequence total system biomass of suspension feeders will be relatively low; (3) flocculation processes involve both inorganic and organic particles, and will tend to co-precipitate algae with SPM, thus increasing net sinking speeds of algae; (4) resuspension–resedimentation cycles will be more intense; and (5) highest above-sediment concentrations of algae may be expected at places that have the strongest sedimentation rates, i.e. those with relatively low current velocities. This pattern was indeed apparent in the Westerschelde data on the tidal flat "Molenplaat" (Figure 6). The highest biomass of suspension feeders (in fact, nearly all suspension feeders found on the flat) was in the centre of the flat, where maximum tidal current velocity was minimal, as was the maximum bottom shear stress over a tidal period.

Consideration of food depletion sets a current-dependent upper limit to the biomass of suspension feeders, as it essentially describes a negative density-dependent effect that is spatially variable. If food depletion were the major factor governing the spatial distribution of suspension feeders, they would be predicted to follow an ideal-free distribution (Sutherland, 1996), i.e. they would be distributed such that everywhere a similar C' is reached. This would lead to a linear dependence between biomass and current velocity. Food depletion theory does not describe an upper limit of current velocity for suspension feeders. Although the direct negative physiological effects of too high currents have been described (Wildish and Kristmanson, 1993), such a description is not available for many species of suspension feeders and its generality, therefore, remains to be tested. It seems logical, however, to assume that stability of the sediment bed, which becomes more vulnerable to resuspension as current velocity increases, sets the upper limit. In fact the critical parameter for the resuspension of particles is the bottom shear stress (Hall, 1994), which increases with the square of the current speed (eqn (6)).

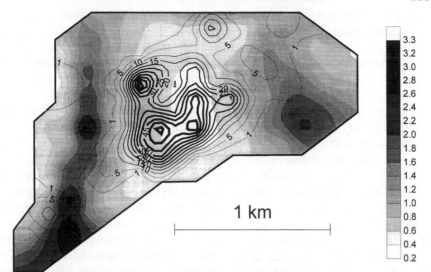

Fig. 6. Spatial distribution of suspension feeders (mainly *Cerastoderma edule* and *Mya arenaria*) on the Molenplaat, Westerschelde, as a function of bottom shear stress. Grey shades depict contours of modelled maximum bottom shear stress during a tidal cycle (Pa), isolines depict biomass (g AFDW m^{-2}) of suspension feeders. Suspension feeders are concentrated in areas of low shear stress on this flat.

C. Seston Depletion and Non-linear Density–current Relations

Is food depletion theory, combined with erosion of animals from the bed, setting an upper limit to current velocity, sufficient to predict the spatial distribution of benthic suspension feeders? Most field validations have been made on reef-building species (e.g. mussels and oysters). Extensive reviews of these studies are given in Wildish and Kristmanson (1997) and Dame (1993, 1996). These animals typically occur in dense patches, up to more than 1 kg of AFDW per square metre, which naturally leads to measurable food depletion in nature, but the occurrence of these patches is at odds with predictions from seston depletion equations. In fact, maximum density is predicted to vary linearly with average current velocity, which in estuarine systems has a range of about one order of magnitude only. This poses a serious problem of interpretation of field results. Wildish and Kristmanson (1993), for example, found an empirical relationship between the logarithm of suspension feeder production and average water column current velocity, which implies an order-of-magnitude increase of production for an increase of current velocity by only 0.2 m s^{-1}.

Beukema and Cadée (1997) showed that the increase of benthic biomass with increased primary production in the Balgzand area did not show a homogeneous spatial distribution. The highest increase in benthic biomass was realized in plots with an intermediate current velocity (measured as an intermediate silt content of the sediment), whereas in low current velocity plots (very fine sediments) or in high current velocity plots (very coarse sediments) no real increase of the (already small) biomass was recorded. This suggests a highly non-linear relation between current velocity and growth potential for the populations.

In order to resolve this problem of non-linearity of the biomass–velocity relation, it may first be noted that suspension-feeding animals may considerably increase the roughness of the sediment surface. Green *et al.* (1998) describe how horse mussels (*Atrina zelandica*) may increase the drag coefficient C_{100} by almost one order of magnitude. This increase, however, is not proportional to the density of the animals: at higher densities, skimming flow develops over the ensemble of the horse mussels, with a lower drag coefficient compared to the situation with sparsely distributed mussels. Not only the physical presence of the animals, but also the water jets from their siphons (O'Riordan *et al.*, 1995), can increase the roughness of the bed. The jets act to the incoming flow as "solid objects" protruding several centimetres from the bed. On an actively filtering mussel bed in a flume, L. van Duren (unpublished findings) measured an increase of bottom roughness of two orders of magnitude compared with a flat bed. This increase was significantly affected by filtration of the animals. Hence, these animals can significantly affect their food supply through affecting their physical environment. The spatial distribution of the animals in clumps and boulders with empty or sparse patches in between can further enhance the bottom roughness (Fréchette *et al.*, 1989). More detailed measurements and modelling should clarify the importance of these structural characteristics for vertical turbulence and hence food supply.

The disadvantages of a clumped spatial distribution for food gathering may be offset by a number of biological processes which may be advantageous or simply inevitable. Protection against (bird) predation by oversaturating predators could be such a mechanism. Several bird species foraging on benthic suspension feeders have interference mechanisms that depress their intake rate when foraging on favourable beds with high density of food (e.g. Ens and Goss-Custard, 1984). Clumped distributions will then lead to a decrease in mortality risk per prey individual. Reducing the risk of erosion by currents may also be an important factor. At least in reef-building species such as mussels or oysters, the critical value for resuspension may depend on animal density. Densely covered mussel beds with their sediment structure full of empty shells are much more resistant to erosion than sand without animals or shells. (Nehls and Thiel, 1993).

By affecting physical conditions and predation pressure, suspension feeders may induce positive feedback between enhanced suspension feeder density and enhanced survival or growth. Positive feedbacks are considered a major influence in structuring ecological communities (Wilson and Agnew,

1992). They may invoke alternative stable states (Van de Koppel *et al.*, 1997). When this concept is applied to the benthic environment, one of these states is characterized by a dense bed of suspension feeders, whereas in the other state suspension feeders are absent. Disturbance of suspension feeder density in the dense state may lead to a permanent switch to the other state, namely collapse of suspension feeder populations. This could explain the slow return rate of natural mussel banks, once they have been destroyed by fisheries (Piersma and Koolhaas, 1997).

The mechanism proposed here differs from the classical concept of "trophic group amensalism" (Rhoads and Young, 1970; see Snelgrove and Butman, 1994, for a critical discussion and review). That theory critically depends on the destabilization of the sediment by deposit feeders as a means of preventing the establishment of suspension feeders, whereas suspension feeders in turn could prevent the establishment of deposit feeders. We do not assume a very strong linkage between the two groups, and we do not predict a spatial separation of deposit-feeding and suspension-feeding communities. In fact, the data shown in Figure 2 do not suggest such a mutual exclusion, but rather the existence of mechanisms causing the absence of suspension feeders at many places, and their concentration at a few places, whereas other governing processes must be at play in deposit-feeding species.

For a number of infaunal species (notably the bivalves *Macoma balthica*, *Cerastoderma edule* and *Mya arenaria*), detailed analyses of recruitment processes have been conducted (Beukema and de Vlas, 1989; Günther, 1991, 1992; Armonies and Hellwig-Armonies, 1992; Beukema, 1993; Armonies, 1996). Primary recruitment of the pelagic larvae of these species into the sediment is dependent on: (1) availability of larvae in the water column; (2) suitable conditions for larval settlement, implying relatively low current speeds; and (3) the ability of young post-larvae to maintain themselves in the sediment once settled, (i.e. a relatively low bottom shear stress). In a transect from a gully to an intertidal flat, recruitment usually peaks at the deepest places with sufficiently calm current conditions. In the Wadden Sea this is the lower intertidal, but in a recent study in the Westerschelde this zone occurs only in the high intertidal, excluding elevation itself as a major factor (M. Duiker *et al.*, H. Bouma *et al.*, unpublished results). After this first phase of primary recruitment, several cycles of motility, transport and resettlement follow. This may lead to a substantial redistribution of the animals, but traces of the original settlement distribution may remain even after 1 year (H. Bouma *et al.*, unpublished results). In any event, the animals lose their motility before they grow to a size where their joint filtration capacity may lead to significant seston depletion problems.

In summary, the use of boundary layer physics as a framework to study and understand the spatial distribution of suspension feeders should be carefully re-evaluated by taking into consideration flocculation, enhanced sinking and fast sedimentation–erosion cycles, particularly in turbid estuaries. This framework should further be critically evaluated for its ability to explain non-linear correlations between animal density and current velocity. More study is needed of the physical maintenance of animals in currents and waves. Finally, recruitment processes and other biological interactions should be explicitly incorporated into the models of spatial distribution.

The relations described in this chapter are highly relevant for measures taken in the light of management policies for estuarine habitats. Dredging, land reclamation or sea level change may lead to changes, within an estuary, in the relative availability of areas with a particular current regime. It is predicted, for example, that further dredging in the Westerschelde estuary will lead to increased steepness of the banks of intertidal flats, and to a general increase in the sediment dynamics on these flats. The question is whether this loss of most suitable habitat will lead to increased production of benthic suspension feeders on the remaining suitable parts (so that the fraction of total system production consumed by benthos will remain constant), or whether any loss of habitat will translate into a loss of fauna, including fish and birds. From the general distribution theory based on seston depletion problems, increased production seems likely. However, more detailed studies of the biological effects related to recruitment and positive density-dependent mechanisms in dense patches are needed before these questions can be answered.

IV. SPATIAL DISTRIBUTION AND DYNAMICS OF DEPOSIT FEEDERS

The development of predictive models for the spatial distribution of deposit feeders is hampered by the lack of insight into the basic processes determining the relation of these animals to the particulate organic carbon (POC) in sediments. Both the flux of carbon into the sediment, and the availability of particulate organic carbon in the sediment, are determined by complex processes, including abiotic and biotic factors. Attempts to relate biomass of deposit feeders directly to POC concentration in the sediment have generally failed, and with respect to POC flux (as, for example, in problems of anthropogenic organic loading) there is a paradox of enrichment. The very similar "succession after disturbance series" or "distance to pollution source series" of Rhoads *et al.* (1978) and Pearson and Rosenberg (1978) founded a solid base of empirical descriptions on how benthic communities typically react to varying loading with organic wastes, and how their structure may change in time or space as the pressure is changed. This evidence has been the basis for a number of methods allowing fast assessment of anthropogenic (organic) stress on benthic communities (Heip, 1995). Due to the generally poor quantification of the organic fluxes and concomitant biogeochemical changes in the sediments, it remains problematic to use this conceptual model as a basis for a process-oriented understanding of how deposit-feeding assemblages react to enhanced POC fluxes.

A. The Sediment as a Habitat

1. Sediment Texture

Sediments comprise a variety of particles and interstitial water in which various components are dissolved. Porosity (ϕ), the fraction of volume occupied by water, typically ranges from 0.9 in the surface layer of muddy sediments to 0.4 in sandy sediments, and usually decreases exponentially with depth due to compaction (Berner, 1980). As a consequence, the solid volume fraction of sediments ($\phi_s = 1 - \phi$) exhibits a pronounced gradient in the upper few centimetres.

Sediments can be classified according to their grain size as clay (< 2 μm), silt (2–63 μm), sand (63–2000 μm) or gravel (> 2 mm). Sedimentary grain-size distributions are the result of sorting processes during deposition, erosion and transport (Allen, 1985). Sediment particles are transported via bedload (also known as traction load and including rolling and saltation) and in suspension. Particles may settle individually according to Stokes' law, or form aggregates due to ionic strength (salinity) changes or biofilm formation. Moreover, suspension feeders may actively remove particles from the water column to the sediments (biodeposition). The physical stability of the sediment surface layer may, in various ways, be influenced by biological processes (Paterson, this volume). Kranck *et al.* (1996a,b) have proposed a grain-size distribution-based partitioning approach that allows distinction between sediments that have settled from suspension with no subsequent reworking and those that have been reworked a number of times. The net result of numerous deposition–erosion cycles is the occurrence of gravel and cobbles in estuarine channels and the dominance of fine sands in tidal flats. The sediment grain-size distribution has important consequences for sediment permeability (a proportionality factor between water pressure gradient and water flow) because voids between larger particles may be clogged by the presence of smaller particles. It also determines the living space for micro-organisms and meiofauna which themselves are not able to displace the sedimentary particles.

2. Sedimentary Organic Matter

The quantity, quality and spatial distribution of particulate organic matter in sediments and the biomass, vertical distribution and composition of benthic communities are related by a number of processes. On the one hand, animals affect the carbon flux to the sediment by biodeposition and via transport from the frequently resuspended surface layer to the deeper sediment layers. On the other hand, particulate organic matter constitutes the food and may (directly or indirectly) control the number, size and diversity of benthic animals.

Sedimentary organic matter concentrations vary from less than 0.05 wt% in sandy sediments to more than 10 wt% in fine-grained sediments, the majority of the values lying between 0.1 and 5 wt% (Hedges and Oades, 1997). For ecological studies, it is more appropriate to express the amount of organic matter available on a volume or area basis because this accounts for differences in water contents and porosity. For instance, organic carbon concentrations in intertidal sediments in the Westerschelde estuary expressed on a dry weight basis varied from 4.6 wt% in the freshwater area to 1.4 wt% in the marine section. There was no such gradient when expressed on a volume or area basis (1.1–1.8 mmol cm^{-3}; Middelburg *et al.*, 1996). Similar arguments in favour of volume- or area-based numbers have been put forward for bacterial densities (Schmidt *et al.*, 1998).

Organic matter is usually concentrated in the surface mixed layer of the sediment, typically the top 10 cm (Mayer, 1993). This is due to deposition or formation of organic matter at the sediment–water interface and subsequent burial and mixing downward of organic matter while being subject to degradation. The relative surface enrichment depends mainly on the rate of mixing (due to moving animals) and the lability of the organic material. Benthic animals such as *Arenicola marina* and *Heteromastus filiformis* can also enrich the surface layer via selective ingestion of organically enriched particles at depth and defaecation at the sediment–water interface (Neira and Höpner, 1994; Grossmann and Reichardt, 1991).

Organic matter comprises living biomass (macrofauna, meiofauna and microbial components), detritus (non-living biomass and its alteration products) and black carbon, a highly condensed carbonaceous residue from incomplete combustion processes on land, e.g. charcoal and soot (Middelburg *et al.*, 1999). The contribution of living biomass to bulk sedimentary organic matter has not often been determined, but is usually less than 10%. Given a typical estuarine macrofauna biomass of 0.6 mol C m^{-2} (i.e. 15 g AFDW m^{-2}) and a typical carbon stock of 200 mol C m^{-2} (1 wt%, porosity of 0.52 and integrated over 20 cm), it is clear that macrofauna biomass contributes little (< 1%) to sedimentary carbon. Meiofauna biomass values in estuaries cluster around 1 g C m^{-2} (Heip *et al.*, 1995; approximately 0.1 mol C m^{-2}) and do not contribute significantly to sediment carbon. The number of bacteria in sediments is rather invariant at about 10^9 to 10^{10} cm^{-3} (Schallenberg and Kalf, 1993; Schmidt *et al.*, 1998). Integration over 20 cm and assuming 2×10^{-13} g C per cell (Lee and Fuhrmann, 1987), this relates to about 3–33 mol C m^{-2} (1.5–15% of the total organic carbon). This is consistent with the 5% (range 1–20%) contribution of bacterial biomass to lacustrine sedimentary organic matter reported by Schallenberg and Kalf (1993).

The majority of sedimentary organic particulate matter is strongly associated with the mineral matrix, but about 10–30% may occur as discrete low-density particles (Mayer et al., 1993). These low-density particles usually have a high nutritional value and organisms may selectively feed on these particles in sandy sediments (e.g. *Callianassa;* Stamhuis et al., 1998). There is a close correlation between organic matter on the one hand and grain size or surface area on the other (Mayer, 1994; Hedges and Oades, 1997), because of (1) the tight association of organic matter with mineral particles and (2) similarities in the hydrodynamic behaviour of fine particles and discrete organic matter. The nature of organic matter–mineral interactions is not well known, but ionic as well as weaker interactions due to hydrogen bonding and van der Waals forces are involved. The majority of estuarine sediments below the bioturbated zone have a fixed amount of organic carbon per unit surface area of the grains (40–80 μmol C m^{-2}; 0.5–1.0 mg C m^{-2}), the amount being equivalent to a monolayer of organic material covering mineral surfaces (Mayer, 1994). This adsorbed organic matter is in dynamic equilibrium with the pore-water (Keil et al., 1994, 1997) and sorption sites may compete with enzymes for labile organic matter. Sorption of organic matter to surfaces in pores smaller than 10 nm (accounting for > 50% of the total surface area) may constitute a mechanism for preservation of intrinsically labile compounds because microbial exo-enzymes are too large to enter these pores. One may speculate whether macrobenthos may perhaps profit from this sorbed material if the conditions in their digestive systems enhance desorption from these sites.

The quality of organic matter is a loosely defined proportionality factor between mineralization and food availability on the one hand and organic carbon quantity on the other. This term has been introduced to reconcile the orders of magnitude variability in mineralization rates with the invariance of sedimentary organic matter concentrations. The quality depends on the accessibility/availability and the composition of the organic material. In a simple approximation, it can be characterized by the first-order degradation rate of the organic material. This is based on the degradation model

$$\frac{dC}{dt} = -kC \qquad (12)$$

where C is the concentration of organic matter, t is time and k (time^{-1}) is the first-order degradation rate. The higher k is, the faster degradation proceeds, hence its use as a quality parameter.

The composition of the organic matter depends on its source and subsequent degradation history because of preferential consumption of more labile compounds and consequent selective accumulation of more refractory compounds (de Leeuw and Largeau, 1993; Cowie and Hedges, 1994; Dauwe and Middelburg, 1998). To the extent that animals are more dependent for their metabolism on the freshly arriving matter than on the matter present in the sediment (e.g. Tsutsumi *et al.*, 1990), this quality difference between flux and stock may be responsible for the lack of any apparent correlation between growth of the animals and POC present in the sediment.

Organic geochemical approaches to constrain the nutritional value of organic matter are hampered by a number of factors. First, only the fraction of organic matter that remains after extensive utilization can be analysed, not the fraction that has been respired or assimilated. Second, despite significant improvements in analytical techniques, a large fraction (> 40%) of the sedimentary organic matter remains biochemically uncharacterized (Wakeham *et al.*, 1997; Dauwe and Middelburg, 1998). Third, chemical characterization usually involves hydrolysation with strong acids (e.g. for amino acids, amino sugars and carbohydrates) or extraction with organic solvents (e.g. for fatty acids) and there is no simple straightforward relation between these chemically defined compounds and enzymatically available substrates (Mayer *et al.*, 1995; Dauwe *et al.*, 1999a). Despite these difficulties, considerable progress has been made in the use of chemical measures for organic matter quality. The ratio between chlorophyll and total organic carbon may be a useful measure (Hargrave and Phillips, 1989) of the nutritional value because chlorophyll is readily degraded ($k = 0.02–0.04$ day^{-1}; Sun *et al.*, 1991). The contribution of amino acids to the total organic carbon (Ittekkot, 1988; Cowie and Hedges, 1994), the molecular composition of amino acids (Dauwe and Middelburg, 1998) and the enzymatically available amino acid pool (Mayer *et al.*, 1995; Dauwe *et al.*, 1999a) are also useful quality measures at time-scales relevant to estuarine sediments.

Molar C/N ratios and sedimentary nitrogen contents have also been used quite extensively because organic nitrogen is an essential and limiting nutrient in deposit feeders diets (Tenore, 1988). However, the use of C/N ratios is not straightforward because organic matter derived from different sources may have variable C/N ratios and different pathways of nitrogen mineralization. Molar C/N ratios of allochtonous marine and autochthonous estuarine material are in the range of 6–8 and increase during degradation because of preferential mineralization of nitrogen-rich material (e.g. Burdige, 1991; Rosenfeld, 1981). Molar C/N ratios of terrestrial and higher (marsh) plant-derived matter are generally higher than 12 but decrease because of nitrogen enrichment as a consequence of bacterial processing. Bacterial processing of higher plant material is a well-known requirement before heterotrophic organisms can assimilate terrestrial organic matter with a low quality (see section IV). However, nitrogen enrichment cannot always be equated with increasing nutritional value to benthic detritivores because part of the nitrogen is incorporated in recalcitrant compounds (Rice, 1982). Bacterial processing of marine and estuarine algal material also occurs, but is not obligatory because bacteria and algae are rather similar in terms of their amino acid and carbohydrate composition, and nutritional value to macrofauna (Cowie and Hedges, 1994; Dauwe and Middelburg, 1998; Dauwe et al., 1999a).

3. Sediment Biogeochemistry

Early diagenesis comprises all processes occurring in the surface layers of sediments subsequent to deposition. The main factor driving these biogeo-chemical processes is the degradation of organic matter, which causes consumption of oxidants and production of metabolites (including nutrients and hydrogen sulfide). Recycling of nutrients is a prerequisite for production by estuarine autotrophs. The production of metabolites such as hydrogen sulfide, however, may toxify benthic fauna. In estuarine sediments, the major electron acceptors for oxidation of organic matter are oxygen, nitrate, manganese oxide, iron oxides and sulfate, and the oxygen bound in organic matter itself (Heip et al., 1995). These oxidants are generally utilized sequen-tially with a distinct biogeochemical zonation pattern and pronounced pore-water–composition–depth gradients as a result (Aller, 1982). However, recent studies have clearly shown that many of the oxidation pathways occur in the same depth interval (e.g. Canfield and Des Marais, 1993; Canfield et al., 1993), partly due to heterogeneity induced by macrobenthic activity.

Early diagenetic models, describing transport and reaction in aquatic sedi-ments, have been used extensively to predict and understand the changes in sediments due to organic matter degradation (Berner, 1980; Soetaert et al., 1996a; Boudreau, 1997). They are almost always restricted to one dimension, (but see, for example, Aller, 1980a; Boudreau and Marinelli, 1994), because the data available are usually related only to depth and because of mathematical complexity. For derivation of equations and their analytical and numerical solution the reader is referred to the excellent treatment by Boudreau (1997). Basically, partial differential equations specifying advection, diffusion and reaction are solved, given appropriate boundary and initial conditions.

Application of early diagenetic models to estuarine and intertidal sediments may require a number of adaptations to capture the salient features of these systems. In sandy and intertidal sediments the formulation of pore-water advection should include lateral water flows (in the Brinkman layer; Svensson and Rahm, 1991; Khalili et al., 1997) and falling water levels during air exposure (Rocha, 1998). These flows enhance solute transport rates and may bring oxygen to much greater depth in the sediments than would be possible by diffusion alone (Hüttel and Gust, 1992; Lohse et al., 1996; Boudreau and Jørgensen, 1999; Güss, 1998).

Steady-state assumptions may not be adequate since seasonal changes in temperature, bioturbation and carbon input markedly influence the system (Aller, 1980b; Klump and Martens, 1989; Soetaert et al., 1996b).

Bioturbation, the mixing of sediments due to the activity of moving animals, is usually approximated as being a diffusional type of process, i.e. continuous, symmetrical and over small distances (Boudreau, 1986). The bioturbation coefficient (D_b), an eddy–diffusion-like mixing parameter, is then estimated from concentration or activity versus depth profiles of tracers with known decay rates, for example radionuclides (Berner, 1980) or chlorophyll (Sun *et al.*, 1991). However, it is clear that this diffusion approximation of sediment mixing does not apply when (1) sediment motions are not random, but directed (e.g. conveyer-belt mixing), (2) mixing occurs between two non-adjacent points, (3) mixing is intermittent and (4) higher dimensional processes must be incorporated into the one-dimensional framework (Boudreau, 1997). Non-local transport processes may then be included (Boudreau and Imboden, 1987; Soetaert *et al.*, 1996c; Boon and Duineveld, 1998). Wheatcroft *et al.* (1990) identified a number of different ways in which animals may move sediment particles; clearly many of these types of movement deviate far from random diffusion-like transport.

In the present generation of diagenetic models, macrobenthos (and meiobenthos) figure only as modulators of sediment (and solute) transport without consumption of any organic matter. This is clearly inconsistent with estimates of animal respiration (see section II.B). Moreover, the dynamics of macrobenthos are also not included.

Macrophytes (e.g. seagrasses) or algae often inhabit shallow water and intertidal sediments. These benthic producers supply significant organic matter to sediments and should therefore be included explicitly in these biogeochemical models. Moreover, they affect oxygen inputs and nutrient cycling (Risgaard-Petersen *et al.*, 1994; Risgaard-Petersen and Jensen, 1997). Specification of the upper boundary conditions in intertidal estuarine sediments requires special attention. Besides the need to include carbon production during low tide by microphytobenthos (Heip *et al.*, 1995) and enhanced solute exchange due to ripples and bioroughness (Boudreau and Jørgensen, 1999), it is also necessary to deal with the alternation between air exposure and submergence. During exposure the sediment surface acts as a reflecting boundary for solutes, while gases may pass, and this is best represented with a surface evaporation type of boundary. During submergence both solutes and gases are exchanged between water and sediments. This alternation requires dynamic diagenetic models able to resolve the tidal time-scale. Estuarine sediments are dynamic in terms of particle movement due to moving wave and current ripples (Khalili *et al.*, 1997), with a consequence that particulate organic matter becomes homogenized. In such a case it is better to specify the concentration of organic matter at the sediment–water interface than to specify the flux of organic matter to the sediments (Rice and Rhoads, 1989).

B. Deposit Feeders and their Food

The basic features of the response of benthic deposit-feeding communities to organic enrichment are (Pearson and Rosenberg, 1978): (1) large species are replaced by smaller species, even if within species individuals may grow larger near to the source of organic pollution (Weston, 1990); (2) deep-dwelling bioturbating species are replaced by surface or sub-surface deposit-feeding species; (3) average lifespan decreases and K-selected species are generally replaced by r-selected species; (4) density increases (faster than biomass) up to a certain loading, but crashes when free sulfide reaches the water column; and (6) species diversity decreases, but not in a monotonic way: intermediate high values may be found. It has been shown that many of these features can be summarized in abundance biomass curves (Warwick, 1986), which are cumulative plots of normalized density and biomass by species rank. "Stable" or "equilibrium" or "undisturbed" communities are characterized by a biomass curve well above the abundance curve, because their biomass is dominated by large, relatively rare species, whereas communities dominated by small opportunistic species tend to have their abundance curve above the biomass curve. Intersecting curves are interpreted as indicative of moderate disturbance, but the justification for this is often unclear. Over the past decade, these methods for pollution detection have had numerous new developments, but we concentrate here on the fundamental patterns of difference between benthic communities subject to different organic fluxes.

It has been pointed out by Beukema (1988) and Craeymeersch (1991) that many estuarine samples exhibit abundance biomass curves of the disturbed types, without other obvious signs of eutrophication or human disturbance. The patterns and community types described as a response to eutrophication stress also occur under natural conditions. These conditions can be rather obvious: under extreme physical stress, causing instability of the bed, only a few small, motile species can survive. However, at the other extreme of the physical disturbance scale, in quiet, extremely muddy environments with a high organic content, the benthic assemblage may also typically be composed of small surface or shallow sub-surface deposit feeders, characteristic of "high loading" or "high disturbance" pollution sites. It is an intriguing question as to which processes could be responsible for these patterns.

Dauwe et al. (1998) have considered the relation between the quality of the arriving organic flux and the optimal rate of bioturbation in a sediment. They used an analytical diagenetic model composed of a bioturbated top layer and a non-bioturbated layer, forced by a constant flux of organic carbon and a constant sediment accretion rate. With this model they calculated the mineralization rate (expressed as a fraction of the arriving flux) at the bottom depth of the bioturbated layer, as a function of bioturbation rate and first-order degradation rate of the organic matter.

The result is shown in Figure 7 as a contour graph in the $k - D_b$ plane. For every quality of incoming flux, there is an increase of mineralization at depth with bioturbation. However, this increase is strongly non-linear, and the bioturbation range where the steepest increase can be realized is dependent on quality. For very refractory influx ($k = 0.01$ year^{-1}) there is no real increase for varying D_b. For $k = 0.1$ year^{-1} the increase is between 1 and 20 cm^2 year^{-1}. For $k = 1$ year^{-1} it is between 20 and 100 cm^2 year^{-1}. For very high-quality material ($k = 10$–100 year^{-1}) no realistic values of D_b allow more than traces of organic matter to be brought to the depth ($x = 10$ cm) used here as a reference. Based on these model results, the most intense and deepest bioturbation would be predicted to occur in sediments receiving low

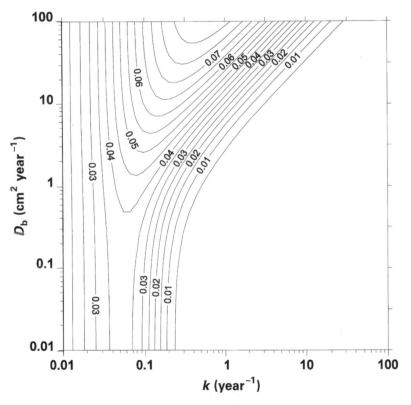

Fig. 7. Isolines of mineralization rate (first-order rate constant $k \times$ concentration of organic matter, mmol C cm^{-3} year^{-1}) at a depth of 10 cm for a two-layer model of organic matter diagenesis in sediments (Dauwe *et al.*, 1998). Parameters of the model (flux of organic matter = 1 mmol C cm^{-2} year^{-1}, sediment accretion rate = 0.5 cm year^{-1}, depth of bioturbated layer = 10 cm k (first-order degradation rate of organic matter, year^{-1}) and D_b (bioturbation coefficient, cm^2 year^{-1})) were varied as shown.

to intermediate quality of organic flux. Sediments with a more refractory flux will show deep but not very intense bioturbation, and in sediments receiving more labile carbon flux, bioturbation depth will become shallower and eventually restricted to the near-surface zone only.

Rice and Rhoads (1989) have argued that it is unrealistic to prescribe a constant-flux upper boundary for sediments in shallow water. The top layer of these sediments is continuously reworked physically, leading to a relatively constant concentration in the upper layers. This boundary specification has important consequences for the effects of bioturbation. If the concentration in the top layer is kept constant, more bioturbation will lead to an increase of the flux into the sediment, since more material is mixed down from the surface layer into the sediment. When applying this boundary condition to the model of Dauwe et al. (1998), the resulting graph is qualitatively similar, but quantitatively different, to the constant flux boundary condition. In particular, the influence of bioturbation on mineralization at depth is predicted to occur at slightly higher quality of the organic matter. Moreover, the influence is steeper and at generally higher bioturbation levels.

All results in this model exercise are normalized to the incoming flux or to the top concentration respectively. The disappearance of deep-deposit feeders upon organic enrichment cannot be explained by a shortage of energy at depth for the bioturbating organisms, since this will increase at least linearly with the magnitude of the flux. A theory for the paradox of benthic enrichment should therefore contain at least one extra factor. Classically, the increased production of sulfide, the shallower depth of the redox potential discontinuity (RPD) layer, depth of oxygen penetration or similar parameters related to increased mineralization in the sediments have been put forward as the main factor contributing to the disappearance of deep-burrowing animals. At least qualitatively, one could argue against these hypotheses that many deep-deposit feeders are known to survive in reasonably anoxic sediments, and that the animals can contribute to the oxygenation of the sediment by flushing the interstitial water (irrigation), which can increase the apparent diffusion coefficient for dissolved species in the interstitial water by an order of magnitude (compared to molecular diffusion).

We extended the constant carbon flux model with the dynamics of a "generic electron acceptor". Positive values of the general electron acceptor denote oxygen concentration, while negative values provide the concentration of reduced substances that can be oxidized by oxygen ("oxygen demand units"; see Soetaert et al., 1996a). The model is specified as follows.

$$0 \leq x \leq L: \qquad \frac{\delta Ox}{\delta t} = \frac{\delta}{\delta x} D_1 \frac{\delta Ox}{\delta x} - kC$$

$x > L$:
$$\frac{\delta Ox}{\delta t} = \frac{\delta}{\delta x} D_2 \frac{\delta Ox}{\delta x} - kC$$

Boundary conditions: $Ox = Ox_0$ at $x = 0$

$$D_1 \left.\frac{\delta Ox}{\delta x}\right|_{x=L}^{\text{upper}} = D_2 \left.\frac{\delta Ox}{\delta x}\right|_{x=L}^{\text{lower}}$$

$$\left.\frac{\delta Ox}{\delta x}\right|_{x=\infty} = 0 \tag{13}$$

where Ox denotes the generalized electron acceptor concentration, x is space, t is time, L is the depth of the bioturbated layer, C is the concentration of organic matter as derived from the model of Dauwe *et al.* (1998), D_1 is the diffusion coefficient of the electron acceptor in the upper (bioturbated) layer, and D_2 is the (molecular) diffusion coefficient of the electron acceptor in the lower (non-bioturbated) layer. D_1 can be higher than D_2 if the animals' activity enhances the diffusion of the electron acceptor.

The results of this model are shown in Figure 8 for different values of the organic matter flux, with a quality (first-order degradation rate) of 1 year^{-1}, a sediment accretion rate of 0.5 cm year^{-1} and a bioturbation depth of 10 cm. When the bioturbation coefficient is held constant (Figure 8A), the concentration of the electron acceptor at depth becomes increasingly negative with the magnitude of the flux. In Figure 8B, the bioturbation coefficient D_b increased proportionally with the flux, from 1 cm^2 year^{-1} at a flux of 1 mol C m^{-2} year^{-1} to 100 cm^2 year^{-1} at a flux of 100 mol C m^{-2} year^{-1}. It can be seen that the result of increased bioturbation is a much more negative concentration of the electron acceptor. Bioturbation shifts the average depth of mineralization to greater depth, thereby making it increasingly difficult for oxygen to reach this depth. Bioturbation in itself therefore results in a shift from aerobic to anaerobic mineralization in sediments, a result also demonstrated by Heip *et al.* (in press) with the aid of the more elaborate diagenetic model of Soetaert *et al.* (1996a). When the diffusion coefficient of the electron acceptor increases with bioturbation (Figure 8C), this effect is counteracted by the increased flushing of the bioturbated layer, and the concentration of the electron acceptor at depth becomes less negative. However, this effect is not unlimited, as the animals cannot influence the diffusion coefficient of the lower layer, and the mineralization taking place there increases with the flux and with the bioturbation (the model effectively leaks organic matter to the lower layer). Thus, no matter how the animals influence the diffusion coefficient of the electron acceptor in the

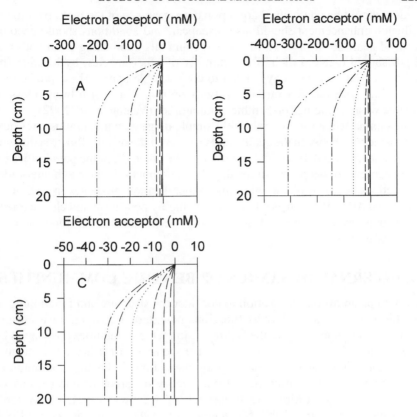

Fig. 8. Concentration profiles of a generalized electron acceptor in a two-layer diagenetic model as specified in the text, for fluxes of organic matter varying in a logarithmic series between 0.1 and 10 mmol C cm^{-2} $year^{-1}$, quality $k = 1$ $year^{-1}$, sediment accretion rate = 0.5 cm $year^{-1}$ and depth of bioturbated layer = 10 cm. Highest fluxes give rise to the highest electron acceptor depletion. (A) Constant D_b = 10 cm^2 $year^{-1}$, $D_1 = D_2$ = molecular diffusion. (B) As in A, but D_b linearly co-varying with magnitude of flux between 1 and 100 cm^2 $year^{-1}$. (C) As in B, but D_1 co-varying with D_b between 1 and 10 times molecular diffusion. Note the differences in scale.

bioturbated layer, an increase in organic flux will always lead to more reduced conditions deeper in the sediment, and bioturbation is a factor enhancing this effect.

From these calculations three qualitative predictions may be made. First, clogging of the sediment, even with inert fine material, should decrease the diffusion enhancement and would therefore be detrimental to deep bioturbating organisms in sandy sediments. Second, the transition between a community dominated by deep bioturbators to a community dominated by surface deposit feeders should be relatively sharp: upon disappearance of the bioturbation, the diffusion enhancement should also disappear, and conditions should become reduced near the surface. Third, the large animals dominating bioturbation at depth should create a niche for a number of other species, by keeping the sediment relatively well oxygenated to considerable depth. These predictions are in line with the general observations on the effects of a broad class of organic waste dumpings on benthic communities (Valente *et al.*, 1992).

In summary, the use of a relatively simple diagenetic model illustrates how the interactions between the quality and quantity of the organic flux to sediments on the one hand, and the animal activities on the other, shape the possibilities for the occurrence of deep-deposit feeding. Even without reference to higher-order interactions between bioturbation and mineralization processes in sediments (Aller and Aller, 1998), these dynamics are highly non-linear and characterized by the presence of strong feedback loops between animals, their food and their chemical environment.

V. INTERNAL DYNAMICS OF BENTHIC COMMUNITIES

It would be an oversimplification to consider the structure and functioning of benthic communities solely as functions of external forcing, be it current regimes or organic flux into the sediment. In fact, communities are shaped by strong internal interaction links as well. Extensive effort has been put into testing the importance of predator–prey interactions in benthic communities (see review in Foreman *et al.*, 1995). Theory on predator control proved successful in explaining the dynamics of freshwater pelagic systems (Carpenter *et al.*, 1985, 1987; McQueen *et al.*, 1986; Brönmark *et al.*, 1992; Strong, 1992). Predatory effects in benthic communities were found to be more complex (Wissinger and McGrady, 1993; Lodge *et al.*, 1994). The discrepancies between the theory of predator control and the results from experimental studies in benthic systems may be due to a high incidence of omnivory (feeding at more than one trophic level) (Ambrose, 1984; Committo and Ambrose, 1985; Polis *et al.*, 1989; Posey and Hines, 1991; Menge *et al.*, 1996). Epibenthic predators such as birds, crabs and fish prey not only on non-predatory infauna, but also on predatory infauna (Ambrose, 1984; Committo and Ambrose, 1985). Most deposit feeders ingest a broad spectrum of potential food sources including detritus, algae, bacteria and protozoa (Lopez and Levinton, 1987; Retraubun *et al.*, 1996). As a consequence, prey in benthic systems may experience not only the direct negative effects of predation by omnivores, but also indirect positive effects because of reduced predation by the omnivores' other prey.

The high incidence of omnivory in benthic marine compared with, for example, terrestrial above-ground communities may be the consequence of physiological difficulties in feeding on different trophic levels (Yodzis, 1984;

Pimm *et al.*, 1991). These difficulties are related to the quality and accessibility of food sources. Consuming terrestrial plants requires a specialized morphology and digestive tract adapted to handling low-quality plant tissues, excluding the adaptations needed to capture often elusive prey. This specialization may explain the low occurrence of omnivory in terrestrial above-ground communities. The requirements needed to catch benthic prey seem less stringent. Most epibenthic predators are able to handle a broad variety of benthic preys (Ambrose, 1984). Many deposit feeders are bulk feeders that ingest both dead organic materials and the micro-organisms that are attached to it (Lopez and Levinton, 1987). It appears that feeding on multiple trophic levels is relatively easy in benthic systems. In fact, it may even be hard to avoid.

Physiological limitations may not be the only factor determining the occurrence of omnivorous links in food webs. Theoretical studies suggest that a high incidence of omnivory is destabilizing to food webs, and therefore that omnivory should be rare in natural communities (Pimm and Lawton, 1978; Pimm, 1982). Early studies on the properties of real food webs indicated that omnivorous interactions are indeed less common than expected by chance alone (Pimm and Lawton, 1978; Pimm *et al.*, 1991), but these findings have been heavily criticized recently by, among others, benthic ecologists. These authors have claimed that omnivory is more common in real, well-documented food webs than in the earlier incomplete food webs analysed (Hall and Raffaelli, 1991, 1993; Polis, 1991). Studies of food web structure in freshwater and desert ecosystems seem to support this view (Vadas, 1990; Polis, 1991). Holt and Polis (1997) made a detailed mathematical study of the dynamics of a system with omnivores. They analysed a three-species system of two predators and a prey, in which one of the predators also preys on the other. Their model predicts that such systems are vulnerable to losing the intermediate predator, especially when production of the bottom prey is high (Figure 9). In benthic communities, however, omnivores seem to be encountered most frequently in more productive environments (Persson *et al.*, 1988, 1996). The discrepancy between theoretical and empirical studies reflects our limited insight in the nature of omnivorous interactions in benthic communities.

To improve insight into the trophic structure of communities, it is important to understand the extent to which predators and omnivores affect their prey and are controlled by their predators (Hairston and Hairston, 1993). If predatory interactions between omnivores and prey on the adjacent trophic level are of only minor importance to their dynamics, it may be more appropriate to regard both species as competitors. If, at the other extreme, prey on the lowest trophic position are only a minor constituent of the food consumed by the omnivore, the system can probably be better regarded as a food chain.

Deposit feeders consume a broad range of potential food sources of which detritus and micro-organisms are the main constituents (Lopez and Levinton, 1987). Deposit feeders can efficiently remove and assimilate

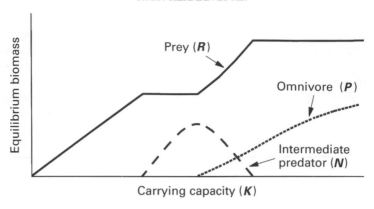

Fig. 9. Relation between prey-carrying capacity (assumed to reflect productivity of the system) on equilibrium biomass of a three-species system of two predators and a prey, in which one of the predators (the omnivore P) also preys on the other predator (N). The dynamics of the system are given by (following Holt and Polis, 1997): $dR/dt = R[r(1 - R/K) - aN - a'P]$; $dN/dt = N(abR - m - \alpha P)$; and $dP/dt = P(b'a'R + \beta\alpha N - m')$. See Holt and Polis (1997) for details. Note that the intermediate predator is lost from the system at high carrying capacity K because of competition and predation by the omnivore.

micro-organisms from the sediment, whereas detritus is in most cases more refractory. The relative importance of detritus and bacteria as a food source for deposit feeders differs among systems. In most marine non-vegetated sediments, bacterial densities seem insufficient to be an important carbon source to the resident stocks of deposit feeders (Kemp, 1987). In mangrove-dominated systems, however, bacteria form a much more important source of carbon for deposit feeders (Odum and Heald, 1975). Mangrove sediments have among the highest recorded bacterial densities in marine systems (Schmidt *et al.*, 1998). In some freshwater sediments bacteria constitute up to 79% of the organic matter consumed by the benthos (Schallenberg and Kalf, 1993), and this is attributed by these authors to the fact that macro-phyte-derived material is often the prime carbon source in lake sediments. The differences between these findings are most likely related to the quality and lability of organic matter. Macrophyte- or mangrove-dominated systems are characterized by low-quality organic matter with high C/N ratios (Giddins *et al.*, 1986; Robertson, 1988). Digestive constraints may limit carbon uptake from detritus by deposit feeders (Alongi and Christoffersen, 1992), and hence deposit feeders rely more strongly on microbial food sources. Despite the high carbon loading of mangrove sediments, deposit feeder densities are generally found to be low (Alongi and Christoffersen, 1992; Sheridan, 1997). This suggests that only a limited part of detrital carbon is eventually incorporated into the production of these animals

(Odum and Heald, 1975; Robertson *et al.*, 1992), which is consistent with the hypothesis that in these systems an additional trophic link (over bacteria) mediates the transfer from detrital carbon to the organisms.

Most marine non-vegetated sediments or sandy beaches obtain detritus from marine algal sources, which provide relatively high-quality organic matter with C/N ratios of about 6–8 (see section IV.A.2). Furthermore, bacterial densities in marine sediments are around 10^9 ml^{-1} (Alongi and Sasekumar, 1992; Schmidt *et al.*, 1998), values that are too low to meet the energy demands of macrofauna (Kemp, 1990). Hence, bacteria are unlikely to be an important food source for macrofaunal detrivores in marine sediments (Kemp, 1987; Lopez and Levinton, 1987). It is furthermore unlikely that deposit feeders remove more than 10–15% of bacterial production in most sediments (Kemp, 1990), unless both bacterial production and deposit feeding are focused in the surface layer. In most non-vegetated marine sediments, the strength of interaction between bacteria and macrofauna is probably weak, and detritus is considered to be the main food source for deposit feeders (Tenore *et al.*, 1982).

Holt and Polis (1997) showed that the intermediate predator in a three-component system with omnivory may be unstable in productive environments. The detritus–bacteria–deposit feeder system may therefore be considered as intrinsically unstable. In benthic soft-bottom communities, however, a number of mechanisms are likely to have a positive influence on the persistence and hence on the stability of omnivorous interactions. Habitat complexity may protect a part of the intermediate predator population by providing a refuge against predation (Diehl, 1992). Some benthic prey may seek refuge under dense root-mats (Peterson, 1982) or by retreating to a depth within the sediment at which predation by epibenthic predators is reduced. Deposit feeders, consuming both detritus and micro-organisms, may depend on micro-organisms as a source of nutrients (Lopez and Levinton, 1987). In mangrove-dominated systems, detritus is a poor source of nutrients such as nitrogen or phosphorous. Most mangrove detrivores prefer ingesting food that is aged and colonized by bacteria and fungi (Alongi, 1998). Furthermore, bacteria may provide an essential source of certain amino acids or vitamins (Lopez and Levinton, 1987). Any deposit-feeding organism that depletes the standing stocks of such micro-organisms may deprive itself from valuable nutrients. Hence, food shortage would effectively prevent deposit-feeding organisms from overexploiting their bacterial food supply, and eliminating omnivorous interactions. This may explain the high densities of micro-organisms found in sediments dominated by low-quality mangrove detritus.

A number of stabilizing factors are typical to benthic environments. Diffusion in aquatic sediments may limit the availability of oxygen to benthic organisms. Many benthic omnivores not only interact directly with their prey, but also have indirect effects on prey by changing environmental conditions. Deposit feeders often disturb the sediment, causing enhanced mixing and transportation of sediment and solutes to deeper layers or to the surface. Enhanced pore-water irrigation due to the activity of deposit feeders results in higher oxygen penetration into the sediment, which may stimulate the growth of micro-organisms (Van de Bund *et al.*, 1994). While the microbial contribution to the decay of labile organic matter was reduced, deposit feeders were found to stimulate bacterial decomposition of relatively refractory organic matter (Anderson and Kristensen, 1992; Kristensen *et al.*, 1992). Some deposit feeders may stimulate the growth of bacteria in their burrow systems, a process called microbial gardening (Hylleberg, 1975; Grossmann and Reichardt, 1991). Growth of micro-organisms may furthermore be facilitated by concentration of organic matter in faecal pellets (Neira and Höpner, 1994). Stimulation of the growth of micro-organisms may compensate for losses due to predation, and prevent deposit feeders from significantly depleting micro-organism densities. When one of the prey species, rather than the top predator, is a deposit-feeding species, predation may decrease pore-water irrigation. In such systems oxygen availability is likely to decrease as predation on the prey species becomes more severe. Studies on the effects of low oxygen availability have shown that benthic predators may become ineffective in low-oxygen environments (Kolar and Rahel, 1993; Sandberg, 1994, 1997). This would prevent the top predator from depleting prey stocks, and hence promote persistence in benthic communities.

A better understanding of the interaction strength between omnivores and their food sources on various trophic levels will be crucial in expanding insight into the dynamics of benthic communities. Recent research has shown that the strength of interaction between organisms is an important determinant of the stability of food webs (De Ruiter *et al.*, 1995; McCann *et al.*, 1998). Little is known about how omnivores such as deposit feeders depend on their prey or influence prey dynamics, and how this interaction is influenced by system-specific properties such as the quantity and quality of the organic supply. Second, the dynamic implications of indirect effects of omnivores on prey, especially effects mediated by the environment, are virtually unknown. The effects of pore-water irrigation on the chemical properties of sediments were illustrated in section IV.B. The ecological consequences of this feedback, however, have mainly been ignored. Incorporating feedback relations between benthic species and their environment into models of species interactions and food webs will significantly improve the predictive power of theory on predator control in benthic environments.

VI. CONCLUSIONS

In shallow, well-mixed estuaries, the coupling between benthic and pelagic systems is intense. Productivity of the system is an important factor limiting average benthic biomass. Respiration rates of benthic animals account for a substantial fraction of the primary production. Within the total sediment metabolism, animal respiration also represents a significant fraction. The dynamics of macrobenthic populations is important to understand estuarine system dynamics.

There are strongly non-linear two-way interactions between macrobenthic animals and their physical and chemical environment, as shown for suspension feeders and flow, and for deposit feeders and the chemistry of the sediment. Moreover, as shown for the deposit-feeding community, food and food quality is determining the type of interactions between the species, and thereby the dynamics of the community. It is not possible to describe macrobenthic populations as passively undergoing the influence of extrinsic factors. The feedback mechanisms by which the animals change their own environment offer a challenging scope for further study of estuarine benthos in the context of the system. As shown by the deposit-feeding community interactions, the macrobenthos does not comply to simple theoretical predictions based on terrestrial and pelagic aquatic environments. The identification of peculiar characteristics of benthic systems leading to this discrepancy can offer more insight into the processes and mechanisms shaping these communities. Eventually, it may also contribute to better and more general concepts in ecology.

ACKNOWLEDGEMENTS

The authors thank their colleagues in the Department of Ecosystem Studies and their partners in the European projects ECOFLAT (Eco-metabolism of a Tidal Flat) and PHASE (Physical Forcing and Biogeochemical Fluxes in Shallow Coastal Ecosystems) for discussion of the ideas expressed in this paper and for use of unpublished results. This research was supported by the Environment and Climate and Marine Science and Technology programmes of the European Union (ENV4-CT96–026 ECOFLAT and MAS3-CT96–0053 PHASE) and by research contracts from Rijkswaterstaat. This is a contribution to the programme ELOISE (European Land–Ocean Interaction Studies) and contribution no. 2501 of the Netherlands Institute of Ecology.

REFERENCES

Allen, J.R.L. (1985). *Principles of Physical Sedimentology*. George Allen and Unwin, London.

Aller, R.C. (1980a). Quantifying solute distributions in the bioturbated zone of marine sediments by defining an average microenvironment. *Geochim. Cosmochim. Acta* **44**, 1955–1965.

Aller, R.C. (1980b). Diagenetic processes near the sediment–water interface of Long Island Sound. 1. Decomposition and nutrient element geochemistry (S,N,P). *Adv. Geophys.* **22**, 238–350.

Aller, R.C. (1982). The effects of macrobenthos on chemical properties of marine sediment and overlying water. In: *Animal–Sediment relations* (Ed. by P.L. McCall and M.J.S. Tevesz), pp. 53–102. Plenum Press, New York.

Aller, R.C. (1988). Benthic fauna and biogeochemical processes in marine sediments: the role of burrow structures. In: *Nitrogen Cycling in Coastal Marine Environments* (Ed. by T.H. Blackburn and J. Sørensen), pp. 301–338. John Wiley, New York.

Aller, R.C. (1994). The sedimentary Mn cycle in Long Island Sound: its role as intermediate oxidant and the influence of bioturbation, O_2, and Corg flux on diagenetic reaction balances. *J. Mar. Res.* **52**, 259–295.

Aller, R.C. and Aller, J.Y. (1998). The effect of biogenic irrigation intensity and solute exchange on diagenetic reaction rates in marine sediments. *J. Mar. Res.* **56**, 905–936.

Alongi, D.M. (1998). *Coastal Ecosystem Processes*. CRC Press, Boca Raton.

Alongi, D.M. and Christoffersen, P. (1992). Benthic infauna and organism–sediment relations in a shallow, tropical coastal area: influence of outwelled mangrove detritus and physical disturbance. *Mar. Ecol. Prog. Ser.* **81**, 229–245.

Alongi, D.M. and Sasekumar, A. (1992). Benthic communities. In: *Tropical Mangrove Ecosystems* (Ed. by A.I. Robertson and D.M. Alongi), pp. 137–171. American Geophysical Union, Washington.

Ambrose, W.G. (1984). Role of predatory infauna in structuring marine soft-bottom communities. *Mar. Ecol. Prog. Ser.* **17**, 109–115.

Anderson, F.Ø. and Kristensen, E. (1992). The importance of benthic macrofauna in decomposition of microalgae in a coastal marine sediment. *Limnol. Oceanogr.* **37**, 1392–1403.

Armonies, W. (1996). Changes in distribution patterns of 0-group bivalves in the Wadden Sea: byssus-drifting releases juveniles from the constraints of hydrography. *J. Sea Res.* **35**, 323–334.

Armonies, W. and Hellwig-Armonies, M. (1992). Passive settlement of *Macoma balthica* spat on tidal flats of the Wadden Sea and subsequent migration of juveniles. *Neth. J. Sea Res.* **29**, 371–378.

Austen, M.C., Buchanan, J.B., Hunt, H.G., Josefson, A.B. and Kendall, M.A. (1991). Comparison of long-term trends in benthic and pelagic communities of the North Sea. *J. Mar. Biol. Ass. UK* **71**, 179–190.

Bacher, C., Duarte, P., Ferreira, J.G., Héral, M. and Raillard, O. (1998). Assessment and comparison of the Marennes-Oléron Bay (France) and Carlingford Lough (Ireland) carrying capacity with ecosystem models. *Aq. Ecol.* **31**, 379–394.

Baird, D. and Milne, H. (1981). Energy flow in the Ythan Estuary, Aberdeenshire, Scotland. *Estuar. Coast. Shelf Sci.* **13**, 455–472.

Banse, K. and Mosher, S. (1980). Adult body mass and annual production/biomass relationships of field populations. *Ecol. Monogr.* **50**, 355–379.

Baretta, J. and Ruardij, P. (1988). *Tidal Flat Estuaries. Simulation and Analysis of the Ems Estuary*. Springer, Berlin.

Berner, R.A. (1980). *Early Diagenesis. A Theoretical Approach*. Princeton University Press, Princeton.

Beukema, J.J. (1988). An evaluation of the ABC-method (abundance/biomass comparison) as applied to macrozoobenthic communities living on tidal flats in the Dutch Wadden Sea. *Mar. Biol.* **99**, 425–433.

Beukema, J.J. (1993). Successive changes in distribution patterns as an adaptive strategy in the bivalve *Macoma balthica* (L.) in the Wadden Sea. *Helgoländer Meeresunters* **47**, 287–304.

Beukema, J.J. and Cadée, G.C. (1997). Local differences in macrozoobenthic response to enhanced food supply caused by mild eutrophication in a Wadden Sea area: food is only locally a limiting factor. *Limnol. Oceanogr.* **42**, 1424–1435.

Beukema, J.J. and de Vlas, J. (1989). Tidal current transport of thread-drifting postlarval juveniles of the bivalve *Macoma balthica* from the Wadden Sea to the North Sea. *Mar. Ecol. Prog. Ser.* **52**, 193–200.

Boon, A.R. and Duineveld, G.C.A. (1998). Chlorophyll *a* as a marker for bioturbation and carbon flux in southern and central North Sea sediments. *Mar. Ecol. Prog. Ser.* **162**, 33–43.

Boudreau, B.P. (1986). Mathematics of tracer mixing in sediments. II. Nonlocal mixing and biological conveyor-belt phenomena. *Am. J. Sci.* **286**, 199–238

Boudreau, B.P. (1997). *Diagenetic Models and their Implementation.* Springer, Berlin.

Boudreau, B.P. and Imboden, D. M. (1987). Mathematics of tracer mixing in sediments. III. The theory of nonlocal mixing within sediments. *Am. J. Sci.* **287**, 693–719.

Boudreau, B.P. and Jørgensen, B.B. (1999). Diagenesis and the benthic boundary layer. In: *Biogeochemistry of the Benthic Boundary Layer* (Ed. by B.P. Boudreau and B.B. Jørgensen). Oxford University Press, Oxford (in press).

Boudreau, B.P. and Marinelli, R.L. (1994). A modelling study of discontinuous irrigation. *J. Mar. Res.* **52**, 947–968,

Brönmark, C., Klowiewski, S.P. and Stein, R.A. (1992). Indirect effects of predation in a freshwater, benthic food chain. *Ecology* **73**, 1662–1674.

Burdige, D.J. (1991). The kinetics of organic matter mineralization in anoxic marine sediments. *J. Mar. Res.* **49**, 727–761.

Caffrey, J.M., Cloern, J.E. and Grenz, C. (1998). Changes in production and respiration during a spring phytoplankton bloom in San Francisco Bay, California, USA: implications for net ecosystem metabolism. *Mar. Ecol. Prog. Ser.* **172**, 1–12.

Canfield, D.E. and Des Marais, D.J. (1993). Biogeochemical cycling of carbon, sulfur and free oxygen in a microbial mat. *Geochim. Cosmochim. Acta* **57**, 3971–3984.

Canfield, D.E., Thamdrup, B. and Hansen, B.W. (1993). The anaerobic degradation of organic matter in Danish coastal sediments: Fe reduction, Mn reduction, and sulfate reduction. *Geochim. Cosmochim. Acta* **57**, 3867–3883.

Carpenter, S.R., Kitchell, J.F. and Hodgson, J.R. (1985). Cascading trophic interactions and lake productivity. *Bioscience* **35**, 634–639.

Carpenter, S.R., Kitchell, J.F., Hodgson, J.R., Cochran, P.A., Elser, J.J., Elser, M.M., Lodge, D.M., Kretchmer, D. and He, X. (1987). Regulation of lake primary productivity by food web structure. *Ecology* **68**, 1863–1876.

Cole, J.J. and Cloern, J.E. (1984). Significance of biomass and light availability to phytoplankton productivity in San Francisco Bay. *Mar. Ecol. Prog. Ser.* **17**, 15–24.

Commito, J.A. and Ambrose, W.G. Jr. (1985). Multiple trophic levels in soft-bottom communities. *Mar. Ecol. Prog. Ser.* **26**, 289–293.

Cowie, G.L. and Hedges, J.I. (1994). Biochemical indicators of diagenetic alteration in natural organic matter mixtures. *Nature* **369**, 304–307.

Craeymeersch, J.A. (1991). Applicability of the abundance/biomass comparison method to detect pollution effects on intertidal macrobenthic communities. *Hydrobiol. Bull.* **24**, 133–140.

Crank, J. (1986). *The Mathematics of Diffusion.* Clarendon Press, Oxford.

Dame, R.F. (ed.) (1993). *Bivalve Filter Feeders in Estuarine and Coastal Ecosystem Processes.* NATO ASI Series, Series G: Ecological sciences Vol. 33. Springer, Berlin.

Dame, R.F. (1996). *Ecology of Marine Bivalves. An Ecosystem Approach.* CRC Press, Boca Raton.

Dame, R.F. and Prins, T.C. (1998). Bivalve carrying capacity in coastal ecosystems. *Aq. Ecol.* **31**, 409–421.

Dauer, D.M. (1993). Biological criteria, environmental health and estuarine macrobenthic community structure. *Mar. Pollut. Bull.* **26**, 249–257.

234 P.M.J. HERMAN *ET AL.*

Dauer, D.M., Maybury, C.A. and Ewing, R.M. (1981). Feeding behavior and general ecology of several spionid polychaetes from the Chesapeake Bay. *J. Exp. Mar. Biol. Ecol.* **54**, 21–38.

Dauwe, B. and Middelburg, J.J. (1998). Amino acids and hexosamines as indicators of organic matter degradation state in North Sea sediments. *Limnol. Oceanogr.* **43**, 782–798.

Dauwe, B., Herman, P.M.J. and Heip, C.H.R. (1998). Community structure and bioturbation potential of macrofauna at four North Sea stations with contrasting food supply. *Mar. Ecol. Prog. Ser.* **173**, 67–83.

Dauwe, B., Middelburg, J.J., van Rijswijk, P., Sinke, J., Herman, P.M.J. and Heip, C.H.R. (1999a). Enzymatically hydrolyzable amino acids in North Sea sediments and their possible implication for sediment nutritional values. *J. Mar. Res.* **57**, 109–134.

Dauwe, B., Middelburg, J.J. and Herman, P.M.J. (1999b). Degradability of organic matter in North Sea and intertidal sediments. *Mar. Ecol. Prog. Ser.* (in press).

de Leeuw, J. W. and C. Largeau. (1993). A review of macromolecular organic compounds that comprise living organisms and their role in kerogen, coal, and petroleum formation. In: *Organic Geochemistry: Principles and Applications* (Ed. by M.H. Engel and S.A. Macko), pp. 23–63. Plenum Press, New York.

De Ruiter, P.C., Neutel, A.M. and Moore, J.C. (1995). Energetics, patterns of interaction strengths, and stability in real ecosystems. *Science* **269**, 1257–1260.

De Vries, I. (1984). The carbon balance of a saline lake (Lake Grevelingen, The Netherlands). *Neth. J. Sea Res.* **19**, 511–528.

Diehl, S. (1992). Fish predation and benthic community structure: the role of omnivory and habitat complexity. *Ecology* **73**, 1646–1661.

Dollar, S.J., Smith, S.V., Vink, S.M., Obrebski, S. and Hollibaugh, J.T. (1991). Annual cycle of benthic nutrient fluxes in Tomales Bay, California, and contribution of the benthos to total ecosystem metabolism. *Mar. Ecol. Prog. Ser.* **79**, 115–125.

Ens, B.J. and Goss-Custard, J.D. (1984). Interference among oystercatchers, *Haematopus ostralegus*, feeding on mussels, *Mytilus edulis*, on the Exe Estuary. *J. Anim. Ecol.* **53**, 217–231.

Fauchald, K. and Jumars, P.A. (1979). The diet of worms: a study of polychaete feeding guilds. *Oceanogr. Mar. Biol. Annu. Rev.* **17**, 193–284.

Foreman, K., Valiela, I. and Sarda, R. (1995). Control of benthic marine food webs. *Sci. Mar.* **59**, 119–128.

Fréchette, C., Butman, C.A. and Geyer, W.R. (1989). The importance of boundary-layer flows in supplying phytoplankton to the benthic suspension feeder, *Mytilus edulis* L. *Limnol. Oceanogr.* **34**, 19–36.

Giddins, R.L., Lucas, J.S., Neilson, M.J. and Richards, G.N. (1986). Feeding ecology of the mangrove crab *Neosarmatium smithi* (Crustacea: Decapoda: Sesarmidae). *Mar. Ecol. Prog. Ser.* **33**, 147–155.

Green, M.O., Hewitt, J.E. and Thrush, S.F. (1998). Seabed drag coefficient over natural beds of horse mussels (*Atrina zelandica*). *J. Mar. Res.* **56**, 613–637.

Grossmann, S. and Reichardt, W. (1991). Impact of *Arenicola marina* on bacteria in intertidal sediments. *Mar. Ecol. Prog. Ser.* **77**, 85–93.

Günther, C.-P. (1991). Settlement of *Macoma balthica* on an intertidal sandflat in the Wadden Sea. *Mar. Ecol. Prog. Ser.* **76**, 73–79.

Günther, C.-P. (1992). Settlement and recruitment of *Mya arenaria* L. in the Wadden Sea. *J. Exp. Mar. Biol. Ecol.* **159**, 203–215.

Güss, S. (1998). Oxygen uptake at the sediment–water interface simultaneously measured using a flux chamber method and microelectrodes: must a diffusive boundary layer exist? *Estuar. Coast. Shelf Sci.* **46**, 143–156.

Hairston, N.G. Jr. and Hairston, N.G. Sr. (1993). Cause–effect relationships in energy flow, trophic structure, and interspecific interactions. *Am. Nat.* **142**, 379–441.

Hall, S.J. (1994). Physical disturbance and marine benthic communities: life in unconsolidated sediments. *Oceanogr. Mar. Biol. Annu. Rev.* **32**, 179–239.

Hall, S.J. and Raffaelli, D. (1991). Food-web patterns: lessons from a species-rich web. *J. Anim. Ecol.* **60**, 823–842.

Hall, S.J. and Raffaelli, D.G. (1993). Food webs: theory and reality. *Adv. Ecol. Res.* **24**, 187–239.

Harding, L.W., Meeson, B.W. and Fisher, T.R. (1986). Phytoplankton production in two east coast estuaries: photosynthesis–light functions and patterns of carbon assimilation in Chesapeake and Delaware Bays. *Est. Coast. Shelf Sci.* **23**, 773–806.

Hargrave, B.T. and Phillips, G.A. (1989). Decay times of organic carbon in sedimented detritus in a macrotidal estuary. *Mar. Ecol. Prog. Ser.* **56**, 217–279.

Hedges, J.I. and Oades, J.M. (1997). Comparative organic geochemistries of soils in marine sediments. *Org. Geochem.* **27**, 319–361.

Heip, C. (1989). The ecology of the estuaries of Rhine, Meuse and Scheldt in the Netherlands. *Sci. Mar.* **53**, 457–463.

Heip, C. (1995). Eutrophication and zoobenthos dynamics. *Ophelia* **41**, 113–136.

Heip, C.H.R., Goosen N.K., Herman, P.M.J., Kromkamp J., Middelburg, J.J. and Soetaert, K. (1995). Production and consumption of biological particles in temperate tidal estuaries. *Oceanogr. Mar. Biol. Annu. Rev.* **33**, 1–150.

Heip, C.H.R., Duineveld, G., Flach, E., Graf, G., Helder, W., Herman, P.M.J., Lavaleye, M., Middelburg, J.J., Pfannkuche, O., Soetaert, K., Soltwedel, T., de Stigter, H., Thomsen, L., Vanaverbeke, J. and de Wilde P. (1999). The role of the benthic biota in sedimentary metabolism and sediment–water exchange processes in the Goban Spur area (NE Atlantic). *Deep Sea Res.* (in press).

Herman, P.M.J. (1993). A set of models to investigate the role of benthic suspension feeders in estuarine ecosystems. In: *Bivalve Filter Feeders in Estuarine and Coastal Ecosystem Processes* (Ed. by R. Dame), pp. 421–454. NATO ASI Series G33. Springer, Berlin.

Herman, P.M.J. and Scholten, H. (1990). Can suspension feeders stabilise estuarine ecosystems? In: *Trophic Relationships in the Marine Environment* (Ed. by M. Barnes and R.N. Gibson), pp. 104–116. Aberdeen University Press, Aberdeen.

Holt, R.D. and Polis, G.A. (1997). A theoretical framework for intraguild predation. *Am. Nat.* **149**, 745–764.

Hüttel, M. and Gust, G. (1992). Solute release mechanisms from confined sediment cores in stirred benthic chambers and flume flows. *Mar. Ecol. Prog. Ser.* **82** 187–197.

Hylleberg, J. (1975). Selective feeding by *Abarenicola pacifica* with notes on *Abarenicola vagabunda* and a concept of gardening in lugworms. *Ophelia* **14**, 113–137.

Ittekkot, V. (1988). Global trends in the nature of organic matter in river suspensions. *Nature* **332**, 436–438.

Josefson, A.B., Jensen, J.N. and Ærtjeberg, G. (1993). The benthos community structure anomaly in the late, 1970s and early, 1980s—a result of a major food pulse? *J. Exp. Mar. Biol. Ecol.* **172**, 31–45.

Keil, R.G., Montluçon, D.B., Prahl, F.G. and Hedges, J.I. (1994). Sorptive preservation of labile organic matter in marine sediments. *Nature* **370**, 549–552.

Keil, R.G., Mayer, L.M., Quay, P.D., Richey, J.E. and Hedges, J.I. (1997). Loss of organic matter from riverine particles in deltas. *Geochim. Cosmochim. Acta* **61**, 1507–1511.

Kemp, P.F. (1987). Potential impact on bacteria of grazing by a macrofaunal depositfeeder, and the fate of bacterial production. *Mar. Ecol. Prog. Ser.* **36**, 151–161.

Kemp, P.F. (1990). The fate of benthic bacterial production. *Rev. Aq. Sci.* **2**, 109–124.

Kemp, W.M., Sampou, P.A., Garber, J., Tuttle, J. and Boynton, W.R. (1992). Seasonal depletion of oxygen from bottom waters of Chesapeake Bay: roles of benthic and planktonic respiration and physical exchange processes. *Mar. Ecol. Prog. Ser.* **85**, 137–152.

Khalili, A., Basu, A.J. and Huettel, M. (1997). A non-Darcy model for recirculating flow through a fluid-sediment interface in a cylindrical container. *Acta Mech.* **123**, 75–87.

Klump, J.V. and Martens, C.S. (1989). The seasonality of nutrient regeneration in an organic-rich coastal sediment: kinetic modeling of changing pore-water nutrient and sulfate distributions. *Limnol. Oceanogr.* **34**, 559–577.

Kolar, C.S. and Rahel, F.J. (1993). Interaction of a biotic factor (predator presence) and an abiotic factor (low oxygen) as an influence on benthic invertebrate communities. *Oecologia* **95**, 210–219.

Koseff, J.R., Holen, J.K., Monismith, S.G. and Cloern, J.E. (1993). Coupled effects of vertical mixing and benthic grazing on phytoplankton populations in shallow, turbid estuaries. *J. Mar. Res.* **51**, 843–868.

Kranck, K., Smith P.C. and Milligan, T.G. (1996a). Grain-size characteristics of fine-grained unflocculated sediments I: one-round distributions. *Sedimentology* **43**, 589–596.

Kranck, K., Smith P.C. and Milligan, T.G. (1996b). Grain-size characteristics of fine-grained unflocculated sediments: multi-round distributions. *Sedimentology* **43**, 597–606.

Kristensen, E. (1993). Seasonal variations in benthic community metabolism and nitrogen dynamics in a shallow, organic-poor Danish lagoon. *Estuar. Coast. Shelf Sci.* **36**, 565–586.

Kristensen, E., Anderson, F.Ø. and Blackburn, T.H. (1992). Effects of benthic macro-fauna and temperature on degradation of macroalgal detritus: the fate of organic carbon. *Limnol. Oceanogr.* **37**, 1404–1419.

Lee, S. and Fuhrmann, J.A. (1987). Relationships between biovolume and biomass of naturally derived marine bacterioplankton. *Appl. Environ. Microbiol.* **53**, 1298–1303.

Lodge, D.M., Kershner, M.W., Aloi, J.E. and Covich, A.P. (1994). Effects of an omnivorous crayfish (*Orconectes rusticus*) on a freshwater littoral food web. *Ecology* **75**, 1265–1281.

Lohse, L., Epping, E.H.G., Helder, W. and van Raaphorst, W. (1996). Oxygen pore-water profiles in continental shelf sediments of the North Sea: turbulent versus molecular diffusion. *Mar. Ecol. Prog. Ser.* **145**, 63–75.

Lopez, G.R. and Levinton, J.S. (1987). Ecology of deposit-feeding animals in marine sediments. *Q. Rev. Biol.* **62**, 235–260.

Lucas, L.V., Cloern, J.E., Koseff, J.R., Monismith, S.G. and Thompson, J.K. (1998). Does the Sverdrup critical depth model explain bloom dynamics in estuaries? *J. Mar. Res.* **56**, 375–415.

McCann, K., Hastings, A. and Huxel, G.R. (1998). Weak trophic interactions and the balance of nature. *Nature* **395**, 794–798.

McQueen, D.J., Post, J.R. and Mills, E.L. (1986). Trophic relationships in freshwater pelagic ecosystems. *Can. J. Fish. Aquat. Sci.* **43**, 1571–1581.

Mayer, L. (1993). Organic matter at the sediment–water interface. In: *Organic Geochemistry: Principles and Applications* (Ed. by M.H. Engel and S.A. Macko), pp. 171–184. Plenum Press, New York.

Mayer, L.M. (1994). Surface area control of organic carbon accumulation in continental shelf sediments. *Geochim. Cosmochim. Acta* **58**, 1271–1284.

Mayer, L.M., Jumars, P.A., Taghon, G.L., Macko, S.A. and Trumbore S. (1993). Low-density particles as potential nitrogenous foods for benthos. *J. Mar. Res.* **51**, 373–389.

Mayer, L.M., Schick, L.L., Sawyer, T., Plante, C.J., Jumars, P.A. and Self, R. L. (1995). Bioavailable amino acids in sediments: a biomimetic, kinetic-based approach. *Limnol. Oceanogr.* **40**, 511–520.

Meire, P.M., Seys, J.J., Ysebaert, T.J. and Coosen, J. (1991). A comparison of the macrobenthic distribution and community structure between two estuaries in SW Netherlands. In: *Estuaries and Coasts: Spatial and Temporal Intercomparisons* (Ed. by M. Elliot and J.P. Ducrotoy), pp. 221–230. Olsen & Olsen, Fredensborg.

Menge, B.A., Daley, B. and Wheeler, P.A. (1996). Control of interaction strength in marine benthic communities. In: *Food Webs. Integration of Patterns and Dynamics* (Ed. by G.A. Polis and K.O. Winemiller), pp. 258–274. Chapman and Hall, New York.

Middelburg, J.J. (1989). A simple rate model for organic matter decomposition in marine sediments. *Geochim. Cosmochim. Acta* **53**, 1577–1581.

Middelburg, J.J., Klaver, G, Nieuwenhuize, J., Wielemaker, A., de Haas, W. and van der Nat, J.F.W.A. (1996). Organic matter mineralization in intertidal sediments along an estuarine gradient. *Mar. Ecol. Prog. Ser.* **132**, 157–168.

Middelburg, J.J., Nieuwenhuize, J. and van Breugel, P. (1999). Black carbon in marine sediments. *Mar. Chem.* (in press).

Nehls, G. and Thiel, M. (1993). Large-scale distribution patterns of the mussel *Mytilus edulis* in the Wadden Sea of Schleswig-Holstein—do storms structure the ecosystem? *Neth. J. Sea Res.* **31**, 181–187.

Neira, C. and Höpner, T. (1994). The role of *Heteromastus filiformis* (Capitellidae, Polychaeta) in organic carbon cycling. *Ophelia* **39**, 55–73.

Nichols, F.H. (1977). Infaunal biomass and production on a mudflat, San Francisco Bay. In: *Ecology of Marine Benthos* (Ed. by B.C. Coull), pp. 339–358. Belle W. Baruch Library in Marine Science. University of South Carolina Press, Columbia.

Nienhuis, P.H. (1992). Eutrophication, water management, and the functioning of Dutch estuaries and coastal lagoons. *Estuaries* **15**, 538–548.

Nixon, S.W. (1981). Remineralization and nutrient cycling in coastal marine ecosystems. In: *Estuaries and Nutrients* (Ed. by B.J. Neilson and L.E. Cronin), pp. 111–138. Humana Press, Clifton, New Jersey.

Odum, W.E. and Heald, E.J. (1975). The detritus-based food web of an estuarine mangrove community. In: *Estuarine Research* (Ed. by L.E. Cronin), pp. 265–286. Academic Press, New York.

Officer, C.B., Smayda, T.J. and Mann, R. (1982). Benthic filter feeding: a natural eutrophication control. *Mar. Ecol. Prog. Ser.* **9**, 203–210.

O'Riordan, C.A., Monismith, S.G. and Koseff, J.R. (1995). The effect of bivalve excurrent jet dynamics on mass transfer in a benthic boundary layer. *Limnol. Oceanogr.* **40**, 330–344.

Parsons, T.R., Takahashi, M. and Hargrave, B. (1977). *Biological Oceanographic Processes,* 2nd edn. Oxford: Pergamon Press.

Pearson, T.H. and Rosenberg, R. (1978). Macrobenthic succession in relation to organic enrichment and pollution of the marine environment. *Oceanogr. Mar. Biol. Ann. Rev.* **16**, 229–311.

Pelegrí, S.P., Nielsen, L.P. and Blackburn, T.H. (1994). Denitrification in estuarine sediment stimulated by the irrigation activity of the amphipod *Corophium volutator. Mar. Ecol. Prog. Ser.* **105**, 285–290.

Persson, L., Anderson, G., Hamrin, S.F. and Johansson, L. (1988). Predator regulation and primary production along the productivity gradient of temperate lake ecosystems. In: *Complex Interactions in Lake Communities* (Ed. by S.R. Carpenter), pp. 45–65. Springer, New York.

Persson, L., Bengtsson, J., Menge, B.A. and Power, M.E. (1996). Productivity and consumer regulation—concepts, patterns, and mechanisms. In: *Food Webs.*

Integration of Patterns and Dynamics (Ed. by G.A. Polis and K.O. Winemiller), pp. 369–434. Chapman and Hall, New York.

Peterson, C.H. (1982). Clam predation by whelks (*Busycon* spp.): experimental tests of the importance of prey size, prey density, and seagrass cover. *Mar. Biol.* **66**, 159–170.

Piersma, T. and Koolhaas, A. (1997). *Shorebirds, Shellfish(eries) and Sediments Around Griend, Western Wadden Sea, 1988–1996.* NIOZ-report, 1997–7. Netherlands Institute for Sea Research. Texel, The Netherlands.

Pimm, S.L. (1982). *Food Webs.* Chapman and Hall, London.

Pimm, S.L. and Lawton, J.H. (1978). On feeding on more than one trophic level. *Nature* **275**, 542–544.

Pimm, S.L., Lawton, J.H. and Cohen, J.E. (1991). Food web patterns and their consequences. *Nature* **350**, 669–674.

Polis, G.A. (1991). Complex trophic interactions in deserts—an empirical critique of food-web theory. *Am. Nat.* **138**, 123–155.

Polis, G.A., Myers, C.A. and Holt, R.D. (1989). The ecology and evolution of intraguild predation: potential competitors that eat each other. *Annu. Rev. Ecol. Syst.* **20**, 297–330.

Posey, M.H. and Hines, A.H. (1991). Complex predator–prey interactions within an estuarine benthic community. *Ecology* **72**, 2155–2169.

Retraubun, A.S.W., Dawson, M. and Evans, S.M. (1996). The role of the burrow funnel in feeding processes in the lugworm *Arenicola marina* (L.). *J. Exp. Mar. Biol. Ecol.* **202**, 107–118.

Rhoads, D.C. (1974). Organism–sediment relations on the muddy sea floor. *Oceanogr. Mar. Biol. Ann. Rev.* **12**, 263–300.

Rhoads, D.C. and Young, D.K. (1970). The influence of deposit-feeding organisms on sediment stability and community trophic structure. *J. Mar. Res.* **28**, 150–178.

Rhoads, D.C., McCall, P.L. and Yingst, J.Y. (1978). Disturbance and production on the estuarine seafloor. *Am. Sci.* **66**, 577–586.

Rice, D.L. (1982). The detritus nitrogen problem: new observations and perspectives from organic geochemistry. *Mar. Ecol. Prog. Ser.* **9**, 153–162.

Rice, D.L. and Rhoads, D.C. (1989). Early diagenesis of organic matter and the nutritional value of sediment. In: *Ecology of Marine Deposit Feeders* (Ed. by G. Lopez, G. Taghon and J. Levinton), pp. 309–317. Springer, Berlin.

Riley, G.A. (1956). Oceanography of Long Island Sound, 1952–1954. II. Physical oceanography. *Bull. Bingham Oceanogr. Collect.* **15**, 15–46.

Risgaard-Petersen, N. and Jensen, K. (1997). Nitrification and denitrification in the rhizosphere of the aquatic macrophyte *Lobelia dortmanna* L. *Limnol. Oceanogr.* **42**, 529–537.

Risgaard-Petersen, N., Rysgaard, S., Nielsen, L.P. and Revsbech, N.P. (1994). Diurnal variation of denitrification and nitrification in sediments colonized by benthic microphytes, *Limnol. Oceanogr.* **39**, 573–579.

Robertson, A.I. (1988). Decomposition of mangrove leaf litter in tropical Australia. *J. Exp. Mar. Biol. Ecol.* **116**, 235–247.

Robertson, A.I., Alongi, D.M. and Boto, K.G. (1992). Food chains and carbon fluxes. In: *Tropical Mangrove Ecosystems* (Ed. by A.I. Robertson and D.M. Alongi), pp. 293–326. American Geophysical Union, Washington.

Rocha, C. (1998). Rhythmic ammonium regeneration and flushing in intertidal sediments of the Sado estuary. *Limnol. Oceanogr.* **43**, 823–831.

Roden, E.E., Tuttle, J.H., Boynton, W.R. and Kemp, W.M. (1995). Carbon cycling in mesohaline Chesapeake Bay sediments. 1: POC deposition rates and mineralization pathways. *J. Mar. Res.* **53**, 799–819.

Rosenfeld, J.K. (1981). Nitrogen diagenesis in Long Island Sound sediments. *Am. J. Sci.* **281**, 436–462.

Rysgaard, S., Christensen, P.B. and Nielsen, L.P. (1995). Seasonal variation in nitrification and denitrification in estuarine sediment colonized by benthic microalgae and bioturbating infauna. *Mar. Ecol. Prog. Ser.* **126**, 111–121.

Sandberg, E. (1994). Does short-term oxygen depletion affect predator–prey relationships in zoobenthos? Experiments with the isopod *Saduria entomon*. *Mar. Ecol. Prog. Ser.* **103**, 73–80.

Sandberg, E. (1997). Does oxygen deficiency modify the functional response of *Saduria entomon* (Isopoda) to *Bathyporeia pilosa* (Amphipoda). *Mar. Biol.* **129**, 499–504.

Schallenberg, M. and Kalf, J. (1993). The ecology of sediment bacteria in lakes and comparisons with other aquatic ecosystems. *Ecology* **74**, 919–934.

Schmidt, J.L., Deming, J.W., Jumars, P.A. and Keil, R.G. (1998). Constancy of bacterial abundance in surficial marine sediments. *Limnol. Oceanogr.* **43**, 976–982

Sheridan, P. (1997). Benthos of adjacent mangrove, seagrass and non-vegetated habitats in rookery bay, Florida, USA. *Estuar. Coast. Shelf Sci.* **44**, 455–469.

Small, L.F., McIntire, C.D., MacDonald, K.B., Lara-Lara, J.R., Frey, B.E., Amspoker, M.C. and Winfield, T. (1990). Primary production, plant and detrital biomass and particle transport in the Columbia River estuary. *Prog. Oceanogr.* **25**, 175–210.

Snelgrove, P.V.R. and Butman, C.A. (1994). Animal–sediment relationships revisited: cause versus effect. *Oceanogr. Mar. Biol. Ann. Rev.* **32**, 111–177.

Soetaert, K., Herman, P.M.J. and Kromkamp, J.C. (1994). Living in the twilight: estimating net phytoplankton growth in the Westerschelde estuary (The Netherlands) by means of a global ecosystem model (MOSES). *J. Plankton Res.* **16**, 1277–1301.

Soetaert, K., Herman, P.M.J. and Middelburg, J.J. (1996a). A model of early diagenetic processes from the shelf to abyssal depths. *Geochim. Cosmochim. Acta* **60**, 1019–1040.

Soetaert, K., Herman, P.M.J. and Middelburg, J.J. (1996b). Dynamic response of deep-sea sediments to seasonal variation: a model. *Limnol. Oceanogr.* **41**, 1651–1668.

Soetaert, K., Herman, P.M.J., Middelburg, J.J., Heip, C., deStigter, H.S., van Weering, T.C.E., Epping, E. and Helder, W. (1996c). Modeling 210Pb-derived mixing activity in ocean margin sediments: diffusive versus nonlocal mixing. *J. Mar. Res.* **54**, 1207–1227.

Stamhuis, E.J., Dauwe, B. and Videler, J.J. (1998). How to bite the dust: morphology, motion pattern and function of the feeding appendages of the deposit-feeding thalassinid shrimp *Callianassa subterranea*. *Mar. Biol.* **132**, 43–58.

Strong, D.R. (1992). Are trophic cascades all wet? Differentiation and donor-control in speciose ecosystems. *Ecology* **73**, 747–754.

Sun, M., Aller, R.C. and Lee C. (1991). Early diagenesis of chlorophyll-*a* in Long Island Sound sediments: a measure of carbon flux and particle reworking. *J. Mar. Res.* **49**, 379–401.

Sun, M.Y., Aller, R.C. and Lee, C. (1984). Spatial and temporal distributions of sedimentary chloropigments as indicators of benthic processes in Long Island Sound. *J. Mar. Res.* **52**, 149–176.

Sutherland, W.J. (1996). *From Individual Behaviour to Population Ecology*. Oxford University Press, Oxford.

Svensson U. and Rahm, L. (1991). Toward a mathematical model of oxygen transfer to and within bottom sediments. *J. Geophys. Res.* **96**, 2777–2783.

Taghon, G.L. and Greene, R.R. (1992). Utilization of deposited and suspended particulate matter by benthic "interface feeders". *Limnol. Oceanogr.* **37**, 1370–1391.

Tenore, K.R. (1988). Nitrogen in benthic food chains. In: *Nitrogen Cycling in Coastal Marine Environments* (Ed. by T.H. Blackburn and J. Sørensen), pp. 191–206. John Wiley, New York.

Tenore, K.R., Cammen, L., Findlay, S.E.G. and Phillips, N. (1982). Perspectives of research on detritus: do factors controlling the availability of detritus to macroconsumers depend on its source? *J. Mar. Res.* **40**, 473–490.

Tsutsumi, H., Fukunaga, S., Fujita, N. and Sumida, M. (1990). Relationship between growth of *Capitella* sp. and organic enrichment of the sediment. *Mar. Ecol. Prog. Ser.* **63**, 157–162.

Vadas, R.L.Jr. (1990). The importance of omnivory and predator regulation of prey in freshwater fish assemblages of North America. *Environ. Biol. Fish.* **27**, 285–302.

Valente, R.M., Rhoads, D.C., Germano, J.D. and Cabelli, V.J. (1992). Mapping of benthic enrichment patterns in Narrangansett Bay, Rhode Island. *Estuaries* **15**, 1–17.

Van de Bund, W.J., Goedkoop, W. and Johnson, R.K. (1994). Effects of deposit-feeder activity on bacterial production and abundance in profundal lake sediment. *J. North Am. Benthol. Soc.* **13**, 532–539.

Van de Koppel, J., Rietkerk, M. and Weissing, F.J. (1997). Catastrophic vegetation shifts and soil degradation in terrestrial grazing systems. *Trends Ecol. Evol.* **12**, 352–356.

Verhagen, J.H.G. (1986). *Tidal Motion, and the Seston Supply to the Benthic Macrofauna in the Oosterschelde.* DHL Report R1310–14. Delft Hydraulics, Delft.

Wakeham, S.G., Lee, C., Hedges, J.I., Hernes, P.J. and Peterson, M.L. (1997). Molecular indicators of diagenetic status in marine organic matter. *Geochim. Cosmochim. Acta* **61**, 5363–5369.

Warwick, R.M. (1986). A new method for detecting pollution effects on marine macrobenthic communities. *Mar. Biol.* **92**, 557–562.

Warwick, R.M. and Price, R. (1975). Macrofauna production in an estuarien mud-flat. *J. Mar. Biol. Assoc. UK* **55**, 1–18.

Weston, D.P. (1990). Quantitative examination of macrobenthic communtiy changes along an organic enrichment gradient. *Mar. Ecol. Prog. Ser.* **61**, 233–244.

Westrich, J.T. and Berner, R.A. (1984). The role of sedimentary organic matter in sulfate reduction: the G-model testes. *Limnol. Oceanogr.* **29**, 236–249.

Wetsteyn, L.P.M.J. and Kromkamp, J.C. (1994). Turbidity, nutrients and phytoplankton primary production in the Oosterschelde (The Netherlands) before, during and after a large-scale coastal engineering project (1980–1990). *Hydrobiologia* **282/283**, 61–78.

Wheatcroft, R.A., Jumars, P.A., Smith, C.R. and Nowell, A.R.M. (1990). A mechanistic view of the particulate biodiffusion coefficient: step lengths, rest periods and transport directions. *J. Mar. Res.* **48**, 177–207.

Wildish, D.J. and Kristmanson, D.D. (1979). Tidal energy and sublittoral macrobenthic animals in estuaries. *J. Fish. Res. Board Can.* **36**, 1197–1206.

Wildish, D.J. and Kristmanson, D.D. (1993). Hydrodynamic control of bivalve filter feeders: a conceptual view. In: *Bivalve Filter Feeders in Estuarine and Coastal Ecosystem Processes* (Ed. by R.F. Dame), pp. 299–324. NATO ASI Ser. Vol. G33. Springer, Berlin.

Wildish, D.J. and Kristmanson, D.D. (1997). *Benthic Suspension Feeders and Flow.* Cambridge University Press. Cambridge.

Wildish, D.J., Peer, D.L. and Greenberg, D.A. (1986). Benthic macrofaunal production in the Bay of Fundy and possible effects of a tidal power barrage at Economy Point-Cape Tenny. *Can. J. Fish. Aquat. Sci.* **43**, 2410–2417.

Wilson, J.B. and Agnew, A.D.Q. (1992). Positive-feedback switches in plant communities. *Adv. Ecol. Res.* **23**, 263–336.

Wissinger, S. and McGrady, J. (1993). Intraguild predation and competition between larval dragonflies: direct and indirect effects of shared prey. *Ecology* **74**, 207–218.

Yodzis, P. (1984). How rare is omnivore? *Ecology* **65**, 321–323.

Integrated Coastal Management: Sustaining Estuarine Natural Resources

S. CROOKS AND R.K. TURNER

I. SUMMARY

Many countries are now recognizing the ecological and economic importance of the resources in their coastal zones. Coastal ecosystems carry out a wide range of processes and functions which the wider environment and human society benefit from and are reliant on. These ecosystems are, however, under mounting severe stress from the associated pressures of rapid human population expansion and climate change-induced sea-level rise. In large part, loss and

ADVANCES IN ECOLOGICAL RESEARCH VOL. 29
ISBN 0–12–013929–4

degradation of these resources has been due to a failure to appreciate, and account for, the full range of goods and services that they provide.

Galvanized by the United Nations Conference on Environment and Development meeting of 1992, there has been a search for methods to build the capacity of coastal nations and communities to manage their coastal and estuarine resources in a sustainable manner. By adopting a systems perspective attention is focused on the fundamental fact that coastal areas should be viewed as linked ecological–socioeconomic systems which are spatially and temporally co-evolving. Management therefore requires an integrated approach to be undertaken, not only by institutions but across scientific disciplines, to provide the required information upon which sound policy decisions are to be based.

II. INTRODUCTION

Coastal and estuarine environments are some of the most productive ecological systems on Earth and recognized to be of extremely high "value" to human society (Costanza et al., 1997). The coast is, however, a difficult place to manage, involving a natural system that is ever changing and is inhabited and pressurized by a relentlessly growing and increasingly demanding human society. Although exact numbers are a matter for debate, because of different spatial definitions of what constitutes a coastal zone, it is currently estimated that in excess of 37% of the human population (more than 2.1 billion people) are concentrated in coastal areas (Vitousek and Mooney, 1997). A large, but unknown, proportion of this population lives on the embanked floodplain shores of estuaries, drawn to the wealth of potential marine resources, and the economic opportunities that have built up around trading ports. In the past, society had unquestionably assumed that marine waters could provide an inexhaustible bounty of resources, could remove and assimilate all land-borne pollutants and waste, that boundless areas of coastal lowlands were available to be occupied and that nature would be unaffected by all this human activity (Kildow, 1997). It is now abundantly clear that there are biogeochemical limits to this development process. The pressures and damaging impacts of human expansion, and inadequate ecological and socioeconomic management, are reflected in over-exploitation of fisheries resources, destruction of coastal ecosystems, and deleterious effects of land-based contamination and eutrophication (Botsford et al., 1997). In the future, coastal areas need to be viewed as the spatial and temporal context for jointly determined socioeconomic and ecological systems on a co-evolutionary development path (Turner et al., 1997a). The overall scale of economic activity will need to be constrained (Daly, 1992).

Management of the coastal zone must contend not only with anthropogenic pressures and impacts but also with the implications of future uncertainty regarding climate change, accelerated sea-level rise and changing storm patterns

(Warrick and Farmer, 1990; Hooligan and Reiners, 1992; Bird, 1993; Warrick *et al.*, 1993; IPCC, 1996; Bijl, 1997; Bijlsma, 1997). Under natural conditions the form of a coastline is an optimal, but ephemeral, morphodynamic response to changing sea level and the impact of wave and tidal energy. The placement of fixed engineering structures (for resource exploitation, sea defence and coastal protection reasons) within this constantly changing system has in many cases reduced the "resilience" of coastlines to respond to the stresses and shocks of environmental change. Loss of intertidal habitats through land-claim and coastal squeeze (the increasing confinement of the intertidal zone between sea defences and rising sea level) is both a symptom of, and factor contributing to, this reduction in coastal resilience, and is associated with a loss of nursery areas for fisheries, biodiversity, coastal defence functions, carbon reservoirs, and buffers to regulate nutrient fluxes (Jickells, 1998).

If the coast and its resources are to be managed in a more sustainable manner, mechanisms must be found to accommodate the pressures from population growth and global environmental change (IPCC, 1996). To do so will require integrated and strategic planning, linking river catchment, estuaries and coastal water management, to prevent development in vulnerable areas, minimization of disruption to environmental processes and, when possible, restoration or creation of coastal habitats and functions to offset losses due to development or sea-level rise. During the past decade a number of countries, such as the UK, with extensive flood-susceptible regions, have begun to consider "coastal realignment" or "managed retreat" (terms describing the landward movement of sea defences), seen as a pragmatic and practicable mechanism by which flood defences in risk-prone areas can be maintained cost effectively whilst at the same time allowing for the restoration of lost intertidal habitats (HOC, 1992, 1998). The feasibility and practicability of a strategy to increase environmental functioning and resilience is one of the most important contemporary issues facing coastal and estuarine managers.

A. Benefits to Society Provided by Coastal and Estuarine Ecosystems

In general terms the cause of ecosystem degradation and loss is often due to a failure to appreciate the full value of beneficial functions provided by such systems. These include: sustaining biodiversity, storage of sediment, flood defence and storm buffering, maintenance of water quality, and support of commercial coastal and marine food chains. Many of the processes and functioning of estuarine systems have been described in previous chapters of this volume, so we will focus on some of the direct management implications of maintaining and restoring intertidal systems.

Estuarine wetlands are known to contribute to the maintenance of water quality (Mitsch and Gosselink, 1993; Reddy and D'Angelo, 1994).

Accreting intertidal deposits are sinks for sediment and also for the metal and organic pollutants bound to them (e.g. Church *et al.*, 1996). Because wetlands have a high rate of biological activity, they are effective in transforming many of the common pollutants found in coastal and estuarine waters into harmless byproducts or essential nutrients which can be utilized for additional biological activity (Kadlec and Knight, 1996; Jickells and Rae, 1997). Estuarine sediments trap nutrients from agricultural runoff and may thus contribute to regulating planktonic blooms (Nedwell *et al.*, this volume; Underwood and Kromkamp, this volume). Eutrophication has become an important management concern over recent years, given the increasing frequency and impact of harmful microalgae and heterotrophic dinoflagellate blooms (Burkholder, 1998). Measures such as coastal realignment to increase the intertidal area and reinstating natural nutrient-buffering systems may contribute to reducing nutrient levels and the severity of phytoplankton blooms. Wetland areas also increase the rate of pathogen die-off in coastal waters by providing a medium that increases exposure to ultraviolet radiation, oxygen, soil–water interactions and the presence of predatory protozoa (Gersberg *et al.*, 1987; Hemond and Benoit, 1988; Breaux *et al.*, 1995). Globally, wetlands may also be involved in global biogeochemical cycles which contribute to atmospheric stability of nitrogen, sulfur, carbon dioxide and methane.

Maintaining intertidal habitats is an important component of an integrated fisheries policy (Done and Reichelt, 1998). In the US, for example, the output of the fisheries industry is valued at 19.8 billion dollars (1994 data), with up to 93% of commercial fisheries species in some regions being dependent at some point in their life cycle on coastal wetlands (Stedman and Hanson, 1997). Saltmarshes and intertidal flats, for example, provide a nursery ground and shelter for many fish and invertebrate species and therefore play an important role in productivity of adjacent waters (Cain and Dean, 1976; Shenker and Dean, 1979; Weinstein and Brooks, 1983; Sogard and Able, 1991; Paterson and Whitfield, 1997; Cattrijsse *et al.*, 1997; Maes *et al.*, 1998; Sarda *et al.*, 1998). As well as providing shelter, marshes also provide a source of carbon for inshore food webs (Kwak and Zedler, 1997; Page, 1997; Paterson and Whitfield, 1997). Knowing that such linkages exist is important for the planning of schemes to restore estuarine habitat (Kwak and Zedler, 1997).

From a flood defence perspective, the value of coastal systems lies in their ability to attenuate wave energy (Frey and Basan, 1985; Brampton, 1992; Asano and Setoguchi, 1996; Moeller *et al.*, 1996; Yang, 1998). In the UK, for example, it was found that at locations where a sufficiently wide swath of marsh exists a simple clay embankment of 3–4 m may be adequate to prevent storm surge flooding. However, as the width of marsh decreases, the need for larger, more robust, flood defences increases, perhaps in exposed areas requiring an 8–12 m high wall protected by rock (Brampton, 1992). This has significant financial implications as at current (1999) prices the smaller wall costs roughly £400 m^{-1} length whereas the larger structure may cost up to £5000 m^{-1} length (King and Lester, 1995). As the UK has over 2000 km of coastal flood defences, the economic consequences of maintaining and upgrading flood defences as intertidal habitat erodes are significant.

B. Pressures on Estuarine Ecosystems

Population growth is often seen as the central cause of environmental degradation and wetland loss, but it is complicated by issues of population migration and settlement, consumption patterns and social deprivation. The United Nations Secretariat (1998) released exploratory long-term projections of possible world-wide population growth up until the year 2150. Based on the medium-fertility scenario (two children per woman), it is estimated that global population will increase gradually towards 10.8 billion over the next 150 years. Such projections, however, are highly sensitive to fertility estimates, with high and low scenarios, which differ by just one child per couple, providing population estimates of between 27 and 3.6 billion persons, respectively, by 2150. Population growth across the globe is projected to be heterogeneous, with an increasing shift in balance towards Asia and Africa. Only in Europe is the population expected to remain static.

Increasingly there is also a trend towards urbanization, with two-thirds of world's largest cities (population exceeding 1.5 million) found in coastal areas. These urban agglomerations are spreading laterally along the coastline in a number of countries. Growth rates in coastal cities are particularly high, supported not only by internal growth but fed by inward economic migration from surrounding rural areas. Urban expansion produces attendant problems such as piecemeal landscape fragmentation and disruption to wildlife and ecological linkages. The sediments derived from construction works, atmospheric fallout from exhaust fumes and urban runoff all carry with them toxic substances that eventually find their way into estuarine sediments (Dakalakis and O'Connor, 1995; Makepiece et al., 1995; Mills, 1997; van Roon, 1997). Development may also reduce estuarine flushing capacity with the consequence that estuaries become more vulnerable to pollutant accumulation and eutrophication. Although a number of nations have passed legislation to limit production of at least some pollutants, many still discharge vast quantities of untreated sewage and industrial waste from expanding coastal cities every day.

Ports have been and, given the continued globalization and trade liberalization trends, will remain very important contributors to national wealth (gross national product) (Kia and O'Neill, 1997; Wang et al., 1998). With increasing globalization there has been a dramatic growth in traffic through ports, particularly with the increase in goods bought from developing countries and south-east Asia. More than 82% of the world's trade, in tonnes, is transported by shipping and therefore at some point passes through port facilities. Over the 12-year period up to 1994, annual world marine freight movements had steadily increased from 46.8 million twenty-foot equivalent units (TEUs) to 125 million TEUs, a compound growth rate of 9% (Kia and O'Neill, 1997). The expansion in shipping volume has been increasingly facilitated by a move towards larger (up to 14 m draft) and faster ships with a greater capacity. These vessels require larger more modern facilities, incorporating more quay length, the dredging of deeper water channels and turning circles, and more land space for storage and transport infrastructure. As well as direct displacement of wetlands through port development and increased risk of pollution, increased global shipping also brings with it the risk of introduction of exotic invasive species which pose threats to local biodiversity and economic productivity (Carlton, 1992; Furlani, 1996; Cohen *et al.*, 1997). Dredging itself may alter sediment accumulation patterns across an estuary, remobilize fine material which, given a change in oxidation state, may release toxic substances, and the dumping of sediment may damage benthic communities.

On a global scale, the impact of climate change on coastal areas is likely to be manifested in a number of ways including: accelerated eustatic sea-level rise, change to ocean currents and wave climates, alteration to storm paths and possible increase in the frequency of extreme events (including El Niño), changes to atmospheric and seawater temperatures, and changes to rainfall patterns with associated impacts on freshwater hydrology and sediment supply (IPCC, 1996). According to IPCC (1996) global sea level has already risen by some 10–25 cm over the last 100 years and, by the end of the next century, is projected to be about a further 50 cm higher (range of uncertainty of 20–86 cm). The net upward movement of relative sea level, a combination of global sea-level rise, regional isostatic and tectonic crustal movements, and local subsidence, poses a problem for coastal settlements because of associated increased vulnerability to flooding and erosion, and landward intrusion of saline waters into freshwater aquifers and other economically important water bodies (e.g. Jelgersma and Tooley, 1992; Chen, 1997).

Coastal marshes in particular are recognized to be under threat from relative sea-level rise (Boorman *et al.*, 1989; Bird, 1993; Pethick, 1993). For these habitats to maintain an elevation sufficient for vegetation survival, and to avoid ecological "drowning", surface accretion must at least keep pace with the rate of sea-level movement (Randerson, 1979; Krone, 1987; Stevenson *et al.*, 1986; Allen, 1990a,b, 1995, 1997; Reed, 1990, 1995; French, 1991, 1993; Churma *et al.*, 1992; Stumpf and Haines, 1998). Disruption to sediment transport pathways by construction works, both in river catchments and along shorelines to prevent soft cliff erosion, has often exacerbated erosion problems in coastal systems (e.g. Sestini, 1992; Kamaludin, 1993; Chen and Zong, 1998). Under conditions of insufficient sediment supply, gradual marsh submergence takes place, and intertidal ecosystems are lost or migrate landward. Loss of intertidal ecosystems is commonplace in most estuarine and deltaic systems. In the Mississippi Delta, Louisiana, for example, where local relative sea level has been rising by about 1 cm per year over the past 50 years, more than 3950 km² of wetlands have been lost (Stumpf and Haines, 1998).

1. Simulation Models of Saltmarsh Development

Geomorphologically, saltmarshes might be defined as high intertidal environments comprised of repeatedly flooded, vegetated platforms, dissected by blind-ending tidal creeks that widen seaward (Allen, 1997). To investigate response to sea-level rise, a range of zero-dimensional, time-stepping models has been constructed describing the vertical growth of minerogenic (sediment dominated) saltmarshes from a number of meso-macrotidal coastlines (Randerson, 1979; Krone, 1987; Allen, 1990a,b, 1995, 1997; French, 1991, 1993) and, less commonly, growth of organogenic (vegetation dominated) microtidal marshes (Chmura *et al.*, 1992). The basis of these models is essentially a mass balance calculation reflecting the change in surface elevation over time, relative to the moving tidal frame in response to mineral and organic sedimentation rates, and the rate of rising sea level. By specifying the sediment supply, tidal and sea-level regimes and initial elevation, the mass balance equation may be integrated over time to reflect the growth of a young marsh from an intertidal flat or the response of a well-established wetland to sea-level fluctuations.

$$\Delta E = \Delta S_{min} + \Delta S_{org} - \Delta M \qquad (1)$$

Over each time step, ΔE is change in surface elevation relative to the tidal frame, ΔS_{min} and ΔS_{org} are the net accretion due to minerogenic and organogenic sedimentation respectively, and ΔM represents net change in relative sea level (Allen, 1997). Points to note are that spatial variability of sedimentation across a marsh surface has yet to be accounted for; predicating the net impact of sea-level rise on organic sedimentation is poorly understood (Reed, 1995); and, although autocompaction is recognized as being not insignificant (Cahoon *et al.*, 1995), models have yet to incorporate this factor specifically.

Although lacking topographical and spatial detail, the outcome from these vertical accretion models is upheld by observations describing sedimentation rate to be strongly related to the hydroperiod (e.g. Cahoon *et al.*, 1995), and to decline rapidly and non-linearly as the marsh surface gains elevation relative to the intertidal frame. A state described as "maturity" is achieved when the marsh surface attains a level some distance lower than the highest astronomical tide (HAT), in equilibrium with the rate of sea-level rise. The models also predict that for minerogenic marshes, given ΔS_{org} is typically a small factor, surface elevation is particularly sensitive to ΔS_{min} and ΔE. On the north Norfolk coastline, French's (1993) model supported corrected empirical data (Pethick, 1980, 1981) describing these marshes as reaching maturity after about 300 years, although at an elevation that was some 60–80 cm below the level of HAT. This elevation deficit was interpreted to reflect an equilibrium point at which local declining sediment yield offset a local relative sea-level rise of approximately 2 mm year^{-1}.

By extending these models it is possible to look at the effect of increased rates of sea-level rise (or reduced sediment influx) on marsh existence. French (1993), for example, re-ran his model to simulate the effect on marsh status of a range of possible eustatic sea-level change scenarios (1.1–15 mm year^{-1}) and found that only under the more extreme predictions did ecological drowning take place by the year 2100. Under more moderate conditions (about 4 mm year^{-1}), the marsh surface stabilized at a lower elevation, with significant implications for the marsh halophyte communities. An additional effect of increased water level, if uncompensated by an adequate supply of sediment, is an increase in wave attack, erosion of intertidal sediments, and movement upward and shoreward in estuarine facies (Reed, 1988; Allen, 1990b). This erosion may be more significant than ecological drowning in many estuaries, particularly those confined by flood embankments.

By contrast to the success of modelling vertical accretion response to sea-level movement, relatively little is known about how the factors governing the morphology of marsh and creek networks (which are fundamental to flux of water sediment and nutrients) will respond to changing hydroperiods. It is, however, recognized that marsh surface morphology can vary widely between regions, although it is often consistent within regions (Pye and French, 1993). Salt-pans may be abundant or absent. Creeks may range in morphology from linear, through dendritic to complex, with varying degrees of density, frequency and regularity in branching. In terms of planiform character, similarities have been drawn between saltmarsh creeks and fluvial networks, and interpreted as reflecting a tidal drainage function (Frey and Basan, 1985; French and Stoddard, 1992). However, unlike fluvial creeks, flow of water through saltmarsh creeks is bidirectional. Maximum flood tide flow velocities often occur as the tide reaches the elevation of the marsh, whilst maximum ebb velocities are often attained at a late stage on the ebbing tide (Bayliss-Smith *et al.*, 1979; French and Stoddard, 1992).

In one study, based on morphometric analysis of 13 marshes from around the coast of England and Wales, a conceptual three-stage model of creek development as marshes progress from a mudflat towards maturity has been suggested (Steel and Pye, 1997). It is proposed that, as the young marsh develops, its primary creek system morphology is inherited from the ebb–flow drainage features of the mudflat. With vertical accretion of the marsh surface, an increase in ebb–flow erosion induces headward extension of the creek network until a maximum creek density is reached associated with the modal frequency of tidal inundation (approximately 280 tides). Finally, as the marsh approaches maturity, reduced flow requirements lead to infilling of lower-order creeks. This study also found a relationship between creek dimensions and the size of the watershed, although in other studies no relationship has been reported (Pethick, 1992). In response to upward sea-level movement, Allen (1997) argued that an increase in hydraulic duty would induce widening of channel cross-sectional area and headward erosion of creek systems. Thus, in confined coastal environments, extension of creek networks may lead to internal breakdown of the marsh surface. In the south-east of England, currently experiencing an uncompensated relative sea-level rise of 3–4 mm year^{-1}, this may be a factor contributing to the high erosion rates.

C. The Failure of Society to Protect Estuarine Ecosystem Functionality

If it is the case that estuarine ecosystems provide so many benefits, it is reasonable to question why loss and degradation continues. Clearly, to some degree the desirability of flat areas, fertile land and ease of access has drawn agriculture, industry and urban settlement to coastal areas, consequently displacing natural areas. When large areas of wetlands exist the marginal cost to society of displacement is relatively low, but this penalty increases as the total area of habitat declines. Some of the land-use changes are in society's best interest, when the returns from that land-use are relatively high. However, wetlands have frequently been lost to activities that result only in a limited gain or even a net cost to society. This loss is attributed by Turner and Jones (1991) to interrelated market and policy intervention failures, through a lack of information and understanding of the full value of the multitude of functions that wetlands provide.

Market failures reflect the inability of the economy properly to value a range of functions from which the ecosystem and society draws direct and indirect benefits but does not pay the full cost. This is further complicated by inappropriate or absent property rights regimes, and the conflict between public and private interests. Many human activities, undertaken to maximize private gain, produce adverse affects elsewhere (such as pollution related to agricultural fertilizer usage, or housing and industrial expansion). Due to the lack of enforceable public rights, no compensation is provided to those adversely affected, resulting in a broad-based public loss (Turner *et al.*, 1997b). In addition, where wetlands are privately owned, financial incentives are often insufficient to convince the land-owner to conserve wetlands (maximizing the benefit to society overall). The result is that land is converted to intensive agriculture, housing, etc., in order to maximize private benefits (Heimlich *et al.*, 1998).

Policy intervention failures stem from government or public policies that induce a detrimental impact on wetlands which society might otherwise have chosen to conserve. Often this failure results from a lack of co-ordination between sectoral interests, for instance subsidies to enhance agricultural productivity at the expense of wetland conservation goals. Across Europe a major underlying cause of wetland loss has been the intervention price paid for agricultural crop production which, as part of the Common Agricultural Policy, has made the drainage of wetlands artificially profitable (Turner, 1992). Similarly, in the USA, up until 1985, farm programme payments were dependent on acreage of crops, so encouraging wetland filling and drainage (Heimlich, 1994).

D. Scoping Frameworks

In recognition of market and intervention failures, and in order to scope the innumerable problems and issues surrounding scientific analysis, economic valuation and management of coastal resources, a number of organizations (e.g. the International Geosphere–Biosphere Programme/Land–Ocean Interactions in the Coastal Zone, the Organization for Economic Co-operation and Development (OECD), the World Bank, European Environment Agency, UK Environment Agency) have chosen to adopt simplified auditing approaches akin to the pressure–state–impact–response (P-S-I-R) framework (Figure 1). The objective of this line of approach is to clarify multisectoral interrelationships and to highlight the dynamic characteristics of ecosystem and socioeconomic changes. The P-S-I-R framework provides a way of identifying the key issues, questions, data and information availability, land-use patterns, proposed developments, existing institutional frameworks, timing and spatial considerations, etc. (Turner *et al.*, 1998).

For any given coastal area (defined to encompass the entire drainage network) there will exist a spatial distribution of socioeconomic activities and related land-uses: urban, industrial mining, agriculture/forestry/aquaculture and fisheries, commercial and transportation. This spatial distribution of human activities reflects the final demand for a variety of goods and services within the defined area and from outside the area. Environmental pressure builds up via these socioeconomic driving forces and, augmented by natural systems variability, causes changes in environmental systems states.

The production and consumption activities result in different types and quantities of residuals, as well as goods and services measured in terms of gross national product (GNP). Thus, the concern might be the role and extent of changes in C, N, P and sediment fluxes as a result of land-use change and other activities.

These state environmental changes impact on human and non-human receptors, resulting in a number of perceived social welfare changes (benefits and costs). Such welfare changes provide the stimulus for management

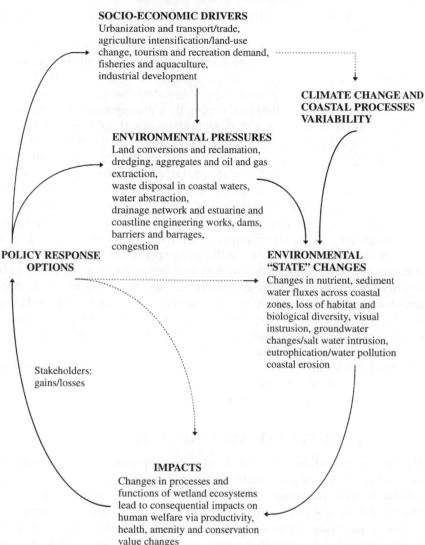

SOCIO-ECONOMIC DRIVERS
Urbanization and transport/trade,
agriculture intensification/land-use
change, tourism and recreation demand,
fisheries and aquaculture,
industrial development

**CLIMATE CHANGE AND
COASTAL PROCESSES
VARIABILITY**

ENVIRONMENTAL PRESSURES
Land conversions and reclamation,
dredging, aggregates and oil and gas
extraction,
waste disposal in coastal waters,
water abstraction,
drainage network and estuarine and
coastline engineering works, dams,
barriers and barrages,
congestion

**POLICY RESPONSE
OPTIONS**

**ENVIRONMENTAL
"STATE" CHANGES**
Changes in nutrient, sediment
water fluxes across coastal
zones, loss of habitat and
biological diversity, visual
instrusion, groundwater
changes/salt water intrusion,
eutrophication/water pollution
coastal erosion

Stakeholders:
gains/losses

IMPACTS
Changes in processes and
functions of wetland ecosystems
lead to consequential impacts on
human welfare via productivity,
health, amenity and conservation
value changes

Fig. 1. Pressure-state-impact-response framework applied to coastal wetlands. From Turner *et al.* (1998).

action which depends on the institutional structure, culture–value system and competing demands for scarce resources, and for other goods and services in the coastal zone. An integrated (modelling) approach will need to encompass within its analytical framework the socioeconomic and biophysical drivers that generate the spatially distributed economic activities and related ambient environmental quality, in order to provide information on future environmental states.

At the core of this interdisciplinary analytical framework is a conceptual model, based on the concept of functional diversity, which links ecosystem processes species, composition, and functions with outputs of goods and services, which can then be assigned monetary economic and/or other values (see Figure 2). A management strategy based on the sustainable utilization of coastal resources should have at its core the objective of maintaining ecosystem integrity, i.e. the maintenance of system components, interactions among them, and the resultant behaviour or dynamic of the system (King, 1993). Functional diversity can then be defined as the variety of responses to environmental change, in particular the variety of spatial and temporal scales with which organisms react to one another and to the environment (Steele, 1991). Marine and terrestrial ecosystems may differ significantly in their functional responses to environmental change and this will have practical implications for management strategies. Thus, although marine systems may be much more sensitive to changes in their environment, they may also be much more resilient (i.e. more adaptable in terms of recovery response to stress and shock). The functional diversity concept encourages analysts to take a wider perspective and examine changes in large-scale ecological processes, together with the relevant environmental and socioeconomic driving forces. The focus is then on the ability of interdependent ecological–economic systems to maintain functionality under a range of stress and shock conditions (Folke *et al.*, 1996).

III. SUSTAINABLE DEVELOPMENT

The advancement of the sustainable development paradigm reflects growing awareness of the need to manage our global environment in a more holistic manner (Boulding, 1966; WCED, 1987; FAO, 1992; OECD, 1997). The literature abounds with definitions of sustainable development, reflecting the broad range of world views and disciplines involved in the debate over what sustainability means and whether it can be achieved. The most widely accepted generalized definition of sustainable development is that given in the Brundtland Report published in 1987 (WCED, 1987) which simply describes sustainable development as "development which meets the needs of the present without compromising the ability of future generations to meet their own needs". It proposes a strategy linking socio-economic and environmental systems such that present key human needs, including the alleviation of poverty, should be met without compromising the basis of future generations to meet their own needs. It is a process of change in which "the exploitation of resources, the direction of investments,

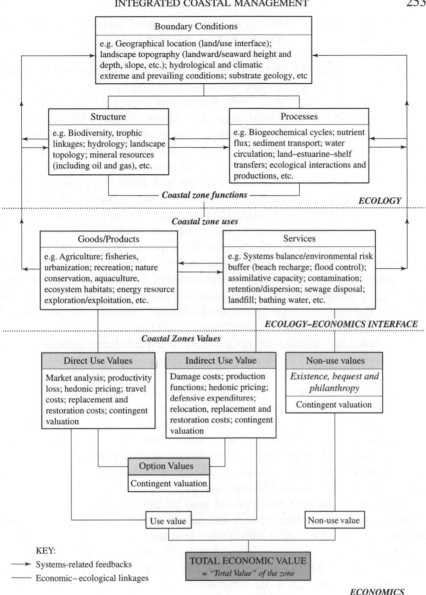

Fig. 2. Coastal zones, environmental functions and associated values. Adapted from Turner *et al.* (1998).

the orientation of technological development, and institutional change are all in harmony and enhance both current and future potential to meet human needs and aspirations" (WCED, 1987, p. 90).

Although existing definitions are extremely broad and can be interpreted liberally, the basic concept underpinning sustainable development is that the foundations of economic growth, and of societal well-being, are ultimately dependent on the health and efficiency of ecological systems and the functions they perform, e.g. resource production, climate stabilization and waste assimilation (Daly and Cobb, 1990; Pearce and Turner, 1990). Interpreting sustainability in this way challenges the current practice of maximizing economic growth by prioritizing maintenance of the environmental resource base (Pearce, 1993). Acceptance of the sustainability objective does not require that economic growth must stop, but rather that compensatory mechanisms be found that maintain the benefits and functions provided by the environment. Along with technological, economic, cultural and regulatory changes, a key mechanism to attaining sustainable use of natural resources is the concept of environmental "substitution" or "compensation". Thus, if economic development is undertaken at the expense of some environmental resources, the strict interpretation of sustainable development demands that an appropriate large compensation investment be directed at protecting, restoring or creating replacement natural ecosystems (Turner, 1993). In addition, the rate at which substitutable natural resources are lost should should not exceed the rate at which they can be replaced.

This indicates a need to re-orientate the traditional approaches in economics and ecology to explore jointly interface issues such as intragenerational and intergenerational equity, the treatment of the long term (50–100 years), potential irreversibility of ecological change, fundamental uncertainty and system complexity, and the processes of endogenous technological change. Sustainable development is also a normative concept, a set of policy objectives that many governments have now signed up to. These social objectives are a matter of judgement based on a range of often competing values and ethical norms. Thus, both weak and strong variants of sustainability have been expounded in the academic literature and in policy circles.

Sustainable economic development may be characterized as a process of change in an economy that ensures that human welfare does not decline over the long term (Pearce et al., 1989). To be consistent with the criterion of intergenrational equity, sustainability from an economic perspective requires a non-declining capital stock over time. The definition of capital used here encompasses natural capital (the structure and functions of ecosystems and the resulting goods and services provision), and manufactured, human and institutional capital (with the latter taken to include an ethical and cultural dimension). Sustainability therefore requires a development process that allows for an increase in the well-being of the current generation, with particular emphasis on the welfare of the poorest members of society, while simultaneously avoiding uncompensated and "significant" costs on future generations. Policy would be based on a long-term perspective, incorporating an equity as well as an efficiency criterion, and would also emphasize the need to maintain a "healthy" global ecological system. While the capital theory itself is criticized by some analysts, who identify problems in aggregating natural and produced capital, the main difference between analysts concerns the problem of the substitutability of produced and natural capital.

The *constant capital* condition for sustainable development can be interpreted in a weak and strong form. The *weak sustainability* condition can be written as

$$K/N = (K_m + K_h + K_n + K_{am})/N$$

K/N should be constant or rising over time. K is total capital, K_m is human-made capital, K_h is human capital, K_n is natural capital, K_{am} is social or moral capital and N is the population.

The *strong sustainability* condition in its environmental form should be: K_n/N must be constant or rising over time, and the conditions for weak sustainability must also hold.

Weak sustainability effectively requires that the total stock of capital, both natural and human-made, be maintained, although allowing unlimited substitution possibilities (via technical progress) between the different forms of capital. However, complete substitution is not always possible and it is questionable whether human-made advances can compensate fully for loss of natural capital systems beyond critical threshold conditions, increasing their vulnerability (reduced integrity and resilience) to stresses and shocks. Hence a more stringent position of strong sustainability requires that the total stock of *natural* capital be non-declining, because of the uncertainties over substitution possibilities and the adaption of a precautionary approach. Natural and human-made capital may be seen as complements rather than substitutes (Goodland and Daly, 1996), making it important that both forms of stock be maintained.

Sustainable coastal development can be described as the proper use and care of the coastal environment on loan from future generations. Although an economic system can be both efficient and sustainable, economic efficiency does not in itself guarantee sustainability (Pearce and Turner, 1990). The "scale" of the socioeconomic activity must be kept within biophysical limits and carrying capacity or ecosystem resilience. Keeping the scale of socioeconomic activities within the resilience capacity of their underpinning resource base, locally, regionally and ultimately globally, is the main challenge for sustainable development (including coastal management) policy. Biodiversity loss, for example, may be related to a loss of system resilience, but the limits are not static and appropriate behaviour, management systems and institutions can serve to delay or postpone their onset. The principle of sustainable utilization of resources will be a key component of any future coastal management strategy.

The objective of coastal zone management is to produce a "socially desirable" mix of coastal zone products and services. This mix is likely to change over time with changing demands, changing knowledge and changing pressures. Fulfilment of this objective will require the mitigation of current market and intervention failures in coastal areas. Market failures are characterized by missing resource prices or by prices that fail to signal real social scarcities, and therefore mislead policy. A number of reasons can be identified to account for such failures, such as high transaction costs (caused by temporal and spatial distances), missing, ill-defined or unprotected property rights for coastal resources, and inadequate or inaccessible information about environmental change effects. Market failure effects are often compounded by inefficient policy interactions, or the lack of any intervention or regulation. Efforts to mitigate these failures must be supplemented with continuous monitoring of management performance and of coastal zone conditions, and the application of research findings over time. The desired social mix of coastal "outputs" can be provided most effectively and efficiently by an integrated approach to coastal zone management (ICM). The analysis for ICM requires overall resource management auditing. It forces decision-makers to ask relevant questions relating to coastal ecosystems and processes, changing trends of pressure on the coastal zone, trade-offs, environmental impacts and various aspects of response strategies, including sources of finance and human or institutional capital resource potential. In macro-policy terms, ICM can be seen as a component of sustainable resource management within a regional or national economic development strategy (Bower and Turner, 1998).

Because the coastal zone is the most biodiverse zone, a strong sustainability strategy would impose a "zero net loss" principle or constraint on resource utilization (affecting habitats, biodiversity and the operation of natural processes); see section VI, B. Wetlands, for example, provide a range of valuable functions and related flows of goods and services. Such systems have also been subject to severe environmental pressures and have suffered extensive degradation and destruction. They may, therefore, be good candidates for a "zero net loss" rule, depending on how critical the functions and systems involved might be. The lost opportunity costs of the wetland conservation policy (i.e. foregone development project net benefits) should be calculated and presented to policy-makers. If the wetland area requires a more proactive management approach (e.g. buffer zone creation, monitoring and enforcement costs, or habitat restoration–creation measures), an aggregate (costs and benefits) valuation calculation will be required. As we will see later, wetland and other habitat mitigation in the United States and Europe might be viewed as moves to impose strong sustainability constraints by demanding that losses be replaced by habitats of equal functional capacity.

IV. INTEGRATING COASTAL MANAGEMENT

In coastal resource management, ecological sustainability requires the safeguarding and transmission to future generations of a quality of natural capital (natural resources) that provides a sustained yield of economic and environmental services (van der Meulen and Udo de Haes, 1996). It is now widely accepted that the key to implementing sustainable development in the coastal zone is an integrated approach across economic sectors, levels of governance, academic disciplines, stakeholders and generations (OECD, 1993a,b; UNEP, 1995, 1997; GESAMP, 1996, Turner and Adger, 1996; World Bank, 1996; Costanza et al., 1998). Under Chapter 17 of Agenda 21 priority is given to the need for integrated coastal management to further the resolution of issues affecting the sustainable use of coastal environments and marine resources. ICM is also supported by the International Panel on Climate Change (IPCC) and the United Nations Framework Convention on Climate Change as the most important vehicle for adapting to climate change while at the same time improving the present situation in coastal areas (IPCC, 1996; Bijlsma, 1997).

For ICM to succeed it must become a continuous, adaptive, day-to-day process which consists of a set of tasks typically carried out by several or many public and private entities. The tasks together produce a mix of products and services from the available coastal resources. ICM involves continuous interaction between human systems and natural systems, among human systems, and among natural systems, as these systems co-evolve over time. The management process must therefore be dynamic and adaptive in order to cope with changing circumstances, changing social tastes, increased knowledge of the behaviour of coastal processes and of human behaviour and "value" of coastal ecosystems, as well as changing technology, changing factory prices and changing governmental policies (Cicin-Sain, 1993; Bower and Turner, 1998).

The ICM process can be broken down into four basic and interrelated elements: problem identification, analysis and planning, detailed design, and implementation. The analysis–planning component must be underpinned by biogeochemical research and data relating to various processes, structures, stocks, flows and quantified relationships between pollutant discharge and related environmental state changes. From an economic perspective it should be based on the cost–benefit economic efficiency criterion and the evaluation method tempered by any relevant equity considerations, other precautionary environmental (e.g. ambient quality) standards and regional economic constraints such as income or employment targets.

The "design" segment of ICM will encompass both preliminary designs of physical biophysical facilities such as effluent treatment facilities, hard and soft sea defence/coastal protection engineered structures, "soft" engineering of restored marsh areas, and subsequent more detailed designs. But the design activity is also not restricted to structures and will include specifying the components and procedures of non-structural measures such as inspection and monitoring systems, charging systems, and land-use planning and implementation provisions.

To be able to manage requires proper resource assessment, involving the evaluation (including costs of, and wherever feasible, monetary evaluation of the benefits) of multiple resources exploitation in the coastal zone and the interactions between and among the competing resources uses. The *first* management problem is that of deciding which of the possible sets of outputs of goods and services can be produced. Various combinations of outputs are possible, involving marine transport, waste disposal, fisheries yield, recreation, national defence, amenity and preservation of unique coastal ecosystems. The different combinations will reflect different trade-offs among the feasible outputs.

Because of the dynamic and "open system" nature of coastal zones, analysis for planning and management must consider at least three areas (multiple foci for ICM).

(1) *Politically designated management area.* The political process in any given country, or in an international setting, will designate the boundaries of the management area, and will assign the management responsibilities to one or more public and private agencies.

(2) *Ecological areas.* A designated coastal management area may be within the boundaries of an identified ecosystem. More likely, the area will encompass, or be encompassed by, several ecosystems or catchment areas.

(3) *Demand areas.* Demand areas are those from which demands are exerted on the resources of the designated coastal area. These demands comprise: demands from within the designated management area; demands from outside the designated management area but within the catchment area; demands from outside the catchment area, with respect to, for example, waste disposal of pollutants transported into the area via atmospheric transport, demands for coastal recreation, including visits to unique marine resources areas; and internationally determined demands, such as for global shipment of crude oil and oil products.

Thus, ICM involves *multiple* regions, the boundaries of which rarely, if ever, coincide. For example, governmental boundaries for countries, states and provinces are rarely contiguous with watershed or ecosystem boundaries. In analysis, explicit consideration must be given to cross-boundary flows in and out, upstream and downstream. However, the management structure established for a designated coastal area is not likely to have jurisdiction over activities beyond its area.

ICM benefits are achieved by reducing damages, mitigating pollution and resource over-exploitation problems, enhancing coastal zone outputs, and preserving unique coastal ecosystems. It is important to recognize that benefits comprise the net effects on coastal zone resources—processes, functions and outputs—linked to a management measure or set of measures. Any given measure often generates multiple effects, not all of which will be positive. Thus a measure that reduces "pollutant" discharges into a coastal water body can reduce damages to recreation, at the same time improving the environment for finfish, but also increasing borer populations which result in increased damage and maintenance costs of wooden structures such as piers.

The benefits of ICM are linked to four environmental impacts/effects categories: direct and indirect productivity effects; human health effects; amenity effects; and existence effects such as loss of biodiversity and/or cultural assets. Different valuation techniques are appropriate for each of the four broad effects categories. Table 1 summarizes both the characteristics of the valuation methods and the valuation options available in each environmental effects category (see section V.A).

V. LINKING ECOLOGY AND ECONOMICS

It is now widely recognized that if, in the face of growing global environmental pressures, natural systems are to be protected with their biodiversity and functionality intact there is a need to create much closer links between ecological and economic understanding (Common and Perrings, 1992; Costanza, 1996) . The last 20–30 years have seen a gradual evolution of strategies aimed at integrated assessment of environmental sciences, technology and policy problems. A particular contribution of socioeconomic research has been the incorporation of evaluation methods and techniques that can be applied to resource damage situations (projects, policies or courses of action which change land-use, alter or modify from point and non-source pollution, etc.) because of material flux changes and related consequences, including loss of habitats and functionality.

The adoption of a systems perspective serves to re-emphasize the obvious but fundamental point that economic systems are underpinned by ecological systems, and not vice versa. There is a dynamic interdependency between economics and ecosystems. The properties of biophysical systems are part of the constraints set that bounds economic activity. The constraints set has its own internal dynamics which react to economic activity that exploits environmental assets (extraction, harvesting, waste disposal, non-consumptive uses). Feedbacks then occur which influence economic and social relationships. The evolution of the economy and of the constraint sets are interdependent; "co-evolution" is thus a pivotal concept (Common and Perring, 1992).

Any integrated assessment framework must include coupled or integrated models (biogeochemical and socioeconomic) but is not just limited to this (Turner et al., 1999). According to Rotmans and Van Asselt (1996) integrated assessment is "an interdisciplinary and participatory process of combining,

Table 1
Valuation methodologies relating to ecosystem functions

Valuation method	Description	Direct use values	Indirect use values	Non-use values
Market analysis	Where market prices of outputs (and inputs) are available. Marginal productivity net of human effort or cost. Could approximate with market price of close substitute. Requires shadow pricing	✓	✓	
Productivity losses	Change in net return from marketed goods: a form of (dose–response) market analysis	✓	✓	
Production functions	Wetlands treated as one input into the production of other goods; based on ecological linkages and market analysis.		✓	
Public pricing	Public investment, for instance via land purchase or monetary incentives, as a surrogate for market transactions.	✓	✓	✓
Hedonic price method (HPM)	Derive an implicit price for an environmental good from analysis of goods for which markets exist and which incorporate particular environmental characteristics	✓	✓	
Travel cost method (TCM)	Costs incurred in reaching a recreation site as a proxy for the value of recreation. Expenses differ between sites (or for the same site over time) with different environmental attributes	✓	✓	
Contingent valuation method (CVM)	Construction of a hypothetical market by direct surveying of a sample of individuals and aggregation to encompass the relevant population. Problems of potential biases	✓	✓	✓
Damage costs avoided	The costs that would be incurred if the wetland function were not present (e.g. flood prevention)		✓	
Defensive expenditures	Costs incurred in mitigating the effects of reduced environmental quality. Represents a minimum value for the environmental function		✓	
Relocation costs	Expenditures involved in relocation of affected agents or facilities; a particular form of defensive expenditure		✓	
Replacement/ substitute costs	Potential expenditures incurred in replacing the function that is lost, for instance by the use of substitute facilities or "shadow projects"	✓	✓	✓
Restoration costs	Costs of returning the degraded wetland to its original state. A total value approach; important ecological, temporal and cultural dimensions	✓	✓	✓

Source: Turner *et al.* (1997b).

interpreting and communicating knowledge from diverse scientific disciplines to achieve better understanding of complex phenomena". The critical importance of making value-laden assumptions transparent in both natural and social scientific components of integrated assessment models needs to be highlighted. Valuation is more than the assignment of monetary values, and includes multicriteria methods and techniques in order to identify practicable trade-offs.

More often than not, governments are reluctant to set environmental policies that constrain economic growth without clearly overwhelming evidence that such, often politically unpopular, action is required. To support environmental decision-making a considerable volume of research has concentrated on the issues of determining methodologies to assess the full value of ecosystems, to provide indicators of the state of the environmental and developing problems, and to develop integrated models to predict impacts of environmental change.

Integrating economic and ecological analysis can be done in a variety of ways. This section discusses advances in, and contraints to the understanding of, integrated methods of resource valuation, environmental indicators, systems modelling and management options evaluation.

A. Methods of Valuation (and Limitations)

Recent years have seen considerable advances in the field of environmental economics towards a classification for economic values as they relate to the natural environment (Figure 3). The total economic value (TEV) of a natural resource can be disaggregated into its constituent parts, consisting of use values and non-use values.

A use value is a value derived from the utilization of a productive function of a natural system. Use values consist of three components: direct use, indirect use, and option. Direct use value refers to the gain from the actual use which may be either: (a) consumptive, in that direct exploitation of resources takes place (fishing, mineral extraction, harvesting, etc.) or (b) non-consumptive, through which an individual benefits by the use of services provided by the coastal resource (aesthetic enjoyment, recreation, education, etc.). Indirect use value refers to the benefits individuals derive from the various ecosystem functions (storm buffering, aquifer protection from salt-water intrusion, species nursery and breeding grounds, waste assimilation, etc.). Option value relates to the value an individual might place on perceived future benefits from the conservation of a resource or one of its components. A number of environmental economists include an additional subdivision of option value, the quasi-option value, which is the value of information gained by delaying a decision to proceed with use of a resource, which may result in an irreversible loss.

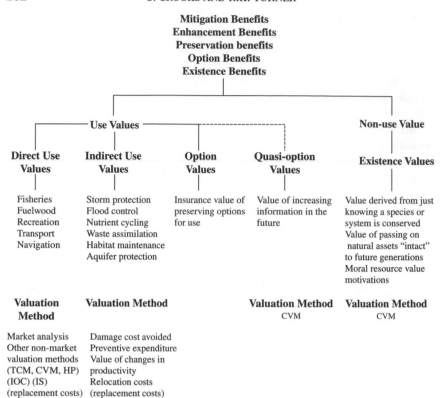

Fig. 3. Methods for valuing coastal resource benefits. HP, hedonic pricing, based on land/property values data; CVM, contingent valuation based on social surveys designed to elicit willingness to pay values; TCM, travel cost method, based on recreationalist expenditure data; IOC, indirect opportunity cost approach, based on options forgone; IS, indirect substitute approach. The benefit categories illustrated do not include the indirect or secondary benefits provided by the coastal zone to the regional economy. From Bower and Turner (1998).

Non-use values are more problematic as they suggest non-instrumental values which are intrinsically real, but unassociated with actual use or even the option to use in the future. In essence, non-use values are associated with benefits derived simply from the knowledge that a resource, such as an individual species or an entire ecosystem, is maintained. It is, by definition, not associated with any use of the resource or any tangible benefit derived from it, although users of a resource might also attribute non-use value to it. Non-use value is closely linked to ethical concerns, often being associated with altruistic preferences and to some degree, self-interest (Crowards, 1997). Non-use values can be subclassified into three main components:

existence, bequest and philanthropic. Existence value is the satisfaction value an individual derives by simply knowing that a feature of the environment continues to exist, whether or not it brings benefits to others (e.g. protection of coral reefs or endangered species). Bequest value relates to the knowledge that a resource will be maintained for future generations, so protecting the opportunity for them to enjoy it. Finally, philanthropic value is associated with the satisfaction an individual derives from ensuring that a resource is maintained and available for contemporaries of his or her generation. Because such non-use values reflect people's preferences, concerns and sympathies, the values derived are anthropocentric, but nevertheless include some recognition of the value of the very existence of certain species or whole ecosystems.

Total economic value, therefore, is equivalent to the sum of these component values derived from ecosystem benefits. However, it has been argued that the full contribution of component species and processes to the aggregate life support service provided by the environment as a whole is not fully encapsulated in economic valuation (Gren *et al.*, 1994). This critique of economic valuation has some validity, not so much in relation to individual species or ecosystems, but to aggregation of the larger scale. This aggregated ecosystem could be said to possess *primary value*. The prior existence of a healthy ecosystem is necessary before the range of use and non-use values, linked to ecosystem structure and function, can be utilized by people. These use and non-use values might then be labelled *secondary values*. It is these various components of the total secondary values, and not primary value, that are encompassed within the TEV.

It is also the case that TEV often fails to capture fully the total range of secondary values. Economic valuation is very much dependent on the understanding of ecosystem functions and linkages which are difficult to analyse scientifically. In particular the indirect use values of ecosystems are shrouded by unquantifiable uncertainty, and the distinction between these values and non-use values is far from clearcut once it is realized how complicated and interrelated natural systems actually are in practice.

A range of valuation techniques exists for assessing the economic value of the functions performed by environmental systems. Valuation methods of coastal zone impacts fall into four main categories (Bower and Turner, 1998).

(1) *Market-oriented benefit valuation.* This is benefit valuation using actual market prices of productive goods and services based on changes in the value output, or loss of earnings (e.g. loss of fisheries output due to pollution, value of productive service or recreational loss, through increased illness caused by reduced water quality).

(2) *Surrogate market benefit valuation.* Environmental surrogates may include market goods, property values, other land values, travel cost of

recreation, wage differentials, compensation payments (e.g. entrance fees to national parks as a proxy for value of visits to protected areas, changes in property value as a result of environmental degradation, compensation for damage of fisheries).

(3) *Cost valuation using actual market prices of environmental protective inputs.* This includes preventive expenditure, replacement costs, shadow projects, cost-effectiveness analysis (e.g. cost of environmental safeguards in designs, cost of replacing damaged resources).

(4) *Survey-oriented (hypothetical) valuation.* This involves contingent valuation or contingent ranking questionnaire-based surveys of individuals to elicit willingness-to-pay or willingness-to-be-compensated.

Many functions of ecosystems result in goods and services that are not traded in financial markets and therefore remain unpriced. It is then necessary to assess the relative economic worth of these goods and services by means of non-market valuation techniques. A detailed review of such techniques is beyond the scope of this chapter but can be found in a number of general texts (Dixon and Hufschmidt, 1986; Braden and Kolstad, 1991; Freeman, 1993; Hanley and Spash, 1993; Turner, 1993; Pearce and Moran, 1994; Bromley, 1995; Turner and Adger, 1996; Turner et al., 1999).

It is important to draw a distinction between alternative valuation techniques, in terms of those that estimate economic benefits directly from those that estimate costs as a proxy for benefits (Turner et al., 1997b). For instance, estimating costs of damage avoided, defensive expenditures, replacement or substitution costs, or restoration costs as part of a valuation study suggests that the costs are a reasonable approximation of the benefits that society attributes to the resource in question. The underlying assumption is that the benefits are at least as great as the costs involved in repairing, avoiding or compensating for damage. These techniques are widely applied as part of the project/policy appraisal process, due to the relative simplicity of the methods and the availability of relevant data. But it is important to be aware of the limitations of such methods. Only where it can be shown that (1) replacement or repair will provide a perfect substitution of the original function, and (2) the costs of doing so are less than the benefits derived from this function will the costs actually represent the economic value associated with the function.

Where market prices exist for resources they may have to be adjusted to provide social or shadow prices, but otherwise they are likely to provide a relatively sound assessment of economic value. Approaches related to market analysis include assessment of productivity losses which may accompany changes to ecosystems, and the incorporation of habitat as just input to the production function of other goods and services. Investment by public agencies to conserve ecosystems may provide a rough proxy for collective individuals' willingness to pay, and hence social value.

In the absence of market prices, other theoretically valid benefits estimation techniques include hedonic pricing and travel cost methods. However, these are based on preferences revealed through observable behaviour and are restricted in their applicability. Contingent valuation, based on surveys that attempt to elicit "stated preferences" is one mechanism whereby it may be possible to include non-use benefits that are not associated with any observed behaviour. The legitimacy of applying contingent valuation and its results is still a matter of intense debate, especially in the context of transferring derived values from one study to another (Brouwer et al., 1998).

B. Environmental Indicators

There has always been interest in a set of indicators that inform decision-makers about existing or emerging environmental problems. Research into this field was galvanized at the Rio Earth Summit in 1992, where most countries committed themselves to developing a set of indicators to help inform government, industry and non-governmental organizations and the public about issues related to sustainable development. Although indicators of economic growth have been used for many years (for example, GNP), they rarely encompass environmental concerns and so fail to provide a measure of sustainable growth. There is, therefore, a need for a set of environmental indicators to clarify the impacts of a changing global environment.

Environmental indicators may be defined as a parameter, or value derived from a parameter, that points to, provides information about or describes the state of a phenomenon, environment or area, with significance extending beyond that directly associated with parameter value (OECD, 1994). Put more simply, the then UK Department of the Environment defined indicators as "quantified information which help to explain how things are changing over time" (DoE, 1996). Thus, in general, environmental indicators are seen as helping to reduce the complexity of environmental problems and to increase the transparency of the possible trade-offs involved in specific policy choices. In addition, indicators should simplify the communication process by which the results of measurements are provided to the public and to those charged with making decisions.

Although simple in concept, the formulation of indicators for sustainable development has been difficult in practice. Nevertheless, such tools and their information are vital to more flexible management processes. The fact that, nowadays, simple descriptive environmental indicators are often referred to as sustainability indicators adds further to the confusion. This confusion arises because indicators of specific environmental issues may fall into one of two groups, either descriptive or normative (Weterings and Opschoor, 1994). Descriptive indicators reflect actual state or condition (for example, changing area of coastal wetlands or levels of fisheries recruitment), whereas normative indicators relate this actual state or condition to a desired state or condition with defined boundary condition (for example, pollutant emission loads relative to estuarine assimilative capacity). Often, indicators change are too descriptive, so fail to qualify as sustainability indicators which, by definition, are normative since they should reflect how far one has drifted from a desired (sustainable) situation or development path.

One approach which is becoming increasingly widely adopted is to consider environmental management beyond individual species and on the geographical scale of landscape elements and ecoregions. An extension to this approach has focused on concepts of ecosystem "health" and/or ecosystem "integrity" incorporating human values with biogeophysical processes (Rapport, 1989; Makarewicz, 1991; Costanza, 1992; Karr, 1993; Woodley et al., 1993; Mageau et al., 1995; Rapport et al., 1998b). There has been difficulty in arriving at a consensus definition of these concepts, but most attempts include the belief that healthy ecosystems are "stable and sustainable", and are able to maintain organization and autonomy over time, displaying resilience to stress (Costanza, 1992). Rapport et al. (1998b) differentiated between health and integrity by classifying ecosystem health to relate to a landscape element that has been impacted by human activity, whilst integrity refers to a reference pristine system evolved in the absence of significant human intervention. Supporters of the ecosystem health and integrity approach argue that indicators should, across a range of spatial and temporal scales, reflect system "organization", "vigour" and "resilience", measuring the systems diversity of components and interactions, their productivity and the system's capacity to maintain structure and function in the presence of stress (Costanza, 1992).

A major challenge to the use of indicators is the existing uncertainty over what they actually measure and whether they encompass the features of the environment that are of interest, as well as changing in some meaningful way relative to environmental change (Landres et al., 1988; Norris and Norris, 1995). There is a trade-off between ease of sampling (or availability of data) and depth of information gathered (Noss, 1990). Data on individual species (Kerr and Dickley, 1984) and composites of species (Karr, 1991), to provide some indication of species richness, are relatively simple to collect but such information provides limited insight into system processes and functioning (Simberloff, 1998). Measures of parameters directed at system organization, vigour, functioning and to some extent resilience have been suggested (Rapport, 1989; Costanza, 1992; Mageau et al., 1995). Resilience, in particular, is an enigmatic parameter to determine because it implies the ability to predict the dynamic interactions between ecosystem component under stress, and identify boundary conditions, before the system flips to an alternative quasi-stable state.

Critics of the ecosystem health and integrity approach to defining management goals point to the fact that ecosystems are constantly changing and do not exist in a "stable state", and as such it is not possible to define an optimum condition for ecosystem preservation (Wicklum and Davies, 1995). Some conservationists are also doubtful about a management approach that permits the loss of individual species as long as given ecosystem functions are not greatly altered, arguing that monitoring of certain key species or measures of species diversity, independent of species identity, is needed to reflect adequately the health of an ecosystem (Tilman and Downing, 1994; Simberloff, 1998). Nevertheless, the concept of ecosystem health or integrity, interpreted broadly, is useful in that it helps to focus attention on larger natural systems and away from the specific interests of individuals and groups (Common and Perrings, 1992).

As we begin to see the disparity in ways of viewing the environment, the reasons why developing indicators is so problematic becomes apparent. The real test of indicator usefulness comes when trying to integrate ecology with socioeconomics. Rapport *et al.* (1998a) argue that "ecosystem health assessments require analysis of linkages between human pressures on ecosystem and landscapes, altered ecosystem structure and function, alterations in ecosystem services, and societal response. Effective diagnosis requires exploring and identifying the most critical of these links." There is therefore a need to incorporate environmental indicators into an organizational framework such as the P-S-I-R (Brouwer *et al.*, 1998). There have also been attempts to aggregate water quality and biological measurements into a framework to derive integrated measurements of coastal ecosystem health (Hameedi, 1997). However, more progress is required before these indicators can be utilized routinely to provide informed and objective assessment of the impacts of anthropogenic changes in the coastal zone (Hameedi, 1997).

C. Integrated Modelling

Integrated ecological–economic modelling provides one mechanism by which the requirements for sustainability can be scrutinized. Considering the coastal zone, integration can be achieved by bringing together the various elements from different disciplines. For instance, a systems dynamic model may be designed to capture all the important cause–effect relationships affecting the state and change of coastal ecosystems. These may include natural (biophysical, ecological) and cultural–socioeconomic (decisions by individuals and organizations to utilize resources, change the landscape and to create pollutants) aspects, including institutional processes (policies, regulation and public management of nature). But information about environmental effects (physical, chemical or biological) and about the value of ecosystem functioning needs to be better integrated than has been the case so far. Quantitative information about increases or decreases in an ecosystem service can be a necessary prerequisite to valuation, but values may also determine which effects are chosen for measurement in the first place. Information about environmental effects requires a systems perspective, a focus on appropriate models, the adoption of relevant measurement end-points and appropriate temporal and spatial scales (Bingham *et al.*, 1995).

While the field of ecology and toxicology have advanced to a relatively high degree of accuracy in predicting the effect of some actions on particular ecosystem attributes, the sheer complexity of ecosystems still poses an immense challenge to analysts seeking to predict system alterations resulting from human actions. The variety of ecosystems and their attributes have not been competely catalogued and not all crucial relationships are known. Bingham *et al.* (1995) concluded that joint modelling exercises are required to review what information existing ecological models accept and in what form information is generated, and to explore ways in which the ecological models need to be adapted better to serve policy or valuation purposes.

Russell (1995) has argued that an ecological–economic modelling project undertaken in the 1970s (the Delaware estuary study: Spofford *et al.*, 1975) has a number of important lessons to offer. Real collaboration requires substantial investment in time, which can be facilitated by the use of a mathematical model of the problem to be investigated. Bockstael *et al.* (1995) reported progress in such a collaborative effort investigating ecosystem functioning and change in the Patuxent River basin in Maryland, USA.

The linking of individual discipline models is very much restricted by the model type (Turner *et al.*, 1997b). If the theories and concepts of ecological and socioeconomic models fit a general systems frame, they may be incorporated within a unified model structure, where modules might represent the various original models, and the derived outputs of each model feed into the next. However, given the different modes of thinking (across temporal and spatial scales) typical of different disciplines, it is not always easy to link models directly. For instance, if both the ecological and economic models systems are represented in the form of programming or optimization models then several options are available: look for a new aggregate objective; adopt a multiobjective or conflict analysis framework; or, when possible, derive multiple sets of optimality conditions and solve these simultaneously. However, when the ecological and economic systems are represented by different model types, it is difficult to suggest how they can be linked to one another. Where economic models have an optimization or programming format and ecosystem models a descriptive format, direct technical integration seems feasible, otherwise heuristic approaches are needed (Turner *et al.*, 1997b).

Model classification can be described in many ways. For instance, Costanza *et al.* (1993) distinguished between economic, ecological and integrated approaches on the basis of whether they optimize:

- *generality*, characterized by simple theoretical or conceptual models that aggregate, caricature and exaggerate;
- *precision*, characterized by statistical, short-term, partial static or linear models with one element examined in much detail; and
- *realism*, characterized by causal, non-linear–dynamic–evolutionary, and complex models.

These three criteria are usually conflicting and trade-offs are inevitable. Ecological studies focus on interactions between living organisms and communities with their biotic and abiotic ecosystem processes, food webs and nutrient fluxes (Hollings *et al.*, 1995). Economics studies the behaviour, costs and benefits associated with the allocation of scarce resources at the micro-level of economic agents (households, firms), and multisectoral developments and costs and benefits of public regulation at macro-levels (Kahn, 1998). Often precision is striven for in natural sciences descriptive levels, whereas generality and realism are sought in the description of socioeconomic value levels. The integration of the two may imply the need for a qualitative approach. Interdisciplinary work, which may be the separating line between ecological–economic analysis and environmental economics or ecology, may involve ecologists or economists transferring elements, or even theories and models, from one discipline to another and transforming them for their specific purpose. This may require action such as reduction, simplification or summation (Turner *et al.*, 1997b). For instance, considering estuarine wetlands, it would be desirable to derive a simple dynamic model summarizing and simplifying some of the statistical and causal relationships of the spatial models of wetland hydrology and statistical models of vegetation community structure. The outcomes could then be linked to a simplified economic interaction and values model.

D. Evaluation and Decision-making

Once environmental values have been elicited, indicators have provided "clues" to the state of the environment, and models have quantified to some extent the consequences of various possible management scenarios, there is a need to evaluate the different options and feed this information into a decision-making process for coastal areas.

Environmental appraisal is a generic term relating to the identification, measurement and assessment of environmental impacts. Assessment refers to a procedure for determining the importance of any impact. In cases where large projects or policies are being assessed it is not possible to trace out all the linkages between economic, social and environmental sectors via a quantified "model" of the working of the ecological and economic systems. More pragmatically, judgemental assessments of the likely consequences are all that are possible, often accompanied by a degree of residual uncertainty (DETR, 1998).

Appraisal techniques all try in different ways to order information so that gains and losses can be placed in a comparative evaluative framework. Environmental impact assessment and cost-effectiveness analysis are fundamental to all coastal resources management issues. No appraisal can take place unless the environmental impacts are known with some degree of accuracy, and any appraisal ought, in principle, to seek "value for money" (i.e. the costs of a policy are no higher than they need be to be related to any given stated goal or objective). Risk analysis is also relevant in a number of coastal contexts (e.g. flood defence and coastal protection policies) and is often combined within an overall cost–benefit analysis (CBA) (Penning-Rowsell *et al.*, 1992; Glassen *et al.*, 1994). In the context of regional or area economic development, the objective of coastal management, interpreted using cost–benefit analysis, can be expressed as follows. Maximize the present value of:

$$GRP - C_p - C_{cm} - D + B - C_a$$

where GRP is gross regional product, C_p is normal production costs, C_{cm} is net coastal management costs (e.g. discharge reduction costs/beach replenishment costs/coastal protection costs), D is remaining damages, B is benefits from improved environmental quality, and C_a is administrative costs of ICM.

Given an integrated approach to coastal management the CBA economic efficiency criterion will have to be supplemented by other conditions such as equity, regional income or employment targets, etc. (Bower and Turner, 1998). Multi-criteria analysis (MCA) methods seek to offer appraisal guidance when policy-makers are striving to trade-off gains and losses against multiple policy objectives (Janssen, 1994).

A number of MCA models have been developed since the 1970s to address different policy options in a range of contexts (see Janssen, 1994; Giaoutzi and Nijkamp, 1995). Such non-monetary evaluation techniques originated in operations research, and were developed partly in response to criticism of monetary valuation methods within CBA. The limitations of CBA in environmental decision-making include the fact that the technique provides only a partial value for policy analysis. The determination of the value of essential environmental attributes has proved particularly problematic (Gren *et al.*, 1994). CBA is also limited by its inability to encompass qualitative information and as such provides a rigorous analysis of an incomplete data set. Finally, the decision-making process is relatively automated and criticism has been levelled at the implied passive role of the decision-maker who is often led to the final outcome (Voogd, 1983).

Multi-criteria techniques attempt to incorporate a range of quantitative and qualitative information forms, and aim to provide a method for systematically appraising a number of alternative projects involving a series of criteria, or individuals, or groups in different ways. Multi-criteria models are particularly useful in a situation where no outright solution which simultaneously optimizes all the criteria exists and, therefore, the decision-maker is presented with trade-offs and has to identify compromise solutions. Furthermore, MCA allows the sensitivity of both social and physical data to be tested for robustness, and makes explicit the trade-offs between competing impacts and stakeholders.

The decision process has been well defined in management texts as having three separate stages: problem definition, development of possible courses of action, and selection of courses of action from the choices available (Janssen, 1994). This means that MCA must: effectively generate information on the decision problem from available data and ideas, effectively generate alternative solutions to a decision problem, and provide a good understanding of the structure and content of a decision problem.

When issues such as social implications, ecological and environmental conservation or biophysical impacts of decisions are also important to decision-makers, then MCA may be an appropriate analytical tool. MCA is particularly useful as a tool in projects where there are conflicting objectives or priorities of different stakeholders, and has been widely applied to a number of problems including land-use planning (Makowski et al., 1996; Joubert et al., 1997; Malczewski et al., 1997).

The first step to a MCA is to define as far as possible the actual problem being faced, such as use or degradation of resources, ideally in discrete measures of the environmental impact (i.e. size of area impacted, number of species held within). A set of possible scenarios for improving the site quality are identified and compiled. The model then requires that the predicted effects of each scenario be described. Before this can be done a set of management objectives must be developed which should reflect the different socioeconomic drivers for the relevant area, and ideally should be grouped to represent the various interest groups. Each scenario is then measured, or valued, in terms of the listed objectives.

To determine a rank ordering of alternative scenarios, the relative importance of the criteria must be distinguished. This is achieved by weighting the criteria, both within each criterion (e.g. different types of economic impact, net costs versus employment impact) and between criteria (e.g. economic versus ecological impact). One of the major problems facing MCA is the derivation of a set of weights that reflects stakeholder opinions. Often, studies using MCA have sent questionnaires to individuals or organizations to elicit weighting vectors (i.e. Chung and Poon, 1996). Alternatively, to avoid the difficulty of deriving a representative set of weights, Maimone (1985) used a set of extreme weights to identify radical outcomes for a multi-criteria evaluation applied to waste management. Others, such as Powell (1996), have manipulated weights to test the sensitivity of the MCA to the major parameters.

VI. CONSERVING ESTUARINE NATURAL RESOURCES

A. Habitat Protection and Restoration

Drawing conservation plans and setting biodiversity goals on a hierarchy of geographical scales, from international to national, is now common practice (Bibby, 1998). Research in the field of landscape ecology (focusing on interrelationships between the spatial patterns of landscape components and their functioning) tells us that heterogeneity in landscape structure, over varying spatial and temporal scales, influences how organisms, populations and whole assemblages interact with the environment (Wiens, 1976; Kotliar and Wiens, 1990; Irlandi, 1994; Naveh, 1994; Kupfer, 1995; With and Crist, 1995; Gustafson and Gardner, 1996; Bell et al., 1997; Ehrenfeld and Toth, 1997). To take coastal fisheries as an example, ecosystem heterogeneity may enhance biodiversity by supporting recruitment and maintaining the numbers of species that require multiple resources (Parrish, 1989; Irlandi and Crawford, 1997; Primavera, 1998). It may also act to influence biological interactions of competition and predation (Coen et al., 1981; Danielson, 1991), as well as foraging behaviour and preditor avoidance (Mittelbach, 1986; Werner and Hall, 1988; Cattrijsse et al., 1997). Development in the coastal zone acts to fragment these landscape linkages and damage ecosystem functioning in a way that is often not yet possible to predict. The maintenance of biodiversity and ecosystem integrity requires planning that extends beyond individual sites, but considers the wider regional biogeography and landscape patterns, perhaps even lowering the priority of local concerns (Noss, 1983).

Maintaining the ecosystem in itself is not enough if current levels of exploitation are to be met. Rees and Wackernagal (1994) introduced the concept of "ecological footprints" to reflect the land area necessary to sustain a level of resource consumption and waste discharge by a given human population (see also Wackernagel and Rees, 1996), and calculated that to support a global population of some six billion people, each with the standard of living of an average American, would require a trebling of the global resource base. Ecological footprints, and the limited capacity of marine and coastal ecosystems to produce resources, has not been taken into account adequately in economic development strategies (Folke et al., 1998).

The achievement of the sustainability objective requires that such landscape-scale market failures are overcome (Gottfried et al., 1996). This will mean planning to protect critical habitats that concentrate specific key functions (e.g. nursery grounds or ecological corridors), and restoration of habitat to substitute for those displaced (see below). Adaptive management may mean that, for the wider public good, governments should consider purchasing the land or the development rights for that land, or more stringently instigating financial incentives to maintain landscape ecological functions within private ownership (Gottfried et al., 1996).

Taking a systems approach to coastal management requires that *all* land-use changes, including habitat restoration projects, are placed in the contexts of larger landscape processes, within which geomorphological, ecological and socioeconomic considerations should be incorporated. When planning to restore habitat within an estuary a number of immediate questions should be asked. Are ecological linkages and functional benefits maximized on the landscape scale? Does the restoration project further ecoregion biodiversity goals? What will be the effect of the restoration project on the morphodynamics of the estuary? How is land-use anticipated to change, and will this affect the future integrity of the site? Current practice is not to ask these questions and, because of poor understanding of landcape parameters, restoration objectives have rarely been set beyond a site-specific context.

B. No-net-loss Policies

In the face of ongoing habitat degradation and loss, and the realization that wetland functions provide valuable (if not fully quantified) goods and services to society, a number of countries have, during the 1990s, adopted "no-net-loss" of designated habitats and species as a central environmental policy goal (e.g. Government of Canada, 1991; White House Office, 1993; NSW Government, 1996). The basic rationale underlying no-net-loss is that preservation of habitat within reserves is not in itself sufficient to maintain biodiversity and ecosystem functioning, but that across wider geographical scales habitat losses must be offset by habitat gains. Pivotal to this approach is the "weak" sustainability belief that "equivalent" habitats can be restored, recreated or enhanced to substitute for those destroyed by development, thus maintaining levels of ecosystem assets and functions. This assumption is controversial, to say the least, and many have argued that it has legitimized and smoothed the path for developers to destroy habitats (Lynch-Stewart, 1992; Roberts, 1993; Race and Fonseca, 1996; Cowell, 1997).

Perhaps two of the more overt examples of no-net-loss policies, enforced by legislation, are those in place in the United States and the European Union. In the United States no-net-loss of wetlands has been adopted as a policy goal both by the Bush and Clinton Administrations (White House Office, 1991, 1993). Although supported by a number of existing legislative measures enacted over the past two decades (Veltman, 1995), it is the Clean Water Act (33 U.S.C. 1344 (1972)), principally section 404, that has provided the legislative basis for no-net-loss. Section 404 specifically regulates the placement of dredge and fill material into all waters of the United States. The term "all areas of water" has been interpreted in its widest sense, and it is under this section of the Act that development in wetland areas is regulated. Under the Act, a permitting system for development projects exists, based on a hierarchical sequence of avoidance, minimization and compensation. First, the permit applicant must show that there is no *practicable alternative* to the activity which would result in less adverse impacts on the wetland. If avoidance is not possible, minimization of disturbance must be achieved. Finally, if a permit is to be granted compensation is required, involving recreation of habitat and functions, or, less favourably, restoration of degraded wetlands as close as possible to the permitted activity. In this context the process of restoring, creating or enhancing wetlands as compensation for wetlands destruction is termed mitigation. An important goal of the restoration process is that the functions of the displaced habitat should be restored at least to the level of those lost (NWPF, 1988).

For European Member States, obligatory compensation for displacement of designated habitat within protected sites is a recent concept, driven by EU legislation, specifically the Habitats Directive (Council Directive 92/43/EEC on the Conservation of Natural Habitat and Wild Flora and Fauna, adopted 1992). This Directive also incorporates the previously existing Birds Directive (Council Directive 79/49/EEC on the Protection of Wild Birds, adopted in 1979) and together they provide a legislative framework intended to protect wildlife and habitat designated as of international importance (DG XI.D, 1996). Like the US legislation to protect wetlands, the Habitats Directive is effectively a no-net-loss policy enacted to maintain and protect habitat and species, although only within designated Natura 2000 sites. Under the Directive, Member States are required to maintain habitat and species levels at a *favourable conservation status*, defined in terms of a natural range that is stable or increasing, and containing structures necessary for the long-term maintenance of the site. Under Articles 6(3) and 6(4) if a *plan or project* is identified to have an adverse affect on the integrity of the site, the developer must clearly show that (1) no alternative solutions exist, that (2) the project is of overriding public interest, and that (3) the Member State must adopt *compensatory measures* which may include habitat restoration or recreation of the same habitat type on the same site or nearby. As such, the Habitats Directive, which has subsequently been adopted into national law, is the most important piece of conservation legislation to influence Member States.

Under both the Clean Water Act and the Habitats Directive, once the developer has shown that an impact on habitat cannot be avoided, and that all efforts have been made to minimize the disturbance, compensatory habitat restoration or replacement is required. It is then an important ecological issue as to whether the functions restored, created or enhanced should directly replace those lost, or represent improvements in functions that once existed (Zedler, 1996a). By and large, conservationists so far have preferred to follow a precautionary approach and demand the replacement of like-for-like habitats as close as possible to the site of loss, as this has the greatest potential for minimizing disruption to functions between landscape elements (Race and Fonseca, 1996). If equivalent habitat replacement is achieved, then the objectives of the no-net-loss are met. However, this approach might not further conservation goals based in biogeographical, or ecoregion, requirements. For instance, in disturbed settings where substantial habitat fragmentation has already occurred it could be argued that like-for-like replacement does not optimize ecological benefits and that out-of-kind and off-site mitigation may provide a mechanism to increase overall functioning (Race and Fonseca, 1996). Under such circumstances it may prove to be wiser to consider net environmental requirements across a wider areas (regional, national or even international) and undertake appropriate trade-offs.

In the USA, where mitigation has become common practice, the adoption of a no-net-loss policy has contributed to the reduction in the rate of wetland loss, but so far a net balance has not been achieved. The mitigation process itself has been very heavily criticized as being developer-friendly, with a failure by regulators to provide high-value wetlands with adequate protection, to ensure that replacement habitats meet mitigatory requirements or even, in a number of cases, to ensure that restoration action was undertaken (Race and Cristie, 1982; Race, 1985; Zedler, 1987; O'Donnell, 1988; Kusler and Kentula, 1990; NRC, 1992; Roberts, 1993; Race and Fonseca, 1996).

Beyond these regulatory failures, a major ecological limitation of site-by-site mitigation is the risk and uncertainty of attempting to restore habitat functions after the original habitat has been lost (Roberts, 1993; Zedler, 1996a). As Kusler and Kentula (1990) pointed out in a review of restoration science, total duplication of natural wetlands "is impossible due to the complexity and variations in natural wetlands as well as created or restored systems and the subtle relationship of hydrology, solid, vegetation, animal life, and nutrients which may have developed over thousands of years in natural systems". These problems are particularly apparent in terrestrial habitats, such as mature forests and peat systems, so the use of mitigation should be tightly controlled. However, with particular relevance to coastal management, Kusler and Kentula (1990) concluded that a relatively high degree of success has been achieved in restoring and recreating wetlands in estuarine, coastal and freshwater marshes, in that order. This higher success rate in recreating tidal wetlands, they argued, reflects the dynamic and self-regulatory nature of coastal systems. Obtaining a self-regulatory system, with no inputs of energy or material from a manager, is the key to success in ecological restoration (Jackson et al., 1995).

Kusler and Kentula's conclusion is supported by a number of more recent experimental projects to restore intertidal habitats both within Europe and the USA (e.g. Broome et al., 1988; Lewis, 1994; Sacco et al., 1994; Burdick et al., 1997; Environment Agency, 1998; Esselink et al., 1998; Lee et al., 1998). Further evidence is provided by the restoration across southern Britain of salt-marshes without management intervention by historic (post-Mediaeval) storm breaching of flood embankments (Pye and French, 1993). Such evidence provides a glimmer of hope that, whereas ecosystem restoration in terrestrial settings is still beset by complex problems, restoration of at least intertidal habitats in the coastal zone may be a feasible option in an no-net-loss strategy and, with adequate landscape planning, may serve to increase coastal resilience. What has also become apparent is that, whether in the coastal zone or elsewhere, provision of a fully functioning replacement habitat prior to loss is an important component of an effective no-net-loss policy. For the more complex substitutable habitats, this may require planning and action years in advance of anticipated losses. This requires managers to take a proactive approach to no-net-loss initiatives, rather than reacting after actions have resulted in habitat loss (Huggett, 1998).

A complementary approach known as mitigation banking (also known as land banking or conservation banking) is a compensatory mechanism gaining favour in the USA, whereby habitat is restored in advance of any proposed development (EPA, 1995; Marsh *et al.*, 1996). As defined by the Association of Wetland Managers, a mitigation bank is a moderately sized to large wetland restoration, creation or enhancement project undertaken by a single developer (public or private) or a consortium of developers not only to compensate for wetland impacts from a particular project but to act as a "bank" with credits to compensate for *future* wetland projects and impacts (Veltman, 1995). Thus, the basis of the banking approach is the setting up of a *credit market*, with restored wetland values and functions being quantified as credits, which are deposited within an account, and later purchased by developers when regulators require compensation for authorized losses of habitat functions (Scodari and Shabman, 1995).

A number of key environmental, economic and regulatory benefits of banking over project-specific mitigation actions have been cited (USFWS, 1988; ELI, 1993; Etchard, 1995; Veltman, 1995; Gardner, 1996; Gilman, 1997; Crooks and Ledoux, 1999). Habitat is created in advance of mitigation requirements, rather than as a compensatory response after the wetland has been lost. This reduces the ecological consequences of a time lag between loss and replacement. It also enables clearer evaluation of restored habitat form and function, so providing greater assurance that restoration will meet mitigation requirements. This is of particular importance if attempting to provide a site for an endangered species whose habitat requirements might not be understood (Zedler, 1996b).

Environmental and financial economies of scale favour the production of larger sites over smaller sites. Larger sites may be more self-sustaining and resilient because they can provide habitats for more species (particularly territorial species and those dependent on multiple landscape elements), more sustainable food chains, which in turn can better accommodate ecosystem succession, migration and change (Kusler, 1992). Given landform, it may be possible to include a range of habitat types and the linkages between them, which would not be provided by smaller, and often isolated, sites. A large bank also provides the opportunity to demand compensation for smaller losses, which individually might not have a significant effect on a site but cumulatively lead to long-term degradation of protected habitats.

Environmentalists in the USA are reported to be supportive of a banking approach if credits are granted only once mitigation efforts are judged successful, thus preventing situations where habitat is destroyed and the ensuing mitigation efforts fail (Veltman, 1995). Reciprocally, when mitigation takes place in advance, banks provide developers with a predictable measure of the costs involved, allowing them to make an economically rational decision based on the full costs and benefits. With adequate landscape planning, it should be possible to create a bank to maximize local or national environmental and biodiversity needs. This may mean restoring habitat, for example, to create ecological corridors or buffer zones, to provide linkages between habitat types, or to support local morphodynamic and sedimentary regimes.

Mitigation banking is, however, not a panacea for coastal wetland management, and for the bank owner there are costs and uncertainties involved. The most significant of these is the financial burden on the bank owner who must fund habitat restoration and perhaps not see a return for many years. A trade-off therefore exists between allowing the sale of credits before full establishment of functions, and the ecological risk that the bank will fail. The bank owner also runs the risk that, over time, conservation regulations change, upsetting the investment decisions made by the bank on the basis of previous rules (Heimlich et al., 1998). To maximize return and provide optimal ecological benefits, the bank owner must predict some time in advance the form of habitat that will be lost to development in the future. It is most likely, given continuing research, that in the future the price of a credit will be based on the functional value of a wetland rather than on acreage. If the price of the credit can be related to the quality of the habitat this will encourage bank owners to create more ecologically desirable habitats. As ecosystem functioning is related to the linkage between landscape elements, it may encourage the creation of banks in the more ecologically "profitable" settings rather than simply a random placement. A recent economic analysis of wetlands mitigation banks, using a simulation model, has indicated that restoration of a whole wetland site will occur when there is a reduction in restoration costs, an increase in biological uncertainty or an increase in the value of the wetlands credits. Continued restoration is more difficult to justify economically as interest rates rise (Fernandez and Karp, 1998).

VII. CONCLUDING REMARKS

Spanning the transition between terrestrial and oceanic environments, coastal and estuarine ecosystems are subject to stress from multiple competing socioeconomic demands, while simultaneously needing to retain their functional diversity and resilience in the face of a highly variable environment. It is now clear that the traditional uncontrolled *ad hoc* and sectoral approaches to resource management based on incomplete economic judgements, shrouded by market and policy intervention failures, has led to the long-term decline of coastal ecosystems and a failure to safeguard them for future generations. A more integrated, systems approach to management in the coastal zone is required to take into account the multiple resource demands and the variety of stakeholders, and their interests and values as well as natural variability.

Moving towards sustainable use of coastal resources means that we must take stock of the resources we have, determine the full range of costs and benefits that management options provide, and develop flexible policies accordingly. To do so requires a transdisciplinary approach, and ecology, because of its linkages to other disciplines (from soil science to economics), is well placed to contribute to this goal.

As drawn out in this review, a number of broad ecological contributions are required for the management of our coastal and estuarine ecosystems. Further research is required to elucidate and predict the impacts of anthropogenic and natural change pressures on coastal resources. For this, we need to understand the basic functions that ecosystems provide and translate these into the various socioeconomic and cultural (monetary and non-monetary) values to society. In addition, it will be necessary to take into account and quantify the importance of landscape heterogeneity in maintaining ecological functioning and the provision of goods and services to society.

Many of the problems being faced in the coastal zone are due to a lack of awareness by the public and policy-makers as to the impacts of their actions. Further work is therefore required to enable the formation of indicators for normative, environmental sustainability. Increased capabilities are also required in the field of integrated modelling to clarify the linkages between human pressures and welfare impacts, and ecological environmental state changes.

Under pressures of global environmental change, sustainable management of coastal and estuarine environments will require the safeguarding of existing ecosystem functions and restoration of degraded systems. Continued research is required into the restoration and enhancement of degraded ecosystems in order to compensate for those that will be lost through growing environmental pressures. Linked to this, more research is required into the possibilities and limitations of resource substitution across a range of spatial scales, and into mechanisms to mitigate the effects of uncertainties and risks to ecosystem integrity and resilience. Current environmental regulations such as the Habitats Directive seem to contain a rather static interpretation of ecosystem integrity and its maintenance. In cases where policy-makers are trying to manage multiple-use wetlands, a more dynamic interpretation will be required to facilitate trade-offs among competing demands for the system services. A combined ecological–economic analysis can provide insight and guidance on the trade-off option set and the relevant ecological, economic and political constraints to the process of substitution.

REFERENCES

Allen, J.R.L. (1990a). Salt-marsh growth and stratification: a numerical model with special reference to the Severn estuary, southwest Britain. *Mar. Geol.* **95**, 77–96.

Allen, J.R.L. (1990b). Constraints on the measurement of sea-level movement from salt-marsh accretion rates. *J. Geol. Soc. Lond.* **147**, 5–7.

Allen, J.R.L. (1995). Salt-marsh growth and fluctuating sea-level: implications of a simulation model for Flandrian coastal stratigraphy and peat-based sea-level curves. *Sed. Geol.* **100**, 21–45.

Allen, J.R.L. (1997). Simulation models of salt-marsh morphodynamics: some implications for high-intertidal sediment couplets related to sea-level change. *Sed. Geol.* **113**, 211–223.

Asano, T. and Setoguchi, Y. (1996) Hydrodynamic effects of coastal vegetation on wave damping. In: *Hydrodynamics* (Ed. by A.T. Chwang, J.H.W. Lee and D.Y.C. Leung), pp. 1051–1056. Balkema, Rotterdam.

Bayliss-Smith, T.P., Healey, R.G., Lailey, R., Spencer, T. and Stoddart, D.R. (1979). Tidal flows in saltmarsh creeks. *Estuar. Coast. Mar. Sci.* **9**, 235–255.

Bell, S.S., Fonseca, M.S. and Motten, L.B. (1997). Linking restoration and landscape ecology. *Rest. Ecol.* **5**, 318–323.

Bibby, C.J. (1998). Selecting areas for conservation. In: *Conservation Science and Action* (Ed. by W.J. Sutherland), pp. 176–201, Blackwell, Oxford.

Bijl, W. (1997). Impact of wind climate on the surge in the southern North Sea. *Climate Res.* **8**, 45–59.

Bijlsma, L. (1997). Climate change and the management of coastal resources. *Climate Res.* **9**, 47–56.

Bingham, G., Bishop, R., Brody, M., Bromley, D., Clark, E., Cooper, W., Costanza, R., Hale, T., Haydon, G., Kellert, S., Norgaard, R., Norton, B., Payne, J., Russel, C. and Suter, G. (1995). Issues in ecosystem valuation: improving information for decision-making. *Ecol. Econ.* **14**, 73–90.

Bird, E.C.F. (1993). *Submerging Coasts: The Effect of Rising Sea-Level on Coastal Environments.* John Wiley, Chichester.

Bockstael, N., Costanza, R., Strand, I., Boynton, W. Bell, K. and Wainger, L. (1995). Ecological modeling and valuation of ecosystems. *Ecol. Econ.* **14**, 143–159.

Boorman, L.A., Goss-Custard, J. and McGorty, S. (1989). *Climate Change, Rising Sea Level and the British Coast.* Institute of Terrestrial Ecology, Research Publication 1. HMSO, London.

Botsford, L.W., Castilla, J.C. and Peterson, C.H. (1997). The management of fisheries and marine ecosystems. *Science* **277**, 509–515.

Boulding, K. (1966). The economics of the coming Spaceship Earth. In: *Environmental Quality in a Growing Economy* (Ed. by H. Jarrett), pp. 3–14. John Hopkins University Press, Baltimore.

Bower, B.T. and Turner, R.K. (1998). Characterising and analysing benefits from integrated coastal management (ICM). *Ocean Coast. Manag.* **38**, 41–66.

Braden, J.B. and Kolstad, C.D. (eds) (1991). *Measuring the Demand for Environmental Quality.* North Holland, Amsterdam.

Brampton, A.H. (1992). Engineering significance of British saltmarshes. In: *Saltmarshes: Morphodynamcs, Conservation and Engineering Significance* (Ed. by J.R.L. Allen and K. Pye), pp. 115–122. Cambridge University Press, Cambridge.

Breaux, A. Farber, S. and Day, S. (1995). Using natural coastal wetland systems for wastewater treatment: an economic benefit analysis. *J. Environ. Man.* **44**, 285–291.

Bromley, D.W. (ed.) (1995). *The Handbook of Environmental Economics*. Blackwell, Oxford.

Broome, S.W., Seneca, E.D. and Woodhouse, W.W., Jr. (1988). Tidal salt marsh restoration. *Aquat. Botany* **32**, 1–22.

Brouwer, R., Crooks, S. and Turner, R.K. (1998). Towards an integrated framework for wetlands ecosystem indicators. *CSERGE Working Paper GEC 98–27*. Centre for Social and Economic Research on the Global Environment, University of East Anglia, Norwich.

Burdick, D.M., Dionne, M., Boumans, R.M. and Short, F.T. (1997). Ecological response to tidal restoration of two northern New England salt marshes. *Wetlands Ecol. Man.* **4**, 129–144.

Burkholder, J.M. (1998). Implications of harmful microalgae and heterotrophic dinoflagellates in management of sustainable marine fisheries. *Ecol. Appl.* **8** (supplement), S37–S62.

Cahoon, D.R., Reed, D.J. and Day, J.W. (1995). Estimating shallow subsidence in microtidal salt marshes of the southeastern United States. Kaye and Barghoon revisited. *Mar. Geol.* **128**, 1–9.

Cain, R. and Dean, J.M. (1976). Annual occurrence, abundance and diversity of fish in a South Carolina intertidal creeks. *Mar. Biol.* **36**, 369–379.

Carlton, J.T. (1992). Introduced marine and estuarine mollusks of North America: an end of the 20th century perspective. *J. Shellfish Res.* **11**, 489–505.

Cattrijsse, A., Hederick, R. and Dankwa, J.M. (1997). Nursery function of an estuarine tidal marsh for the brown shrimp *Crangon crangon*. *J. Sea Res.* **38**, 109–121.

Chen, J. (1997). The impact of sea level rise on China's coastal areas and its disaster hazard evaluation. *J. Coast. Res.* **13**, 925–930.

Chen, X. and Zong, Y. (1998). Coastal erosion along the Changjiang deltaic shoreline, China: history and prospective. *Estuar. Coast. Shelf Sci.* **46**, 733–742.

Chung, S.S. and Poon, C.S. (1996). Evaluating waste management alternatives by the multicriteria approach. *Resources, Conservation and Recycling* **17**, 189–210.

Church, T.M., Sarin, M.M. Fleisher, M.Q. and Ferdelman, T.G. (1996). Salt marshes: an important coastal sink for dissolved uranium. *Geochim. Cosmochim. Acta* **60**, 3879–3887.

Churma, G., Constanza, R. and Kosters, E.C. (1992). Modelling coastal marsh stability in response to sea level rise: a case study in Louisiana, USA. *Ecol. Model.* **64**, 47–64.

Cicin-Sain, B. (1993). Integrated coastal management. *Ocean Coast Manag.* **21**, 1–3.

Coen, L., Heck, K.L. Jr and Abel, L.G. (1981). Experiments on competition and predation among shrimps of seagrass meadows. *Ecology* **62**, 1484–1493.

Cohen, J.E., Small, C., Mellinger, A., Gallup, J. and Sachs, J. (1997). Estimates of coastal populations. *Science* **278**, 1211–1212.

Common, M. and Perrings, C. (1992). Towards and ecological economics and sustainability. *Ecol. Econ.* **6**, 7–34.

Costanza, R. (1992). Toward an operational definition of health. In: *Ecosystem Health—New Goals for Environmental Management* (Ed. by R. Costanza, B. Norton and B. Haskell), pp. 239–256. Island Press, Washington, D.C.

Costanza, R. (1996). Ecological economics: reintegrating the study of human and nature. *Ecol. Appl.* **6**, 978–990.

Costanza, R., Wainger, L., Folke, C. and Maler, K.-G. (1993). Modelling complex ecological economic systems: towards and evolutionary, dynamic understanding of people and nature. *BioScience* **43**, 545–555.

Costanza, R., d'Arge, R., de Groot, R., Farber, S., Grasso, M., Hannon, B., Limburg, K., Naeem, S., O'Neill, R.V., Paruelo, J., Raskin, R.G., Sutton, P. and van der Belt, M. (1997). The value of the world's ecosystem services and natural capital. *Nature* **387**, 253–260.

Costanza, R., Andrade, F., Antunes, P., van der Belt, M., Boersma, D., Boesch, D.F., Catarino, F., Hanna, S., Limburg, K., Low, B., Molitor, M., Pereira, J.G., Raynor, S., Santos, R., Wilson, J. and Young, M. (1998). Principles of sustainable governance of the oceans. *Science* **281**, 198–199.

Cowell, R. (1997). Stretching the limits: environmental compensation, habitat creation and sustainable development. *Trans. Inst. Br. Geogr.* **22**, 292–306.

Crooks, S. and Ledoux, L. (1999). Mitigation banking as a tool for strategic coastal zone management. *CSERGE Working Paper GEC 99–02*. Centre for Social and Economic Research on the Global Environment, University of East Anglia, Norwich.

Crowards, T.M. (1997). Nonuse values and the environment: economics and ethical motivations. *Environ. Values* **6**, 143–167.

Dakalakis, K.D. and O'Connor, T.P. (1995). Distribution of chemical concerntrations in US coastal and estuarine sediments. *Mar. Environ. Res.* **40**, 381–398.

Daly, H. (1992). Allocation, distribution and scale: towards and economics which is efficient, just, and sustainable. *Ecol. Econ.* **6**, 185–194.

Daly, H. and Cobb, J. (1990). *For the Common Good*. Greenprint, London.

Danielson, B.J. (1991). Communities in a landscape: the influence of landscape heterogeneity on the interactions between species. *Am. Nat.* **138**, 1105–1120.

Department of the Environment (1996). *Indicators of Sustainable Development for the UK*. HMSO, London.

DETR (1998). *Review of Technical Guidance on Environmental Appraisal*. Department of Environment, Transport and the Regions, London.

Dixon, J.A. and Hufschmidt, M.M. (1986). *Economic Valuation Techniques for the Environment*. John Hopkins Press, Baltimore.

Done, T.J. and Reichelt, R.E. (1998). Integrated coastal zone and fisheries ecosystem management: generic goals and performance indices. *Ecol. Appl.* **8** (supplement), S110–S118.

Ehrenfeld, J.G. and Toth, L.A. (1997). Restoration ecology and the ecosystem perspective. *Restor. Ecol.* **5**, 307–317.

ELI (1993). *Wetland Mitigation Banking*. Environmental Law Institute, Washington, DC.

Environment Agency (1998). *Results of Post Breach Monitoring of Orplands Coastal Realignment Site, August 1997–March 1998*. Environment Agency, Bristol.

EPA (1995). *Federal Guidance for the Establishment, Use and Operation of Mitigation Banks*. Notice Federal Register: November 28, 1995 (Vol. 60, pp. 58605–58614). US Environment Protection Agency.

Esselink, P., Dijkema, K.S., Reents, S. and Hageman, G. (1998). Vertical accretion and profile changes in abandoned man-made tidal marshes in the Dollard Estuary, the Netherlands. *J. Coast. Res.* **14**, 570–582.

Etchard, G. (1995). Mitigation banks: a strategy for sustainable development. *Coast. Manag.* **23**, 233–237.

FAO (1992). *Sustainable Development and the Environment*. Food and Agriculture Organization. Rome.

Fernandez, L. and Karp, L. (1998). Restoring wetlands through wetlands mitigation banks. *Environ. Resource Econ.* **12**, 323–344.

Folke, C., Holling, C. and Perrings, C. (1996). Biological diversity, ecosystems and the human scale. *Ecol. Appl.* **6**, 1018–1024.

Folke, C., Kautsky, N., Berg, H., Jansson, Å. and Troell, M. (1998). The ecological footprint concept for sustainable seafood production: a review. *Ecol. Appl.* **8** (supplement), S63–S71.

Freeman, A.M.I. (1993). *The Measurement of Environmental Resource Values.* Resources for the Future, Washington, DC.

French, J.R. (1991) Eustatic and neotectonic controls on salt-marsh sedimentation. In: *Coastal Sediments '91* (Ed. by N.C. Kraus, K.J. Gingerich, and J.R. Kriebel), pp. 1223–1236. American Society of Civil Engineers, New York.

French, J.R. (1993). Numerical simulation of vertical marsh growth and adjustment to accelerated sea-level rise, North Norfolk, U.K. *Earth Surf. Proc. Landf.* **18**, 63–81.

French, J.R. and Stoddard, D.R. (1992). Hydrodynamics of saltmarsh creek systems: implications for marsh morphological development and material exchange. *Earth Surf. Proc. Landf.* **17**, 235–252.

Frey, R.W. and Basan, P.B. (1985). Coastal salt marshes. In: *Coastal Sedimentary Environments* (ed. by R.A. Davis), pp. 101–159. Springer, New York.

Furlani, D.M. (1996). *A Guide to the Introduced Marine Species in Australian Waters.* CSIRO Division of Fisheries Technical Report No. 5. Centre for Research on Introduced Marine Pests, Holbart, Tasmania.

Gardner, R.C. (1996). Banking on entrepreneurs: wetlands mitigation banking, and takings. *Iowa Law Rev.* **81**, 527–587.

Gersberg, R.M., Brenner, R., Lyon, S.R. and Elkins, R.V. (1987). Survival of bacteria and viruses in municipal wastewaters applied to artificial wetlands. In: *Aquatic Plants for Water Treatment and Resource Recovery* (Ed. by K.R. Reddy and W.H. Smith), pp. 237–245. Magnolia Publishing, Orlando, Florida.

GESAMP (Groups of Experts on Scientific Aspects of Marine Pollution) (1996). *The Contribution of Science to Coastal Zone Management.* Reports/Studies, vol. 61. p. 61.

Giaoutzi, M. and Nijkamp, P. (1995). *Decision Support Models for Regional Sustainable Development.* Avebury, Aldershot.

Gilman, E.L. (1997). A method to investigate wetland mitigation banking for Saipan, Commonwealth of Northern Mariana Islands. *Ocean Coast. Manag.* **34**,117–152.

Glassen, J., Threivel, R. and Chadwick, A. (1994). *Introduction to Environmental Impact Assessment.* UCL Press, London.

Goodland, R. and Daly, H. (1996). Environmental sustainability: universal and non-negotiable. *Ecol. Appl.* **6**, 1002–1017.

Gottfried, R., Wear, D. and Lee, R. (1996). Institutional solutions to market failure on the landscape scale. *Ecol. Econ.* **18**, 133–140.

Government of Canada (1991). *The Federal Policy of Wetland Conservation.* Ministry of Supply and Services, Canada.

Gren, I.-G., Folke, C., Turner, R.K. and Bateman, I. (1994). Primary and secondary values of wetland ecosystems. *Environ. Resource Econ.* **4**, 55–74.

Gustafson, E.J. and Gardner, R.H. (1996). The effect of landscape heterogeneity on the probability of patch colonization. *Ecology* **77**, 94–107.

Hameedi, M.J. (1997). Strategy for monitoring the environment in the coastal zone. In: *Coastal Zone Management Imperative for Maritime Developing Nations* (Ed. by B.U. Haq, S.M. Haq, G. Kulerberg and J.H. Stel), pp. 111–142. Kluwer Academic Publishers, London.

Hanley, N. and Spash, C.L. (1993). *Cost–Benefit Analysis and the Environment.* Edward Elgar, Vermont.

Heimlich, R.E. (1994). Cost of an agricultural wetland reserve. *Land Econ.* **70**, 234–246.

Heimlich, R.E., Wiebe, K.D., Classen, R., Gatsby, D. and House, R.M. (1998) *Wetlands and Agriculture: Private Interests and Public Benefits*. Resource Economics Decision, US Department of Agriculture, Report No. 765, Washington, DC.

Hemond, H.F. and Benoit, J. (1988). Cumulative impacts on water quality functions of wetlands. *Environ. Manag.* **12**, 639–653.

HOC (House of Commons) (1992). *Coastal Zone Protection and Planning*. Environment Committee Second Report. HMSO, London.

HOC (House of Commons) (1998). *Flood and Coastal Defence*. Agricultural Committee Sixth Report. HMSO, London.

Hollings, C.S., Schindler, D.W., Walker, B.W. and Roughgarden, J. (1995). Biodiversity in the functioning of ecosystems: an ecological synthesis. In: *Biodiversity Loss: Economic and Ecological Issues* (Ed. by C. Perrings, K.-G. Maler, C. Folke, C.S. Hollings and B.O. Jansson), pp. 44–83. Kluwer Academic, Dordrecht.

Hooligan, P.M. and Reiners, W.A. (1992). Predicting the response of the coastal zone to global change. *Adv. Ecol. Res.* **22**, 211–255.

Huggett, D. (1998). Coasts and wetlands: developing a no net loss approach. In: *Marine Environmental Management: Review of Events in 1997 and Future Trends* (Ed. by R. Earll), pp. 95–100. Earll, Kempley, UK.

IPCC (Intergovernmental Panel on Climate Change) (1996*). Second Assessment Report; Climate Change: The Science of Climate Change*. Cambridge University Press, Cambridge.

Irlandi, E.A. (1994). Large and small scale effects of habitat structure on rates of predation: how percent cover of seagrass affects rates of predation and siphon nipping on an infaunal bivalve. *Oecologia* **98**, 176–183.

Irlandi, E.A. and Crawford, M.K. (1997). Habitat linkages: the effect of intertidal saltmarshes and adjacent subtidal habitats on abundance, movement, and growth of an estuarine fish. *Oecologia* **110**, 222–230.

Jackson, L.L., Lopoukhine, N. and Hillyard, D. (1995). Ecological restoration: a definition and comments. *Restor. Ecol.* **3**, 71–76.

Janssen, R. (1994). *Multiobjective Decision Support for Environmental Management*. Kluwer, Dordrecht.

Jickells, T. (1998). Nutrient biogeochemistry of the coastal zone. *Science* **281**, 217–222.

Jickells, T. and Rae, J. (eds) (1997). *Biogeochemistry of Intertidal Sediments*. Cambridge University Press, Cambridge.

Joubert, A.R., Leiman, A., deKlerk, H.M., Katua, S. and Aggrenbach, L.C. (1997). Fynbos (fine bush) vegetation and the supply of water: a comparison of multi-criteria analysis and cost–benefit analysis. *Ecol. Econ.* **22**, 123–140.

Kadlec, R.H. and Knight, R.L. (1996). *Treatment Wetlands*. Lewis, London.

Kahn, J.R. (1998). *The Economic Approach to Environmental and Natural Resources*, 2nd edn. The Dryden Press, Harcourt Brace College Publishers, Orlando.

Kamaludin, B.H. (1993). The changing mangrove shoreline in Kuala Kuran, Peninsular Malaysia. *Sed. Geol.* **83**, 187–193.

Karr, J.R. (1991). Biological integrity: a long neglected aspect of water resource management. *Ecol. Appl.* **1**, 66–84.

Karr, J.R. (1993). Measuring biological integrity: lessons from streams. In: *Ecological Integrity and the Management of Ecosystems* (Ed. by S. Woodley, J. Kay and G. Francis). St Lucie Press, Delray Beach, Florida.

Kerr, S.R. and Dickley, L.M. (1984). Measuring the health of aquatic ecosystems. In: *Contaminant Effects on Fisheries* (Ed. by V.W. Carins, Hodgson, P.V. and J.O. Nriangu), pp. 279–284. J. Wiley, New York.

Kia, M.P. and O'Niell, D. (1997). Large and fast container ships and the perspective of Australian ports. *Proceedings of the 13th Combined Australasian Coastal Engineering and Ports Conference*, Christchurch, New Zealand, 7–11 September 1997, Centre for Advanced Engineering, pp. 769–774.

Kildow, J. (1997). The roots and context of coastal zone management. *Coast. Man.* **25**, 231–263.

King, A.W. (1993). Considerations of scale and hierarchy. In: *Ecological Integrity and the Management of Ecosystems* (Ed. by S. Woodley, J. Kay and G. Francis), pp. 19–46. St Lucie Press, Ottawa.

King, S.E. and Lester, J.N. (1995) The value of salt marsh as a sea defence. *Mar Poll. Bull.* **30**(3), 180–185.

Kotliar, N.B. and Wiens, J.A. (1990). Multiple scales of patchiness and patch structure: a hierarchical framework for the study of heterogeneity. *Oikos* **59**, 253–260.

Krone, R.B. (1987). A method for simulating historic marsh elevations. In: *Coastal Sediments '87* (Ed. by N.C. Kraus), pp. 316–323. American Society of Civil Engineers, New York.

Kupfer, J.A. (1995). Landscape ecology and biogeography. *Prog. Phys. Geogr.* **19**, 18–34.

Kusler, J.A. (1992). Mitigation banks and the replacement of wetland functions and values. In: *Effective Mitigation Banks and Joint Projects in the Context of Wetland Mitigation Plans*. Proceedings from a National Symposium, Association of Wetland Managers, Palm Beach Gardens, Florida, 1992 pp. 51–56.

Kusler, J.A. and Kentula, M.E. (eds) (1990). *Wetland Creation and Restoration: The Status of the Science*. Island Press, Washington, DC.

Kwak, J. and Zedler, J.B. (1997). Food web analysis of southern California coastal wetlands using multiple stable isotopes. *Oecologia* **110**, 262–277.

Landres, P.B., Verner, J. and Thomas, J.W. (1988). Ecological uses of vertebrate indicator species: a critique. *Conserv. Biol.* **2**, 316–328.

Lee, J.G., Nishijima, W., Mukai, T., Takimot, K., Seiki, T., Hiraoka, K. and Okada, M. (1998). Factors determining the functions and structures in natural and constructed tidal flats. *Water Res.* **32**, 2601–2606.

Lewis, R.R. (1994). Enhancement, restoration and creation of coastal wetlands. In: *Applied Wetlands Science and Technology* (Ed. by D.M. Kent), pp. 167–191. Baco Raton, London.

Lynch-Stewart, P. (1992). *No Net Loss: Implimentating "No Net Loss" Goals to Conserving Wetlands in Canada*. Sustaining Wetlands Issue Paper No. 1992–2. North American Wetlands Conservation Council, Ottawa, Ontario, Canada.

Maes, J., Tailieu, A., Van Damme, P.A., Cottenie, K. and Ollevier, F. (1998). Seasonal patterns in the fish and crustacean community of a turbid temperate estuary (Zeeschelde estuary, Belgium). *Estuar. Coast. Shelf Sci.* **47**, 143–151.

Mageau, M.T., Costanza, R. and Ulanowicz, R.E. (1995). The development and initial testing of a quantitative assessment of ecosystem health. *Ecosystem Health* **1**, 201–213.

Maimone, M. (1985). An application of multicriterial evaluation in assessing municipal solid waste treatment and disposal system. *Waste Man. Res.* **3**, 217–231.

Makarewicz, J.C. (1991). Photosynthetic parameters as indicators of ecosystem health. *J. Great Lakes Res.* **17**, 333–343.

Makepeace, D.K., Smith, D.W. and Stanley, S.J. (1995). Urban storm water quality: summary of contaminant data. *Crit. Rev. Environ. Sci. Technol.* **25**, 93–139.

Makowski, M., Somyody, L. and Watkins, D. (1996). Multiple criteria analysis for water quality management in the Nitra Basin. *Water Res. Bull.* **32**, 937–951.

Malczewski, J., Moreno-Sanchez, R., Bojórquez-Tapia, L.A. and Ongay-Delhumeau, E. (1997). Multicriteria group decision-making for environmental conflict analysis in the Cape Region, Mexico. *J. Environ. Plan. Manag.* **40**, 349–374.

Marsh, L.L., Porter, D.R. and Salvesen, D.A. (eds) (1996). *Mitigation Banking: Theory and Practice.* Island Press, Washington, DC.

Mills, G.N. (1997). Organic contamination in New Zealand coastal sediments. *Proceedings of the 13th Combined Australasian Coastal Engineering and Ports Conference,* Christchurch, New Zealand, 7–11 September 1997, Centre for Advanced Engineering, pp. 581–586.

Mitsch, W.J. and Gosselink, J.G. (1993). *Wetlands.* Van Nostrand Reinhold, New York.

Mittelbach, G.G. (1986) Predator-mediated habitat use: some consequences for species interactions. *Environ. Biol. Fish.* **16**, 159–169.

Moeller, I., Spencer, T. and French, J.R. (1996). Wind wave attenuation over salt-marsh surfaces: preliminary results from Norfolk, England. *J. Coast. Res.* **12**, 1009–1016.

Naveh, Z. (1994). From biodiversity to ecodiversity: a landscape-ecology approach to conservation and restoration. *Restor. Ecol.* **2**, 180–189.

Norris, R.H. and Norris, K.R. (1995). The need for biological assessment of water quality: Australian perspective. *Aust. J. Ecol.* **20**, 1–6.

Noss, R.F. (1983). A regional landscape approach to maintain biodiversity. *Bioscience* **33**, 700–706.

Noss, R.F. (1990). Indicators for monitoring biodiversity: a hierarchical approach. *Conserv. Biol.* **4**, 355–364.

NRC (National Research Council) (1992). *Restoration of Aquatic Ecosystems: Science, Technology, and Public Policy.* National Academic Press, Washington, DC.

NWPF (National Wetlands Policy Forum) (1988). *Protecting America's Wetlands: An Action Agenda.* The Conservation Foundation, Washington, DC.

NSW Government (1996). *The NSW Wetlands Management Policy.* Department of Land and Water Conservation, Sydney, Australia.

O'Donnell, A. (1988). The policy implications of wetland. In: *Increasing our Wetland Resources* (Ed. by J. Zelanzy and J.S. Feierbbend), pp. 86–110. National Wildlife Foundation. Washington, DC.

OECD (1993a). *Integrated Policies for Integrated Coastal Management.* Organization for Economic Co-operation and Development, Paris.

OECD (1993b). *Coastal Zone Management: Selected Case Studies.* Organization for Economic Co-operation and Development, Paris.

OECD (1994). *Environmental Indicators.* Organization for Economic Co-operation and Development, Paris.

OECD (1997). *Sustainable Development: OECD Policy Approaches for the 21st Century.* Organization for Economic Co-operation and Development, Paris.

Page, H.M. (1997). Importance of vascular plant and algal production to macro-invertebrate consumers in a southern California salt marsh. *Estuar. Coast. Shelf Sci.* **45**, 823–834.

Parrish, J.D. (1989). Fish communities of interacting shallow-water habitats in tropical oceanic regions. *Mar. Ecol. Prog. Ser.* **58**, 143–160.

Paterson, A.W. and Whitfield, A.K. (1997). A stable isotope study of the food-web in a freshwater-deprived South African estuary, with a particular emphasis on the Ichthyofauna. *Estuar. Coast. Shelf Sci.* **45**, 705–715.

Pearce, D.W. (1993). *Blueprint 3: Measuring Sustainable Development.* Earthscan, London.

Pearce, D.W. and Moran, D. (1994). *The Economic Value of Biodiversity.* Earthscan in association with IUCN, London.

Pearce, D.W. and Turner, R.K. (1990). *Economics of Natural Resources and the Environment*. Harvester Wheatsheaf, Hemel Hemstead, UK.

Pearce, D.W., Markandya, A. and Barbier, E. (1989). *Blueprint for a Green Economy*. Earthscan, London.

Penning-Rowsell, E., Green, C., Thompson, P., Coker, A., Tunstall, S., Richards, C. and Parker, D. (1992). *The Economics of Coastal Management: A Manual of Benefit Assessment Techniques*. Belhaven Press, London.

Pethick, J.S. (1980). Salt marsh initiation during the Holocene transgression: the example of the north Norfolk marshes. *J. Biogcogr.* **7**, 1–9.

Pethick, J.S. (1981). Long-term accretion rates on tidal salt marshes. *J. Sediment Petrol.* **51**, 571–577.

Pethick, J.S. (1992). Saltmarsh geomorphology. In: *Saltmarshes: Morphodynamics, Conservation and Engineering Significance* (Ed. by J.R.L. Allen and K. Pye), pp. 41–62. Cambridge University Press, Cambridge.

Pethick, J.S. (1993). Shoreline adjustment and coastal management: physical and biological processes under accelerated sea-level rise. *Geogr. J.* **159**, 162–168.

Powell, J.C. (1996). The evaluation of waste management options. *Waste Manag. Res.* **14**, 515–526.

Primavera, J.H. (1998). Mangroves as nurseries: shrimp populations in mangrove and non-mangrove habitat. *Estuar. Coast. Shelf Sci.* **46**, 457–464.

Pye, K. and French, P.W. (1993). *Erosion and Accretion Processes on British Saltmarshes*. Final report to the Ministry of Agriculture, Fisheries and Food. Vols 1–5. Cambridge Environmental Research Consultants, Cambridge.

Race, M.S. (1985). Critique of present wetland mitigation policies in the United States based on analysis of past restoration projects in San Francisco Bay. *Environ. Manag.* **9**, 71–82.

Race, M.S. and Cristie, D.R. (1982). Coastal zone development: mitigation, marsh creation and decision-making. *Environ. Manag.* **6**, 317–328.

Race, M.S. and Fonseca, M.S. (1996). Fixing compensatory mitigation—what will it take? *Ecol. Appl.* **6**, 94–101.

Randerson, P.F. (1979). A simulation model for salt-marsh development and plant ecology. In: *Estuarine and Coastal Land Reclamation and Water Storage* (Ed. by B. Knight and A.J. Phillips), pp. 48–67. Saxon House, Farnborough, UK.

Rapport, D.J. (1989). Symptoms of pathology in the Gulf of Bothnia (Baltic Sea): ecosystem response to stress from human activity. *Biol. J. Linn. Soc.* **37**, 33–49.

Rapport, D.J., Costanza, R. and McMichael, A.J. (1998a). Assessing ecosystem health. *Tree* **13**, 397–402.

Rapport, D.J., Gaudet, C., Karr, J.R., Baron, J.S., Bohlen, C., Jackson, W., Jones, B., Naiman, R.J., Norton, B. and Pollock, M.M. (1998b). Evaluating landscape health: integrating societal goals and biophysical process. *J. Environ. Manag.* **53**, 1–15.

Reddy, K.R. and D'Angelo, E.M. (1994). Soil processes regulating water quality in wetlands. In: *Global Wetlands: Old World and New* (Ed. by W.J. Mitsch), pp. 309–324. Elsevier, Amsterdam.

Reed, D.J. (1988). Sediment dynamics and deposition in a retreating coastal salt marsh. *Estuar. Coast. Mar. Sci.* **26**, 67–79.

Reed, D.J. (1990). The impact of sea-level rise on coastal salt marshes. *Prog. Phys. Geogr.* **14**, 24–40.

Reed, D.J. (1995). The response of coastal marshes to sea-level rise: survival or submergence? *Earth Surf. Proc. Landf.* **20**, 39–48.

Rees, W.E. and Wackernagal, M. (1994). Ecological footprints and appropriate carrying capacity. In: *Investing in Natural Capital: The Ecological Economics*

Approach to Sustainability (Ed. by A.M. Jansson, M. Hammer, C. Folke and R. Costanza), pp. 362–390. Island Press, Washington, DC.

Roberts, L. (1993). Wetland trading is a losers game, say ecologists. *Science* **260**, 1890–1892.

Rotmans, J. and Van Asselt, M. (1996). Integrated assessment: a growing child on its way to maturity—an editorial essay. *Clim. Change* **34**, 327–336.

Russell, C. (1995). Are we lost in the vale of ignorance or on the mountain of principle? *Ecol. Econ.* **14**, 91–99.

Sacco, J.N., Seneca, E.D. and Wentworth, T.R. (1994). Infaunal community development of artificially established salt marshes in North Carolina. *Estuaries* **17**, 489–500.

Sarda, R. Foreman, K., Werne, C.E. and Valiela, I. (1998). The impact of epifaunal predation on the structure of macroinfaunal invertebrate communities of tidal saltmarsh creeks. *Estuar. Coast. Shelf Sci.* **46**, 657–669.

Scodari, P. and Shabman, L. (1995). *National Wetland Mitigation Banking Study, Commercial Wetland Mitigation Credit Markets: Theory and Practice.* Institute for Water Resources Report, Alexandria, Virginia, 95-WMB-7.

Sestini, G. (1992). Implications of climate changes for the Po Delta and Venice Lagoon. In: *Climatic Change and the Mediterranean* (Ed. by L. Jeftic, J.F. Milliman and G. Sestini), pp. 428–494. Edward Arnold, London.

Shenker, J.M. and Dean, J.M. (1979). The utilisation of an intertidal saltmarsh creek by larval and juvenile fish: abundance, diversity and temporal variation. *Estuaries* **2**, 154–163.

Simberloff, D. (1998). Flagships, umbrellas, and keystones: is single species management passé in the landscape era? *Biol. Conserv.* **83**, 245–257.

Sogard, S.M. and Able, K.W. (1991). A comparison of eelgrass, sea lettuce, microalgae and marsh creeks as habitats for epibenthic fishes and decapodes. *Estuar. Coast. Shelf Sci.* **33**, 501–519.

Spofford, W.O., Jr., Russell, C.S. and Kelly, R.A. (1975). Operational problems in large-scale residual management models. In: *Economic Analysis of Environmental Problems* (Ed. by E.S. Mills), pp. 171–238. National Bureau of Economics Research, New York.

Stedman, S.-M. and Hanson, J. (1997). *Habitat Connections.* Vol. 1, Nos 1–5. US Department of Commerce, National Oceanic and Atmospheric Administration, Silverspring, Maryland.

Steel, T.J. and Pye, K. (1997). The development of saltmarsh tidal creek networks: evidence from the UK. *Proceedings of the Canadian Coastal Conference*, pp. 227–280. Guelph University, Guelph, Ontario.

Steele, J.H. (1991). Marine functional diversity. *Bioscience* **41**, 470–474.

Stevenson, J.C., Ward, L.G. and Kearney, M.S. (1986). Vertical accretion in marshes with varying rates of sea-level rise. In: *Estuarine Variability* (Ed. by D.A. Wolfe), pp. 241–260. Academic Press, Orlando.

Stumpf, R. P. and Haines, J.W. (1998). Variations in the gulf of Mexico and implications for tidal wetlands. *Estuar. Coast. Shelf Sci.* **46**, 165–173.

Tilman, D. and Downing, J.A. (1994). Biodiversity and stability in grasslands. *Nature* **367**, 363–365.

Tooley, M.J. and Jelgersma, S. (eds) (1992). *Impacts of a Future Sea-Level Rise on European Coastal Lowlands.* Blackwell, Oxford.

Turner, R.K. (1992). Policy failures in managing wetlands. In: *Market and Government Failures in Environmental Management: Wetlands and Forests.* pp. 10–43, OECD, Paris.

Turner, R.K. (ed.) (1993). *Sustainable Environmental Economics and Management: Principles and Practice.* Belhaven Press, London.

Turner, R.K. and Adger, W.N. (1996). *Coastal Zone Resource Assessment Guidelines.* LOICZ/R&S/96–4. LOICZ, Texel, The Netherlands.

Turner, R.K. and Jones, T. (1991) *Wetlands, Market and Intervention Failures. Four Case Studies.* Earthscan, London.

Turner, R.K., Perrings, C. and Folke, C. (1997a). Ecological economics: paradigm or perspective. In: *Economy and Ecosystems in Change* (Ed. by J. van den Bergh and J. van der Straaten), pp. 25–49. Edward Elgar, Cheltenham, UK.

Turner, R.K., van den Bergh, J.C.J.M., Barendregt, A. and Maltby, E. (1997b). Ecological–economic analysis of wetlands: science and social science integration. In: *Wetlands: Landscape and Institutional Perspectives, Proceedings of the Fourth Workshop of Global Wetlands Economics Network (GWEN)* (Ed. by Tore Söderqvist), Beijer Occasional Paper Series, pp. 293–326. Beijer International Institute of Ecological Economics, Stockholm.

Turner, R.K., Lorenzoni, I., Beaumont, N., Bateman, I.J., Langford, I.H. and McDonald, A.L. (1998). Coastal management for sustainable development: analysing environmental and socioeconomic changes on the UK coast. *Geogr. J.* **164**, 269–281.

Turner, R.K., Adger, W.N., Crooks, S., Lorenzoni, I. and Ledoux, L. (1999). Sustainable coastal resources management: principles and practice. *Natural Resource Forum* (in press).

UNEP (1995). *Guidelines for the Integrated Management of Coastal and Marine Areas with Special Reference to the Mediterranean Basin.* UNEP Regional Seas Reports and Studies 161, PAP/RAC (MAP-UNEP). United Nations Environment Programme, Split, Croatia.

UNEP (United Nations Environment Programme) (1997). *Conservation and Sustainable Use of Marine and Coastal Biological Diversity.* Convention on Biological Diversity, Montreal. UNEP/CBD/SBSTTA/3/4.

United Nations Secretariat (1998). *World Population Projections to 2150.* Population Division of the Department of Economic and Social Affairs at the United Nations Secretariat, United Nations, New York.

USFWS (US Fish and Wildlife Survive) (1988). Mitigation banking. *Biol. Rep.* **88**, 1–103.

van der Meulen, F. and Udo de Haes, H.A. (1996). Nature conservation and integrated coastal zone management in Europe: present and future. *Landscape and Urban Planning* **34**, 401–410.

van Roon M. (1997). The effect of transportation infrastructure upon harbour edges. *Proceedings of the 13th Combined Australasian Coastal Engineering and Ports Conference,* Christchurch, New Zealand, 7–11 September 1997. Centre for Advanced Engineering, pp. 859–864.

Veltman, V.C. (1995). Banking on the future of wetlands using federal law. *Northwest. University Law Rev.* **89**, 654–689.

Vitousek, P.M and Mooney, H.A. (1997). Estimates of coastal populations. *Science* **278**, 1211–1212.

Voogd, H. (1983). *Multicriteria Evaluation for Urban and Regional Planning.* Pion, London.

Wackernagal, M. and Rees, W. (1996). *Our Ecological Footprint: Reducing Human Impacts on the Earth.* New Society, Gabriola Island, British Columbia.

Wang, Y., Luo, Z. and Zhu, D. (1998). Economic development and integrated management issues in coastal China. In: *Coastal Management Imperative for Maritime Developing Nations* (Ed. by B.U. Haq, S.M. Haq, G. Kullenberg and J.H. Stel), pp. 371–384. Kluwer, London.

Warrick, R.A. and Farmer, G. (1990). The greenhouse effect, climate change and rising sea level—implications for development. *Trans. Inst. Br. Geogr.* **15**, 5–20.

Warrick, R.A., Barrow, E.M. and Wigley, T.M.L. (1993). *Climate and Sea-Level Change*. Cambridge University Press, Cambridge.

WCED (World Commission on Environment and Development) (1987). *Our Common Future*. Oxford University Press, Oxford.

Weinstein, M.P. and Brooks, H.A. (1983). Comparative ecology of the nekton residing in a tidal creek and adjacent sea grass meadow: community composition and structure. *Mar. Ecol. Prog. Ser.* **12**, 15–27.

Werner, E.E. and Hall, D.J. (1988). Ontogenetic habitat shifts in bluegill: the foraging rate–predator risk trade-off. *Ecology* **69**, 1352–1366.

Weterings, R. and Opschoor, J.B. (1994). Towards environmental performance indicators based on the notion of environmental space. Advisory Council for Research on Nature and Environment, The Netherlands.

White House Office (1991). *Fact Sheet: Protecting America's Wetlands*. Press release dated 9/8/91, Washington, DC.

White House Office (1993). *Protecting America's Wetlands: A Fair, Flexible and Effective Approach*. White House Office on Environmental Policy, Washington, DC.

Wicklum, D. and Davies, R.W. (1995). Ecosystem health and integrity. *Can. J. Botany* **73**, 997–1000.

Wiens, J.A. (1976). Population responses to patchy environments. *Annu. Rev. Ecol. Syst.* **7**, 81–121.

With, K.A. and Crist, T.O. (1995). Critical thresholds in species responses to landscape structure. *Ecology* **76**, 2446–2460.

Woodley, S., Kay, J. and Francis, G. (1993). *Ecological Integrity and the Management of Ecosystems*. St Lucie Press, Delray Beach, Florida.

World Bank (1996). *Guidelines for Integrated Coastal Zone Management*. Environmentally Sustainable Development Studies and Monographs Series No. 9. World Bank, Washington, DC.

Yang, S.L. (1998). The role of *Scirpus* marsh in attenuation of hydrodynamics and retention of fine grained sediment in the Yangtze Estuary. *Estuar. Coast. Shelf Sci.* **47**, 227–233.

Zedler, J.B. (1987). Mitigation problems on the southern Californian coast. *Californian Waterfront Age* **3**, 32–33.

Zedler, J.B. (1996a). Ecological issues in wetland mitigation: an introduction to the forum. *Ecol. Appl.* **6**, 33–37.

Zedler, J.B. (1996b). Coastal management in southern California: the need for a regional restoration strategy. *Ecol. Appl.* **6**, 84–93.

Advances in Ecological Research
Volumes 1–29

Cumulative List of Titles

Aerial heavy metal pollution and terrestrial ecosystems, **11**, 218

Age-related decline in forest productivity: pattern and process, **27**, 213

Analysis of processes involved in the natural control of insects, **2**, 1

Ant-plant-homopteran interactions, **16**, 53

Biological strategies of nutrient cycling in soil systems, **13**, 1

Bray-Curtis ordination: an effective strategy for analysis of multivariate ecological data, **14**, 1

Can a general hypothesis explain population cycles of forest lepidoptera?, **18**, 179

Carbon allocation in trees: a review of concepts for modelling, **25**, 60

Catchment properties and the transport of major elements to estuaries, **29**, 1

A century of evolution in *Spartina anglica*, **21**, 1

The climatic response to greenhouse gases, **22**, 1

Communities of parasitoids associated with leafhoppers and planthoppers in Europe, **17**, 282

Community structure and interaction webs in shallow marine hard-bottom communities: tests of an environmental stress model, **19**, 189

The decomposition of emergent macrophytes in fresh water, **14**, 115

Delays, demography and cycles: a forensic study, **28**, 127

Dendroecology: a tool for evaluating variations in past and present forest environments, **19**, 111

The development of regional climate scenarios and the ecological impact of greenhouse gas warming, **22**, 33

Developments in ecophysiological research on soil invertebrates, **16**, 175

The direct effects of increase in the global atmospheric CO_2 concentration on natural and commercial temperate trees and forests, **19**, 2

The distribution and abundance of lake-dwelling Triclads—towards a hypothesis, **3**, 1

The dynamics of aquatic ecosystems, **6**, 1

The dynamics of field population of the pine looper, *Bupalis piniarius* L. (Lep., Geom.), **3**, 207

Earthworm biotechnology and global biogeochemistry, **15**, 379

Ecological aspects of fishery research, **7**, 114

Ecological conditions affecting the production of wild herbivorous mammals on grasslands, **6**, 137

Ecological implications of dividing plants into groups with distinct photosynthetic production capabilities, **7**, 87

Index